Neuropharmacology of insects

The Ciba Foundation is an international scientific and educational charity. It was established in 1947 by the Swiss chemical and pharmaceutical company of CIBA Limited—now CIBA-GEIGY Limited. The Foundation operates independently in London under English trust law.

The Ciba Foundation exists to promote international cooperation in medical and chemical research. It organizes international multidisciplinary meetings on topics that seem ready for discussion by a small group of research workers. The papers and discussions are published in the Ciba Foundation series.

The Foundation organizes many other meetings, maintains a library which is open to all graduates in science or medicine who are visiting or working in London, and provides an information service for scientists. The Ciba Foundation also functions as a centre where scientists from any part of the world may stay during working visits to London.

Neuropharmacology of insects

Ciba Foundation symposium 88

1982

Pitman
London

ISBN 0 272 79652 2

Published in April 1982 by Pitman Books Ltd, London. Distributed in North America by CIBA Pharmaceutical Company (Medical Education Administration), Summit, NJ 07006, USA

Suggested series entry for library catalogues:
Ciba Foundation symposia.

Ciba Foundation symposium 88
x + 330 pages, 65 figures, 21 tables

British Library Cataloguing in Publication Data

Neuropharmacology of insects.—(Ciba
Foundation symposium; 88)
1. Insects—Physiology—Congresses
2. Pharmacology—Congresses
I. Evered, David II. O'Connor, Maeve
III. Whelan, Julie IV. Series
595.7'0188 QL495

ISBN 0-272-79652-2

Text set in 10/12 pt Linotron 202 Times, printed and bound
in Great Britain at The Pitman Press, Bath

Contents

Symposium on Neuropharmacology of insects, held at the Ciba Foundation, London, 30 June–2 July 1981
Editors: David Evered (Organizer), Maeve O'Connor and Julie Whelan

Participants

E. A. BARNARD Department of Biochemistry, Imperial College of Science and Technology, Exhibition Road, London SW7 2AZ, UK

M. J. BERRIDGE ARC Unit of Invertebrate Chemistry and Physiology, Department of Zoology, University of Cambridge, Downing Street, Cambridge CB2 3EJ, UK

H. BREER Department of Zoophysiology, University of Osnabrück, D-4500 Osnabrück, German Federal Republic

S. G. CULL-CANDY Department of Biophysics, University College London, Gower Street, London WC1E 6BT, UK

J. F. DONNELLAN Sittingbourne Research Centre, Sittingbourne, Kent ME9 8AG, UK

Y. DUDAI Department of Neurobiology, The Weizmann Institute of Science, Rehovot 76100, Israel

A. T. ELDEFRAWI Department of Pharmacology and Experimental Therapeutics, School of Medicine, University of Maryland, Baltimore, Maryland 21201, USA

M. E. ELDEFRAWI Department of Pharmacology and Experimental Therapeutics, School of Medicine, University of Maryland, Baltimore, Maryland 21201, USA

P. D. EVANS ARC Unit of Invertebrate Chemistry and Physiology, Department of Zoology, University of Cambridge, Downing Street, Cambridge CB2 3EJ, UK

K. A. F. GRATION Pfizer Ltd, Sandwich, Kent, CT13 9NJ, UK

L. M. HALL Department of Genetics, Albert Einstein College of Medicine of Yeshiva University, 1300 Morris Park Avenue, Bronx, New York 10461, USA

J. G. HILDEBRAND Department of Biological Sciences, Sherman Fairchild Center for the Life Sciences 913, Columbia University in the City of New York, New York, NY 10027, USA

C. R. HOUSE Department of Veterinary Physiology, Royal (Dick) School of Veterinary Studies, Summerhall, Edinburgh EH9 1QH, UK

Y. N. JAN Department of Physiology, University of California, School of Medicine, San Francisco, California 94143, USA

J. S. KELLY Department of Pharmacology, St George's Hospital Medical School, Jenner Wing, Cranmer Terrace, London SW17 0RE, UK

G. G. LUNT Biochemistry Department 4-West, School of Biological Sciences, University of Bath, Claverton Down, Bath BA2 7AY, UK

W. MORDUE Department of Zoology, University of Aberdeen, Tillydrone Avenue, Aberdeen AB9 2TN, UK

T. NARAHASHI Department of Pharmacology, The Medical and Dental Schools, Northwestern University, 303 East Chicago Avenue, Chicago, Illinois 60611, USA

R. NEUMANN CIBA-GEIGY Ltd, CH-4002 Basel, Switzerland

R. W. OLSEN Division of Biomedical Sciences, University of California, Riverside, California 92521, USA

H. PENZLIN Department of Animal Physiology, Friedrich Schiller University, Erberstrasse 1, DDR-69 Jena, German Democratic Republic

T. PIEK Pharmacology Laboratory, University of Amsterdam, Polderweg 104, 1093 KP Amsterdam, The Netherlands

D. B. SATTELLE ARC Unit of Invertebrate Chemistry and Physiology, Department of Zoology, University of Cambridge, Downing Street, Cambridge CB2 3EJ, UK

P. N. R. USHERWOOD Department of Zoology, The University of Nottingham, School of Biological Sciences, University Park, Nottingham NG7 2RD, UK

Introduction

P. N. R. USHERWOOD

Department of Zoology, Nottingham University, University Park, Nottingham, NG7 2RD, UK

The symposia of the Ciba Foundation are recognized world-wide as major events in science. Looking at the long list of the Foundation's publications, which span a period of over three decades, I am impressed by their timeliness and, above all, by the Foundation's considerable achievement in identifying growth areas that have since become leading fields of scientific endeavour. I believe this is the first Ciba Foundation symposium exclusively concerned with a group of invertebrate animals; indeed, the first phylum-oriented symposium in a list of over 200. Those of us engaged in research on insect neuropharmacology are gratified by the recognition that has been bestowed on our field.

Studies of insect biology, particularly insect physiology, have occupied a special position in science for many decades, arguably because insects are major agricultural pests and vectors of various parasitic diseases. The search for new forms of chemical and biological control of these pests has provided a major impetus for basic research in insect biology and the growing problem of resistance to control measures has catalysed the so-called rational approach to the discovery of new pesticides. Insect neuropharmacology plays a central role in this approach by underpinning the collaborative efforts of pure and applied scientists to identify and exploit target sites within the insect central nervous system. Although I do not deny the importance of this development and the attention it receives from insect neuropharmacologists, I think we would be ill-advised to encourage the view that the relevance of insect neuropharmacology is restricted to insect pests and their control. By all means let us search for neuropharmacological phenomena which are insect-specific because of their potential value to the insecticide industry, but let us not ignore the fact that insect central and peripheral nervous systems do not differ fundamentally from their counterparts in other animals.

There are insect preparations which potentially rival the squid axon and eel electroplax in contributing to our knowledge of the properties which govern

1982 Neuropharmacology of insects. Pitman, London (Ciba Foundation symposium 88) p 1-4

nervous function in general. I am hopeful that some of the contributors to this symposium will take the opportunity to advertise the unique properties of their preparations which make them such exciting prospects for the future. Who would deny the advantages of studying neurohormones in systems where these chemicals are produced by identifiable and accessible single nerve cells and whose target sites are equally accessible to pharmacological and biochemical analysis? Amino acid synapses, which are found in man and probably in all other animals endowed with nervous systems, are arguably best understood in insects, and there is no reason to suppose that the conceptual advances made by studying insect amino acid synapses are not generally applicable to amino acid synapses in other animals. The application of genetics to the dissection and identification of the structures responsible for signalling chemical events in the nervous system provides the neurobiologist with exciting possibilities. Selection of the best preparation on which to develop and use these techniques is clearly the 'name of the game', and insects that can be easily cultured, have short generation times, and are well known genetically provide excellent opportunities for this genetic approach. It can be seen from this brief scenario that insect neuropharmacology has implications not only for the pesticide industry but also for neurobiology as a whole. Vertebrate neuropharmacologists attempting to solve basic problems with preparations that are unlikely to release their secrets readily might be well advised to consider insect material as an alternative.

Having argued the case for the wider relevance of insect studies, I must admit to some sense of disappointment that neurobiologists engaged in research on other animals, particularly vertebrates, do not always share my enthusiasm for insect neuropharmacology. Why is it that an international scientific community which rapidly accepted the squid axon as a tool for basic studies in neurobiology is reluctant to exploit the unique opportunities afforded by insects? Perhaps the fault lies in part with insect neuropharmacologists. Are we too obsessed with the parochial but enormously important goal of generating new chemicals for controlling insect pests? Is it possible that our approach is too often superficial, either because rigour in depth is not essential to our aims or because the background information necessary to substantiate our results is not available? Do we fail to seek opportunities to present our findings to audiences whose views and knowledge of insects extend no further than the pests on their roses? I hope this symposium will go some way towards establishing insect neuropharmacology as an important discipline and perhaps improve the two-way flow of information between neuropharmacologists working with insect preparations and their counterparts who exploit higher (and lower) life-forms to satisfy their curiosity.

We have known for many years that acetylcholine is a transmitter candidate in the insect central nervous system and it is proper that studies on acetylcholine receptors should occupy an important place at this meeting. By using techniques that have been developed primarily for vertebrate excitable tissues it has been possible to study acetylcholine receptors *in situ* and to gain some information on their pharmacological properties. It is arguable that *in situ* studies of acetylcholine synapses in insects have not kept pace with biochemical studies aimed at identifying and characterizing acetylcholine receptors *in vitro*, but there are encouraging signs that this imbalance is being redressed. The opposite is true for the amino acid synapses of the insect peripheral nervous system. *In situ* studies using voltage clamp, noise analysis and, latterly, patch clamp analysis have all contributed to a level of understanding of the basic properties of, for example, L-glutamate receptors which is almost on a par with our knowledge of acetylcholine receptors of vertebrate excitable systems. Unfortunately, technical problems have beset attempts to characterize these receptors *in vitro* using standard biochemical procedures, and there have been similar problems with efforts to investigate presynaptic properties of amino acid synapses using synaptosome preparations. Papers at this symposium should throw some light on possible future developments in this field. The present difficulties clearly underline the importance of technological development for conceptual and factual advancement and I anticipate that the meeting will pinpoint those areas of insect neuropharmacology that can best benefit from a technological boost.

A few years ago I was asked to identify which aspects of insect biology would contribute most to basic science in the next decade or so. Not being a visionary I was reluctant to offer an opinion but when pressed I settled for insect neurosecretion. In retrospect I am well pleased with this choice. The involvement of chemical messengers such as octopamine, dopamine, 5-hydroxytryptamine and peptide neurohormones with membrane cyclases and secondary messengers heralds a new era in insect neuropharmacology and extends the subject into the areas of long-term neuronal modulation, learning and memory. Fortunately, those working with insects have quickly appreciated the problems which arise from attempts to distinguish between neurohormones and neurotransmitters on a functional basis.

If much of what I have said about insect neuropharmacology conveys the impression of qualitative investigations being undertaken by a handful of entomologists, I would argue that for any science emerging from its infancy it is the qualitative side that is most apparent. There is already evidence in the work to be presented that quantitative rigour is entering our field, which must surely augur well for the prospects of those of us who wish to use our insect preparations to influence events in the wider field of general neuropharmacology. I believe that a watershed has been reached in the development of insect neuropharmacology which makes the timing of this meeting particularly apt. It now seems appropriate to ask whether basic research in this field can resolve some of the problems currently besetting the pesticide industry by providing a rational approach to the discovery of new pesticides. It is also a good moment to take stock of our subject and identify the areas most likely to contribute in a major fashion to neuropharmacology in general. I hope that we shall discuss these points.

The Ciba Foundation has given us a unique opportunity to reach a wide audience of scientists through the vehicle of the symposium volume; it has erected a stage on which to present our work; it has assembled an international cast of scientists who are leaders in their field. It is time for me to open the curtains and let the proceedings begin.

Chemical signalling in the insect nervous system

JOHN G. HILDEBRAND

Department of Biological Sciences, Fairchild 913, Columbia University, New York, NY 10027, USA

Abstract Neurons in insects, as in other animals, communicate with each other and with effector cells such as muscle fibres and gland cells principally by means of chemical messengers. Information can be transmitted between particular classes of cells by rapid or slow synaptic transmission mediated by neurotransmitters; by relatively slower and more global communication mediated by neurohormones; by fine-tuning of the activity of target cells mediated by focally released neuromodulators; and by chronic chemical interactions mediated by trophic substances. Because most pharmacological agents that affect the function of the nervous system do so by altering some aspect of chemical-messenger mechanisms, and because of the biological, agricultural and medical importance of insects, it is both timely and important to survey knowledge of the neuropharmacology of insects and to urge new explorations of chemical signalling in the insect nervous system.

Insect neuropharmacology is a relatively recent science born of neurobiology and insect biology and standing where these fields meet. The nervous system of insects and other invertebrates, in contrast to the vertebrate nervous system, offers the investigator the advantage of working with relatively small numbers of experimentally manipulatable and, in many cases, identifiable neurons. Mechanisms of chemical signalling between cells in the nervous system can be probed in individual neurons that can be characterized morphologically, physiologically, biochemically and functionally, and are recognizable from animal to animal. Such preparations are already proving to be of unsurpassed usefulness in the evolution of our understanding of chemical messengers and their receptors in the nervous system.

At the same time, insect neuropharmacology is beginning to advance our understanding of the functional organization of the major, and possibly the most biologically successful, fauna on earth. Constituting more than three-quarters of all known animal species, insects are of enormous interest and

---

1982 Neuropharmacology of insects. Pitman, London (Ciba Foundation symposium 88) p 5-11

importance as agricultural pests, as vectors of infectious and parasitic diseases, and as beneficial pollinators, predators, and elements in food chains. In order to control harmful insects while sparing their beneficial relatives and other living things, we urgently need a sound biochemical and pharmacological basis for understanding the mechanisms of action and selectivity of, and resistance to, insecticides and insect behaviour-modifying substances. Because many of these agents act on synapses, it is especially important to unravel the details of synaptic structure, physiology, biochemistry and pharmacology in insects.

This symposium thus appropriately focuses on the cellular and molecular machinery of synaptic transmission as well as other, related, kinds of chemical signalling in insects. This is one advancing front in a vigorous and rapidly evolving field in which much of the current effort and progress involves diverse invertebrate species as well as vertebrate animals. It is useful, therefore, to begin by reviewing briefly not only the chemical messengers of the insect nervous system but also some principles of chemical signalling that have been uncovered in, and appear to apply to, widely different phyletic groups.

Chemical messengers in the nervous system

Nerve cells communicate with each other and with effector cells such as muscle fibres and gland cells primarily by means of the controlled release of chemical messengers. Information can be conveyed rapidly and specifically between cells, or even between particular parts of cells, by transmission at chemical synapses mediated by *neurotransmitters*. A neurotransmitter is a chemical messenger that is secreted into the intercellular cleft at a morphologically differentiated synapse between a neuron and its follower cell and then acts on a receptive region of the membrane of the postsynaptic cell. The functional specificity of synaptic chemical signalling depends on the distribution of the synapses made by the presynaptic cell.

Synaptic transmission is but one type, although the best understood, of chemical signalling systems and is related, on a continuum, to several other kinds of chemical signalling whose mechanistic and functional characteristics are of increasing interest. A relatively global and much slower and less transient dissemination of information, for example, results from the release of *neurohormones* by neurosecretory cells directly into the circulating blood. Often a neurohormone is released more focally in the vicinity of target cells and serves as a local neurohormone or a *neuromodulator*. Neuromodulators either change the physiological state, such as the spontaneous firing rate or the resting potential, of target cells or alter the functioning of target synapses.

Neurohormones, local neurohormones, and neuromodulators do not necessarily act at or by means of conventional synapses, nor is their chemical signalling confined specifically to an individual, anatomically apposed follower cell, as in the action of a neurotransmitter. Instead, neurosecretory nerve terminals, which may end quite far from target cells, release their chemical messengers into spacious domains. The functional specificity of this kind of chemical signalling depends on the anatomical distribution of cells with appropriate receptors. Finally, in this scheme of chemical signals one could include the extreme case of substances whose role may be primarily nutritive but at the same time can be classified as communicative. These are the *trophic factors* apparently passed between functionally related neurons and target cells and variously required for the survival, differentiation and normal maintenance of the cells involved.

Neurotransmitters and their receptors

At a chemical synapse, depolarization of the presynaptic nerve terminal upon arrival of a propagated action potential or an electrotonic potential triggers the release of a neurotransmitter. This substance diffuses within the synaptic cleft between pre- and postsynaptic cells and binds to specific receptors in the postsynaptic membrane. These receptors are generally believed to be membrane proteins or complexes of proteins and may also include complex lipids of the membrane, such as glycolipids. Combination of the neurotransmitter with the receptor leads to generation of a specific physiological response in the postsynaptic cell. This process clearly involves a 'recognition' component, in which the neurotransmitter binds to an appropriate receptor site, and an 'effector' component, through which the binding produces the physiological change in the cell. A challenge in receptor studies is to discover whether these functions are performed by a single membrane molecule or by a multimolecular receptor complex.

Postsynaptic responses fall into two general categories: 'fast' (milliseconds) and 'slow' (up to seconds or perhaps minutes). In fast responses the binding of the neurotransmitter to its receptor directly results in the opening or closing of selective ion-permeability channels in the membrane. The ions that move, the charges they carry, and the directions of their movement determine whether the postsynaptic effect of transmitter action is excitatory or inhibitory. The nicotinic acetylcholine (ACh) receptor is the prototypic fast receptor, and receptors for amino acids such as glutamate and γ-aminobutyric acid (GABA) also fall into this category.

Contrasting with this mechanism is the slow type of response, in which one or more metabolic steps are interposed between formation of the neurotransmitter–receptor complex and a physiological consequence such as the opening or closing of specific ion channels, an effect on ion pools within the cell or ion pumping across the cell membrane, or modification of the properties of membrane macromolecules. According to the prevailing model for such slow synaptic events, transmitter–receptor interaction leads to activation of an adenylate or guanylate cyclase and thereby to the production of cyclic AMP or cyclic GMP. These 'second messengers' can then activate specific protein kinases that may catalyse the phosphorylation of cell proteins, such as membrane proteins, and thereby give rise to relatively slowly developing and long-lasting postsynaptic effects. Among the increasing examples of substances that are believed to act through such slow mechanisms are ACh at muscarinic receptors and, in many cases, monoamines (such as catecholamines, phenolamines and indolalkylamines) and neuropeptides.

The action of the neurotransmitter is limited by one or more specific mechanisms that serve to remove or to inactivate the transmitter after it has been released into the synaptic cleft. Often the action of the transmitter is terminated by high affinity uptake into the presynaptic cell or into neighbouring glial or Schwann cells. This appears generally to be the case for amino acid and monoamine neurotransmitters. ACh, by contrast, is inactivated by hydrolysis catalysed by extracellular acetylcholinesterase in the synaptic cleft.

Neurotransmitter candidates

To prove that a particular substance is the neurotransmitter at a certain synapse, one must establish that application of the candidate to the postsynaptic membrane—in the same amount and with the same time course as when transmitter release is evoked by stimulation of the presynaptic neuron—elicits a postsynaptic response *identical* to that produced by normal synaptic action. This rigorous standard has not been fulfilled for any synapse, but it has been approached rather satisfactorily in a few cases, most notably the vertebrate skeletal neuromuscular junction (ACh) and the crustacean inhibitory neuromuscular junction (GABA). Thus it is not surprising that conclusive demonstration of a neurotransmitter role of any substance in the insect nervous system is lacking. Abundant evidence indicates, however, that many of the substances believed to function as neurotransmitters in vertebrates also do so in insects and other invertebrates.

The degree to which a candidate substance exhibits the several properties expected of a neurotransmitter can be assessed in insects by the available methods. For example, the presence of appreciable concentrations of various candidates in individual neurons or regions of the insect nervous system can be detected by microanalytical chemical techniques (including high performance liquid chromatography, radioenzymic assays, gas–liquid chromatography combined with mass spectroscopy, and radioimmunoassay) or by histochemical methods (including induced histofluorescence, radiochemical uptake labelling and autoradiography, and immunohistochemistry). Moreover, the presence and pharmacological properties of putative receptors for particular neurotransmitter candidates can be probed by ligand-binding techniques, neuropharmacological methods, and microiontophoresis with intracellular recording.

Among the familiar neurotransmitter candidates that have been found and studied in the insect nervous system are ACh, the amino acids GABA and glutamic acid, the catecholamine dopamine, the phenolamine octopamine, and the indolalkylamine serotonin (5-hydroxytryptamine). ACh is the predominant neurotransmitter in the insect central nervous system (CNS), where it appears to be the major or exclusive transmitter at synapses made by primary mechanosensory and chemosensory afferent axons and where it appears also to be used by some interneurons. Glutamic acid is believed to be the neurotransmitter at excitatory neuromuscular junctions and may also function at some central synapses. Inhibitory synapses in the CNS, like inhibitory neuromuscular junctions, are generally believed to use GABA as neurotransmitter.

In contrast to these transmitters at fast synapses, the monoamines serve more varied chemical signalling functions. Dopamine, octopamine and serotonin may all serve as neurotransmitters (mediating slow synaptic effects), neurohormones, or neuromodulators at different sites in insects. For example, dopamine is believed to be a neurotransmitter at synapses in salivary glands and a transmitter or neuromodulator in visceral muscle and in the CNS. Octopamine is a neuromodulator in skeletal muscle, while it may function as a neurotransmitter as well as a neuromodulator in the CNS and as a neurohormone governing carbohydrate metabolism in a variety of tissues. Finally, serotonin apparently is a neurohormone acting on salivary glands and may also serve as a neuromodulator in visceral muscle as well as a neurotransmitter or neuromodulator in the CNS.

In addition to these relatively well-established neurotransmitters (some of which play other chemical messenger roles as well) in insects, a variety of other candidates have emerged recently and merit attention. Among these substances, known to be present in insect nervous tissue and suspected of being chemical messengers, are histamine, tyramine, noradrenaline and a host of neuropeptides.

Neuropeptides

The recent discovery in vertebrates that many peptides may act as neurotransmitters or neuromodulators has provoked much interest and vigorous research activity. While it has been known for some time that insects possess neuropeptides such as proctolin and adipokinetic hormone, the contemporary explosion of interest in peptides in the nervous system has begun to be felt in insect research. In addition to proteins (such as prothoracicotropic hormone and bursicon) and peptides (such as proctolin, adipokinetic hormone, diuretic and antidiuretic hormones, and eclosion hormone) that were discovered in insects, a growing variety of substances, presumably peptides, with immunoreactivity like that of vertebrate neuropeptides, are known to be present in cells in the insect nervous system. Among these are materials resembling somatostatin, insulin, glucagon, gastrin, α-endorphin, Met- and Leu-enkephalins, substance P and vasopressin. It will be of great interest to learn whether these neuropeptides serve as neurotransmitters or modulators, or as neurohormones, in insects and whether their functions resemble those of neuropeptides in the vertebrate nervous system. Among other intriguing questions, one wonders whether these neuropeptides coexist with monoamines and even other chemical messengers in the same neurons, as has been observed repeatedly in vertebrates.

Epilogue

The anatomists and physiologists of the past who described and named the structures of the nervous system and began to explore their functions endowed them with a daunting aura of complexity. That aura has been penetrated successfully, however, through boldly integrative application of diverse research tools. Nothing short of a revolution has taken place in neuroanatomy as powerful methods such as intracellular staining, computer-aided graphics, axonal transport-mediated labelling, metabolic labelling with radioactive 2-deoxyglucose, and a host of other new techniques have permitted investigators to discern order in what had seemed to be the cellular chaos of the nervous system. Now, as we probe through the old aura of structural complexity, we unexpectedly find new molecular complexities welded to the fine cellular fabric of the nervous system.

In insects as in other animals there now appears to be a plethora of chemical messengers. Why are there so many and what are their different roles and mechanisms of action? By studying experimentally tractable nervous systems, including that of insects, one hopes not only to identify universal neurochemical mechanisms and to solve the riddle of the abundance of chemical messengers, but also to discern peculiarities or specializations that set one phyletic group apart from another. Most pharmacological agents that affect the nervous system, including many insecticides, do so by altering the mechanisms responsible for the biosynthesis, storage, release, or inactivation of chemical messengers such as neurotransmitters and neuromodulators, or by interfering with their interactions with receptors on their target cells. As our understanding of the biochemistry and functional roles of these chemical signalling systems in insects grows, it should be increasingly feasible to work toward new approaches to the chemical modification of insect behaviour and to selective destruction of harmful populations of insect pests. Progress in insect neuropharmacology should help in that effort by providing basic information about the differences between the chemical signalling systems of various insect groups and between insects and vertebrate animals.

Acetylcholine receptors of identified insect neurons

IAN D. HARROW, JONATHAN A. DAVID and DAVID B. SATTELLE

ARC Unit of Invertebrate Chemistry and Physiology, Department of Zoology, University of Cambridge, Downing Street, Cambridge, CB2 3EJ, UK

Abstract Insects provide material well suited to the investigation of the pharmacology of neuronal acetylcholine receptors. Current knowledge of the molecular pharmacology of cholinergic receptors has been enhanced by preparations that are amenable to both biochemical and electrophysiological analysis. Radiolabelled ligand binding, using ^{125}I-labelled α-bungarotoxin, and single-cell electrophysiological techniques have been used to study the properties of a putative nicotinic acetylcholine receptor in the central nervous system of the cockroach, *Periplaneta americana*. A role for this putative receptor in synaptic transmission between cercal afferents and giant interneuron 2 (GI2) in the terminal abdominal ganglion of the cockroach has been established. Because of the inaccessibility of these synaptic membranes, the pharmacological properties of acetylcholine receptors on the cell body membrane of GI2 have also been investigated. These findings have been compared to the results of pharmacological investigations of extra-synaptic acetylcholine receptors on the fast coxal depressor motor neuron (D_f) and dorsal unpaired median (DUM) neurons of the metathoracic ganglion of the same insect. Voltage-clamp techniques have been applied to the cell body membrane of D_f to investigate the ionic currents underlying the acetylcholine-induced depolarization. The differences in sensitivity to cholinergic receptor ligands noted for the neurons investigated has emphasized the need for pharmacological studies on identified cells and pathways in the insect central nervous system.

Acetylcholine (ACh) (Fig. 1) is considered to be an excitatory neurotransmitter at synapses in the insect central nervous system (CNS) and does not appear to be involved in insect neuromuscular transmission. Evidence for this has been summarized in several earlier reviews (Pitman 1971, Gerschenfeld 1973, Callec 1974, Sattelle 1977). The essential components of a cholinergic system have been demonstrated in the CNS of several insect species. For example, ACh, choline acetyltransferase (the acetylating enzyme catalysing the synthesis of ACh), and acetylcholinesterase (the hydrolytic enzyme which

---

1982 Neuropharmacology of insects. Pitman, London (Ciba Foundation symposium 88) p 12-31

terminates the synaptic actions of ACh) are all present in high concentrations (cf. Colhoun 1963). The capacity of insect nervous tissues to synthesize ACh has been clearly demonstrated (Prescott et al 1977). Finally, a high sensitivity of certain insect central neurons to either locally applied or bath-applied ACh has been reported after the inhibition of endogenous acetylcholinesterase (cf. Sattelle 1980).

Nevertheless, the technical difficulties of proving that a putative neurotransmitter exerts a functional role at a particular central synapse are acute. So far, the criteria listed below (cf. Gerschenfeld 1973), which must be satisfied to establish the identity of a neurotransmitter, have not been fulfilled for any particular insect central pathway.

Presynaptic criteria:
(a) The presence and storage of the neurotransmitter candidate and its precursors.
(b) The presence of the synthesizing enzyme and the capacity to synthesize the neurotransmitter candidate from precursors.
(c) Release and collectability of the neurotransmitter candidate when the presynaptic neurons are stimulated.

Postsynaptic criteria:
(d) The presence of an inactivating mechanism.
(e) The ability of the transmitter candidate to mimic the actions of the natural transmitter.
(f) Modification of the actions of the putative transmitter by pharmacological agents should closely resemble the effects on the natural transmitter of the same chemicals.

However, detailed studies on the antennal (olfactory) afferent input to the antennal lobes of the moth *Manduca sexta* (Sanes & Hildebrand 1976, Sanes et al 1977) and the cercal (mechanoreceptor) afferent input to the giant interganglionic interneurons of the abdominal ganglion of the cockroach *Periplaneta americana* (cf. Callec 1974, Sattelle 1980) have in each case fulfilled several of these criteria, leading to the hypothesis that ACh is a major sensory neurotransmitter in the insect CNS. At cholinergic synapses, in response to depolarization of the nerve terminal, ACh molecules are released into the synaptic cleft between the pre- and postsynaptic cells. As a consequence of the interactions of ACh with specific receptor (recognition) sites on the postsynaptic cell, changes in ionic permeability of the postsynaptic membrane are induced. The functional unit that accounts for this regulation of ionic permeability—the ACh receptor—therefore comprises at least two categories of sites (cf. Heidmann & Changeux 1978): (a) the recognition site which binds ACh and related compounds; (b) the ionophore or site of ion translocation. ACh can act on more than one specific recognition site (Dale 1937) and different types of ACh receptors can be distinguished using nicotine and muscarine (see Fig. 1), two pharmacological agents that activate the receptors (agonists). Receptor-active ligands that block ACh receptor-controlled responses (antagonists) enable further characterization of ACh receptors to be made. For example, electrophysiological studies on vertebrates have shown that *d*-tubocurarine is a more effective antagonist than atropine at nicotinic receptors, whereas the reverse is true at muscarinic receptors. It has also been shown that hexamethonium is a potent antagonist at the ganglionic nicotinic receptors of vertebrates whereas

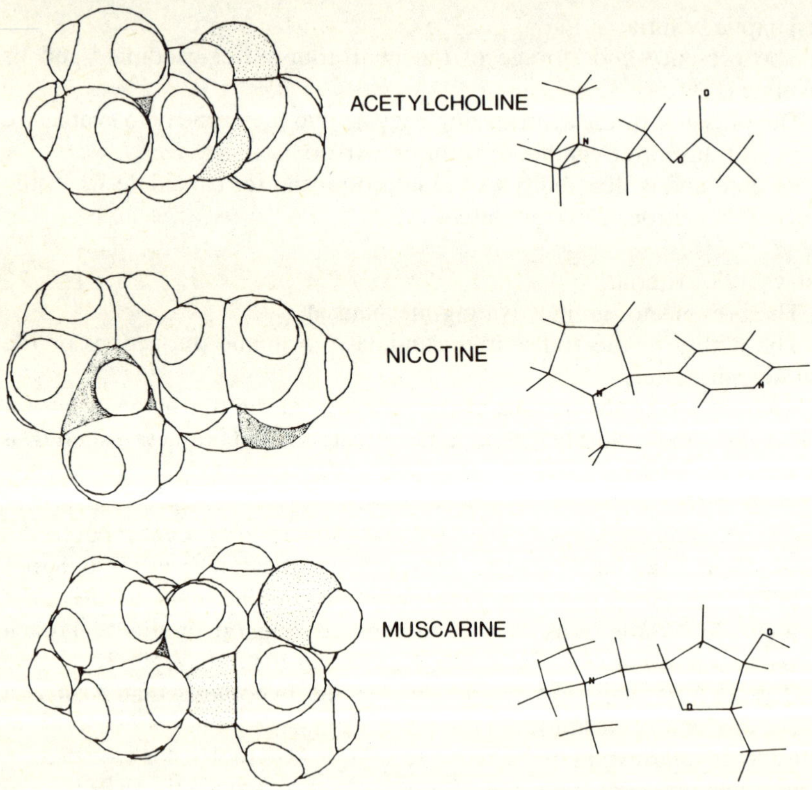

FIG. 1. Molecules of acetylcholine, nicotine and muscarine: 'space-filling' models (left-hand column) and models showing a perspective representation of the orientation of bonds, with the line lengths proportional to interatomic distances (right-hand column). In the space-filling representation, lightly shaded areas indicate the positions of oxygen atoms; densely shaded areas show the positions of nitrogen atoms. Carbŏn atoms with hydrogens attached are also shown (unshaded).

decamethonium and α-bungarotoxin (α-BGTX), effective blocking agents at the vertebrate neuromuscular junction, are less effective at vertebrate ganglionic nicotinic receptors (cf. Triggle & Triggle 1976).

Recently the development of receptor-specific ligands, radiolabelled to a high specific activity, has enabled binding to receptor sites of vertebrates and invertebrates to be studied directly rather than simply being inferred from the pharmacological responses of the tissues (Yamamura et al 1978). By means of this experimental approach the following three putative ACh receptors, each with a distinct pharmacological specificity, have been characterized in extracts prepared from insect central nervous tissues (cf. Sattelle 1980 for review):

(a) A ^{125}I-labelled α-bungarotoxin-binding component, located in low speed (~20–40 000 × *g*) particulate extracts, with the properties of a nicotinic ACh receptor (Schmidt-Nielsen et al 1977, Dudai 1978, Rudloff 1978).
(d) A [^{3}H]quinuclidinyl benzilate-binding component, also located in low speed particulate extracts, with the properties of a muscarinic ACh receptor (Haim et al 1979).
(c) A [^{3}H]decamethonium-binding component, present in high speed (100 000 × *g*) supernatant fractions, exhibiting a mixed pharmacological specificity (Jewess et al 1975, Mansour et al 1977).

The adoption of an integrated experimental approach has proved invaluable in obtaining an understanding of many central nervous phenomena. The work reported here incorporates anatomical, physiological, biochemical and pharmacological techniques in an attempt to characterize the properties and functions of an insect nicotinic ACh receptor. In addition, preliminary attempts have been made to discover whether a functional role can be attributed to putative ACh receptors other than the nicotinic receptor. The CNS of the cockroach *Periplaneta americana* provides experimental material well suited for such investigations. Material in sufficient quantities for radiolabelled ligand-binding studies can readily be obtained and single-cell electrophysiological recording techniques can be applied to investigations of the cholinergic pharmacology of synaptic and extrasynaptic sites on identifiable insect neurons. The cholinergic receptors of insects are of interest not only from a comparative and evolutionary standpoint but also because they provide unique experimental material for the study of genetic (Hall 1980) and developmental (Hildebrand 1980) aspects of receptor biology. In addition there is evidence that certain insecticides may act at insect ACh receptors (see Eldefrawi 1976, Sattelle 1980).

Acetylcholine receptors at synapses in the insect central nervous system

A specific, saturable component of ^{125}I-labelled α-bungarotoxin binding in CNS extracts

Following the demonstration by Hall & Teng (1975) that ^{125}I-labelled α-bungarotoxin (^{125}I-α-BGTX) bound in a non-uniform way to frozen sections of the fruitfly *Drosophila melanogaster*, several laboratories began investigations on the toxin-binding component. Schmidt-Nielsen et al (1977) found that after filtration to remove chitinous material and a low speed (2000 × ***g***) spin to remove pigment granules and other debris, the bulk of the toxin-binding activity remained in the 2000 × ***g*** supernatant fraction. After centrifugation at 20 000 × ***g*** the resultant pellet, which contained the bulk of the toxin-binding sites, was used in subsequent studies. In the same investigation binding was shown to saturate with increasing concentrations of ^{125}I-α-BGTX. By preincubating extracts with different concentrations of test ligands the effectiveness of cholinergic ligands in inhibiting the specific binding of ^{125}I-α-BGTX was tested. Schmidt-Nielsen et al (1977) showed that nicotine, *d*-tubocurarine and ACh were particularly strong inhibitors of α-bungarotoxin binding, whereas drugs such as D,L-muscarine, atropine and pilocarpine only weakly inhibited toxin binding. Similar findings were also noted for *Drosophila melanogaster* by other laboratories (Dudai 1978, Rudloff 1978), so that for this insect there is considerable evidence for the existence of a putative ACh receptor with a nicotinic pharmacological profile. A putative cholinergic receptor with comparable pharmacological properties has been reported in head extracts of the housefly *Musca domestica* (Dudai 1977, Harris et al 1979) and brain extracts of a moth *Manduca sexta* (Sanes et al 1977).

A binding site for ^{125}I-labelled α-bungarotoxin in abdominal nerve cord extracts of Periplaneta americana

The receptor probe ^{125}I-labelled α-bungarotoxin was used to study putative nicotinic receptors in abdominal nerve cord extracts of the cockroach *Periplaneta americana* (Gepner et al 1978). Background binding, determined as the amount of ^{125}I-α-BGTX bound to heat-inactivated extract, was routinely subtracted. Similar levels of background binding were noted in the presence of 1.0×10^{-3} M-nicotine. 100% binding was defined as the amount of ^{125}I-α-BGTX bound in the absence of added ligand. In this way it was shown that nicotine was a very effective inhibitor of toxin binding (Fig. 2). The ID_{50} concentrations for nicotine were estimated to be 3.0×10^{-6} M

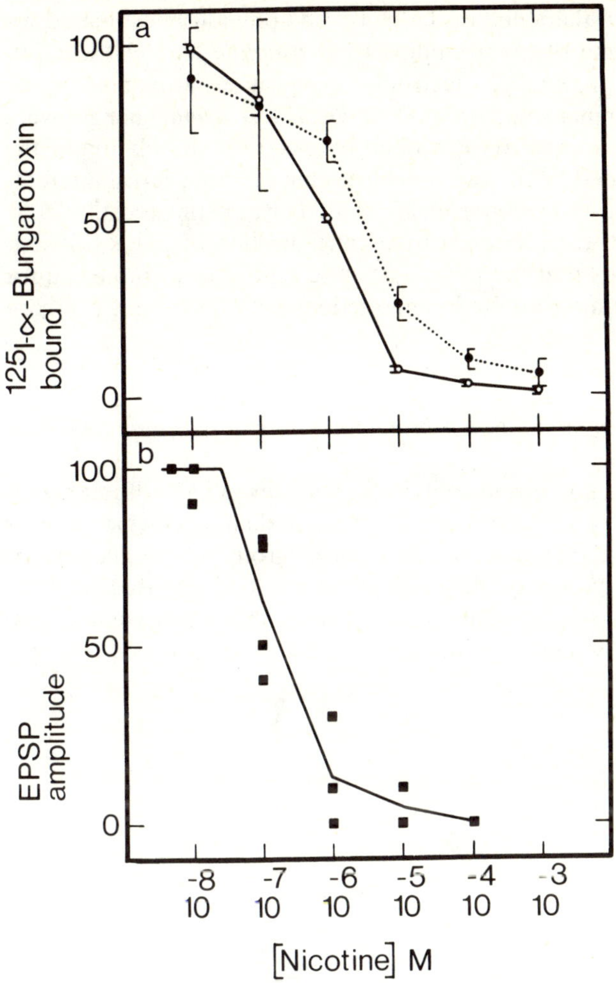

FIG. 2 (a) Concentration dependence of inhibition by nicotine of ^{125}I-labelled α-bungarotoxin binding to extracts of whole flies (*D. melanogaster*) and cockroach abdominal nerve cords (*P. americana*). Results are expressed as a percentage of toxin bound to extract in the absence of nicotine. Concentrations for 50% inhibition of toxin binding are estimated to be 1.0×10^{-6} M (*Drosophila*, ○) and 3.0×10^{-6} M (*Periplaneta*, ●). Each point is the average of three replicates; vertical bars denote 1 SD. (b) Effects of various concentrations of nicotine on the amplitude of the evoked EPSP recorded from cercal-afferent, giant-fibre synapses in the terminal abdominal ganglion of *P. americana* using the sucrose-gap technique (Callec & Sattelle 1973). Concentration for 50% suppression of the EPSP is estimated to be 2.0×10^{-7} M. (Reprinted from Gepner et al 1978, by permission of Macmillan Journals Ltd.)

(*Periplaneta*) and 1.0×10^{-6} M (*Drosophila*). This value for *Drosophila* extract agreed well with the value of 0.5×10^{-6} M previously estimated for the inhibition by nicotine of toxin binding to *Drosophila* extract (Schmidt-Nielsen et al 1977). In parallel electrophysiological experiments on the desheathed sixth abdominal ganglion of *Periplaneta americana,* changes were noted in the amplitude of the EPSP evoked by electrical stimulation of the cercal nerve, and recorded by the sucrose-gap technique from cercal afferent, giant interneuron synapses (Gepner et al 1978). It was estimated that 50% inhibition of the EPSP was produced by a concentration of 2.0×10^{-7} M-nicotine. Thus it appears that the potent synaptic actions of nicotine can for the most part be accounted for by its interactions with an α-bungarotoxin-sensitive ACh receptor.

A role in synaptic transmission for an α-bungarotoxin-binding component

The CNS of *Periplaneta americana* contains at least seven pairs of interganglionic interneurons with giant axons (up to 50 μm in diameter) which appear in sections of the abdominal nerve cord as a ventral group of four giant axons (GI1–4) and a dorsal group of three giant axons (GI5–7). Recently, by back-filling with cobalt single axons dissected from abdominal connectives, selective staining of individual giant interneurons has been achieved (Harrow

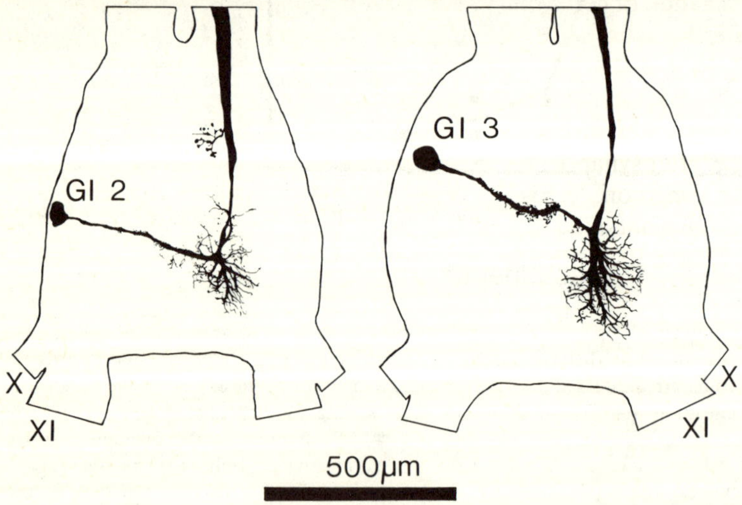

FIG. 3. Morphology of two giant interneurons (GI2 and GI3) of the cockroach (*Periplaneta americana*). Camera lucida representations of silver-intensified, cobalt back-filled, single, giant interneurons in the sixth abdominal ganglion. (Modified from Harrow et al 1980.) The cercal afferent input to both cells via nerve XI is blocked by α-bungarotoxin—see text.

et al 1979, 1980). This work has shown that each of the cockroach giant interneurons GI1–3 has unique morphological features (see Fig. 3 for examples). Cell body position, neurite shape, dendritic branching pattern and the presence or absence of an axon collateral provide anatomical criteria for the identification of these neurons. They receive excitatory monosynaptic input from mechanoreceptor afferents via cercal nerve XI and are highly sensitive to ACh (Callec 1974, Sattelle et al 1976, Sattelle 1978).

The actions of α-BGTX on GI2 and GI3 (Fig. 3) have been investigated in some detail, using the oil-gap, single-fibre technique (Harrow et al 1979, Sattelle et al 1980). By this means, unitary excitatory postsynaptic potentials (EPSPs) corresponding to the activity of a single afferent sensory neuron were recorded in response to mechanical stimulation of a single cercal mechanoreceptor hair (see Fig. 4 for the results of an experiment on GI2). During the same experiment composite EPSPs were also recorded in response to electrical stimulation of many afferents in nerve XI. Bath application of micromolar concentrations of α-BGTX to the desheathed sixth abdominal ganglion resulted in a progressive reduction in the amplitude of the unitary EPSP (Fig. 4). Complete block of the unitary EPSP was noted after a 50-minute exposure to 1.0×10^{-6} M-α-BGTX. No significant change in membrane potential of the giant interneuron was detected during synaptic blockade.

The composite EPSP recorded in the same experiment in response to direct electrical stimulation of cercal nerve XI was also completely blocked by 1.0×10^{-6} M-α-bungarotoxin. After 65 minutes of exposure to toxin, despite increasing the stimulus intensity to recruit further cercal afferent input, no excitatory postsynaptic response could be recorded from the giant interneuron under test. Rebathing the sixth abdominal ganglion in normal saline for 150 minutes after synaptic block by α-BGTX did not result in any recovery of synaptic transmission. Action potentials elicited by direct stimulation of the portion of the giant axon in the sixth abdominal ganglion were unaffected by 1.0×10^{-6} M-α-BGTX. The single-fibre axon back-filling procedure (Harrow et al 1980) was used to establish the identity of the cell under test, giant interneuron 2 (G12), using the morphological criteria outlined above. The time required to completely block synaptic transmission between cercal afferents and GI2 depended on the concentration of α-BGTX perfused over the desheathed sixth abdominal ganglion. For example, concentrations as low as 1.0×10^{-9} M were effective in blocking transmission, provided that the toxin was applied for 200 minutes. Thus nanomolar concentrations of this nicotinic ACh receptor probe are able to block transmission at synapses between cercal afferents and giant interneurons, indicating that the α-BGTX-binding component is a constituent of an ACh receptor that is functional in this monosynaptic pathway.

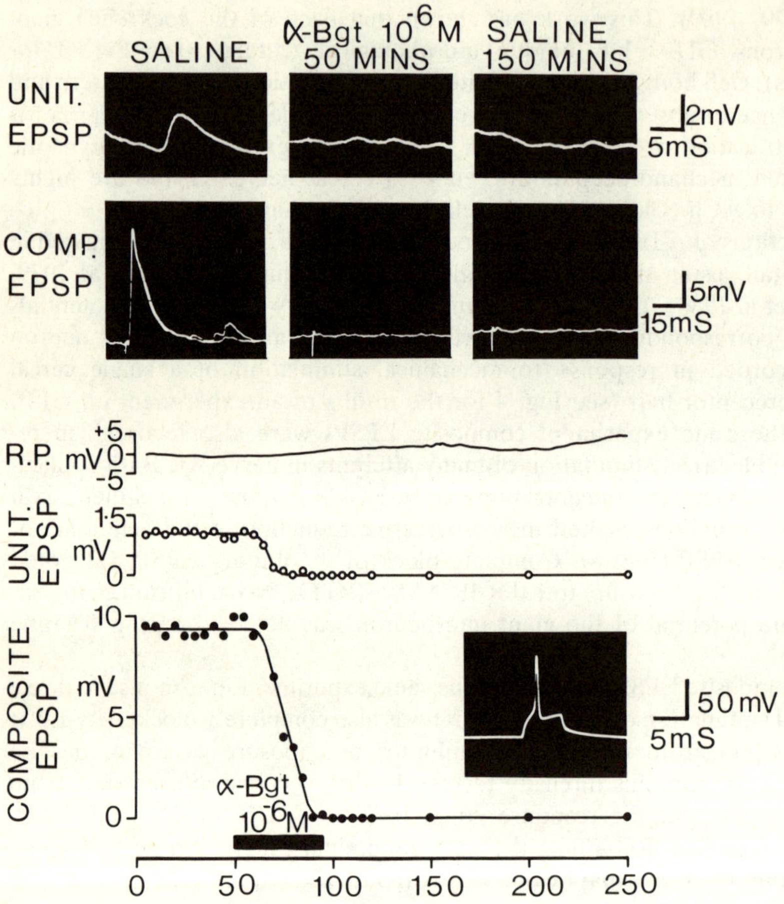

FIG. 4. Actions of α-bungarotoxin (1.0×10^{-6} M) on the cercal afferent input to giant interneuron 2 (GI2). Both unitary and composite excitatory postsynaptic potentials (EPSPs) were blocked by superfusing the desheathed sixth abdominal ganglion with a saline solution containing 1.0×10^{-6} M-α-BGTX. No significant change in the resting potential (R. P.) was noted during the actions of α-BGTX and normal action potentials (see inset) were recorded after complete block of EPSPs.

Extrasynaptic acetylcholine receptors of identified neurons

Limitations of the sucrose-gap method of studying synaptic transmission preclude detailed quantitative pharmacological studies with this technique. For example, the preparation used in the studies on nicotine provides an averaged response derived from a population of giant interneurons and any differences in the receptors between individual cells could be distorted. A further shortcoming of investigations of receptor properties by means of bath application of ligands, that is equally applicable to oil-gap, single-fibre studies, is the possibility of presynaptic or other indirect effects of the drug under test. To overcome some of these limitations we have studied the ability of key antagonists to modify the local (ionophoretic) application of ACh to the cell body (extrasynaptic) membranes of identified insect central neurons.

As shown in Fig. 4, cercal afferent inputs to GI2 in the sixth abdominal ganglion of the cockroach were blocked by micromolar α-BGTX. Prolonged exposures to toxin concentrations as low as nanomolar were also effective (Sattelle et al 1980). In studies on the extrasynaptic receptors of the cell body membrane of GI2 in desheathed ganglia it was demonstrated that the depolarizing response to ionophoretically applied ACh was also blocked by submicromolar concentrations of α-BGTX (Harrow 1981). By contrast, micromolar concentrations of the muscarinic cholinergic antagonist quinuclidinyl benzilate (QNB) were ineffective (Harrow 1981). Thus of these two highly specific cholinergic receptor probes, the nicotinic antagonist was the more effective in blocking the depolarization induced by ionophoretic application of ACh.

The cell body membrane of the fast coxal depressor motor neuron (D_f)

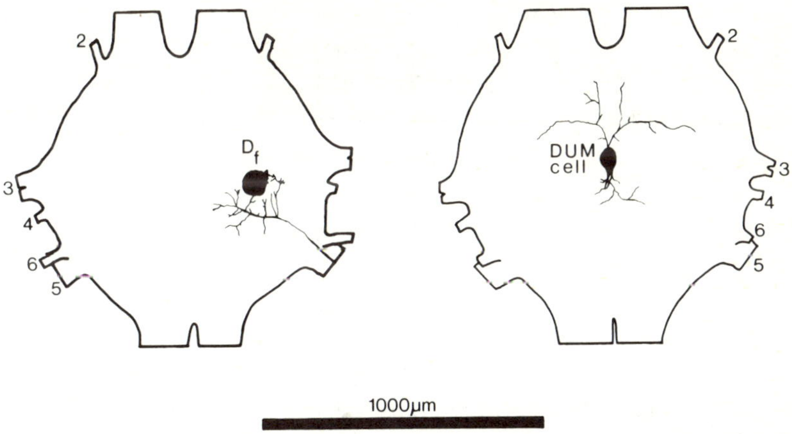

FIG. 5. Morphology of the fast coxal depressor motor neuron (D_f) and an unidentified dorsal unpaired median (DUM) neuron of the metathoracic ganglion of *Periplaneta americana*. Camera lucida representations of cells injected with cobalt via a microelectrode located in the cell body. The nerve trunks are enumerated.

provides an even more accessible cell body membrane (Fig. 5). Situated on the ventral surface of the cockroach metathoracic ganglion, the cell body can be located by visual inspection. Recently, David & Pitman (1982) reported the relatively high sensitivity of this cell body membrane to ionophoretically applied ACh. When α-BGTX at concentrations of 1.0×10^{-8} M and higher was applied to desheathed metathoracic ganglia, the sensitivity of the cell body membrane to ionophoresis of ACh from a micropipette located close to the cell body membrane was reduced (Fig. 6). Complete block of the

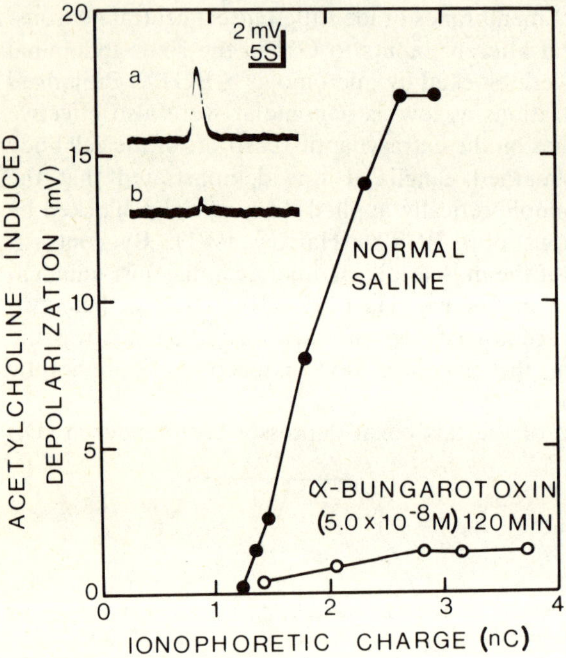

FIG. 6. Actions of α-bungarotoxin (5.0×10^{-8} M) on the depolarization induced by ionophoretic application of ACh to the cell body membrane of the fast coxal depressor motor neuron (D_f) in a desheathed metathoracic ganglion. Dose–response curves are shown in normal saline and after a prolonged exposure to α-BGTX. Insets show responses to ACh: (a) in normal saline and (b) after α-BGTX treatment (5.0×10^{-8} M for 120 min).

transient ACh-induced depolarization was achieved at a concentration of 5.0×10^{-8} M after exposure to the toxin for 120 minutes. By contrast, QNB was without effect on the ACh response at concentrations of 1.0×10^{-5} M (60-minute exposure). At higher concentrations of QNB a significant block was observed, and above 1.0×10^{-3} M complete block of the ACh-induced depolarization was noted (David & Sattelle 1982). Decamethonium was without effect on the local application of ACh to the cell body membrane of D_f after a 60-minute exposure at concentrations up to 1.0×10^{-3} M. At higher concentrations a blocking action was noted (David & Sattelle 1982).

Thus of the three molecular probes used so far in binding studies to characterize putative cholinergic receptors in insects, α-BGTX was the most effective on the ACh receptors of the cell body membrane of D_f. Quinuclidinyl benzilate was three orders of magnitude less effective and decamethonium was five orders of magnitude less effective than α-BGTX. As noted for GI2, nicotinic cholinergic receptors are clearly present on the extrasynaptic (cell body) membrane of D_f. The extremely high concentrations required for an action on D_f of the other receptor probes counsels caution against concluding that either the muscarinic receptor or the decamethonium-binding receptor are present on the cell body membrane of this cell.

One further group of insect central neurons has been investigated in our studies on the cholinergic pharmacology of identified insect neurons. These are the dorsal unpaired median (DUM) cells of the metathoracic ganglion (see Fig. 5). As with GI2 and D_f, when ACh was ionophoretically applied through a microelectrode to the cell body membrane in a desheathed ganglion, a transient depolarizing response ensued (Sattelle et al 1980). Micromolar concentrations of α-BGTX were completely ineffective on these cells, although at higher concentrations (1.0×10^{-5} M) α-BGTX did reduce the sensitivity of DUM neurons to ACh (Sattelle et al 1980). Further studies are needed to explore the basis of these strikingly different sensitivities to α-BGTX between D_f and DUM neurons.

It has been established, therefore, that nicotinic ACh receptors are present both on certain interneurons (GI2, GI3) and on a motor neuron (D_f) (see Table 1). Although the pharmacological profile of the recognition site of this receptor is becoming more clearly defined, less is known of the properties of the receptor ionophore.

Ionic currents induced by acetylcholine in an identified insect neuron

It was recently shown by David & Pitman (1982), working on the cockroach motor neuron (D_f), that the depolarization induced by ionophoretic application of ACh was substantially reduced by a 10-fold reduction in the external sodium ion concentration. Also, the reversal potential of the ACh response (−34 mV), noted in the same study, suggested that sodium ions carry at least part of the ionic current of the ACh response. We have, therefore, applied voltage-clamp techniques to the cell body membrane of D_f in an attempt to measure directly the ionic basis of the currents induced by ACh. The results of such an experiment are shown in Fig. 7. The inward current induced by ACh was practically abolished when the external sodium ions were replaced by either choline or Tris. Thus there is now direct evidence of a role for sodium ions in the ACh-induced potential changes of the fast coxal depressor motor neuron of the cockroach.

TABLE 1 **Actions on insect central neurons of cholinergic ligands used to characterize putative receptors in binding studies**

Neuron	*Concentration (M) required to completely block response (EPSP or ionophoretically induced ACh depolarization)*			
	α-Bungarotoxin	*Quinuclidinyl benzilate*	*Decamethonium*	*Reference*
G12				
synaptic receptors	1.0×10^{-9}	$>1.0 \times 10^{-5}$	Not tested	Sattelle et al 1980 Sattelle, Harrow, Hue, Pelhate, Gepner & Hall (unpublished observations)
extrasynaptic (cell body) receptors	1.0×10^{-7}	No effect at 1.0×10^{-5}	Not tested	Harrow 1981
G13				
synaptic receptors	5.0×10^{-8}	Not tested	Not tested	Harrow et al 1979
D_f				
extrasynaptic (cell body) receptors	1.0×10^{-7}	1.0×10^{-4}	1.0×10^{-2}	David & Sattelle 1982

FIG. 7. Voltage-clamp traces showing currents induced by acetylcholine (ionophoretically applied) on the cell body membrane of the fast coxal depressor motor neuron (D_f) in a desheathed metathoracic ganglion. The effects on the current of replacing normal saline by a sodium-free saline are illustrated. E_m, −60 mV.

Conclusions

The central nervous system of the cockroach (*Periplaneta americana*) has proved amenable to both radiolabelled ligand-binding studies and single-cell electrophysiology of identifiable neurons. Binding studies have enabled the characterization *in vitro* of three putative ACh recognition sites, each with a distinct pharmacological specificity. A physiological role has been established for only one of these putative receptors—the ^{125}I-labelled α-BGTX-binding component. There is strong evidence that this binding site is a constituent of ACh receptors at synapses between cercal mechanoreceptor afferent neurons and giant interneurons (GI2 and GI3) in the sixth abdominal ganglion. This binding site also appears to be a constituent of extrasynaptic receptors on the cell body membranes of GI2 and the fast coxal depressor motor neuron.

Pharmacological differences have emerged between the insect nicotinic cholinergic receptor and the nicotinic receptor of vertebrate muscle and electroplax tissue. The insect receptor is for example much less sensitive to decamethonium and carbamylcholine. Also, the receptors mediating cercal-afferent, giant interneuron synaptic transmission are sensitive to α-BGTX, whereas the nicotinic receptors of vertebrate autonomic ganglia are not (Brown & Fumagalli 1977). Some similarities are noted between the properties of this insect receptor and the nicotinic receptors of the tectum of vertebrates (Barnard et al 1979).

A series of isothiocyanate choline acetyltransferase inhibitors designed as potential insecticides are also active on the insect nicotinic receptor (Sattelle & Callec 1977) and a good agreement has been shown between their ability to modify synaptic transmission and their inhibition of ^{125}I-labelled α-BGTX binding (Gepner et al 1978). Thus the nicotinic ACh receptors of the CNS of insects can be conveniently studied using identifiable neurons, thereby facilitating quantitative pharmacological investigations. In addition, they represent a well-characterized potential target site for the design of new insecticidally active molecules.

Acknowledgements

The authors are deeply indebted to their colleagues J.-J. Callec, J. I. Gepner, L. M. Hall, B. Hue and M. Pelhate, all of whom have contributed to aspects of the work reviewed in this paper. We are particularly grateful to Dr J. E. Treherne for his encouragement of this work.

REFERENCES

Barnard EA, Dolly JO, Lang B, Lo M, Shorr R G 1979 Application of specifically acting toxins to the detection of functional components common to peripheral and central synapses. Adv Cytopharmacol 13:409-435

Brown DA, Fumagalli L 1977 Dissociation of α-bungarotoxin binding and receptor block in the rat superior cervical ganglion. Brain Res 129:165-168

Callec J-J 1974 Synaptic transmission in the central nervous system of insects. In: Treherne JE (ed) Insect neurobiology. North-Holland Biomedical Press/American Elsevier, Amsterdam & New York, p 119-178

Callec J-J, Sattelle DB 1973 A simple technique for monitoring the synaptic actions of pharmacological agents. J Exp Biol 59:725-738

Colhoun EH 1963 The physiological significance of acetylcholine in insects and observations upon other pharmacologically active substances. Adv Insect Physiol 1:1-45

Dale HH 1937 Acetylcholine as a chemical transmitter of the effects of nerve impulses. J Mt Sinai Hosp 4:416-429

David JA, Pitman RM 1982 The effects of axotomy upon the extrasynaptic sensitivity of an identified motoneurone in the cockroach *Periplaneta americana*. J Exp Biol, in press

David JA, Sattelle DB 1982 Actions of cholinergic ligands on the cell body membrane of the fast coxal depressor motoneurone of the cockroach *Periplaneta americana*. J Exp Biol, in press

Dudai Y 1977 Demonstration of an α-bungarotoxin-binding nicotinic receptor in flies. FEBS (Fed Eur Biochem Soc) Lett 76:211-213

Dudai Y 1978 Properties of an α-bungarotoxin-binding cholinergic nicotinic receptor from *Drosophila melanogaster*. Biochim Biophys Acta 539:505-517

Eldefrawi AT 1976 The acetylcholine receptor and its interaction with insecticides. In: Wilkinson CF (ed) Insecticide biochemistry and physiology. Plenum Press, New York, p 297–326

Gepner JI, Hall LM, Sattelle DB 1978 Insect acetylcholine receptors as a site of insecticide action. Nature (Lond) 276:188-190

Gerschenfeld HM 1973 Chemical transmission in invertebrate central nervous systems and neuromuscular junctions. Physiol Rev 53:1-119

Haim N, Nahum S, Dudai Y 1979 Properties of a putative muscarinic cholinergic receptor from *Drosophila melanogaster*. J Neurochem 32:543-552

Hall LM 1980 Biochemical and genetic analysis of an α-bungarotoxin-binding receptor from *Drosophila melanogaster*. In: Sattelle DB et al (eds) Receptors for neurotransmitters, hormones and pheromones in insects. Elsevier/North-Holland Biomedical Press, Amsterdam, p 111-124

Hall LM, Teng NNH 1975 In: McMahon D, Fox CF (eds) Developmental biology—pattern formation—gene regulation. Benjamin, Menlo Park, California (ICN–UCLA Symp Mol Cell Biol vol 2) p 282-289

Harris R, Cattell KJ, Donnellan JF 1979 Identification of a putative nicotinic acetylcholine receptor in fractions from housefly brain. Biochem Soc Trans 7:136-138

Harrow ID 1981 Some cholinergic properties of an identified neuronal pathway. PhD thesis, University of Cambridge

Harrow ID, Hue B, Pelhate M, Sattelle DB 1979 α-Bungarotoxin blocks excitatory postsynaptic potentials in an identified insect interneurone. J Physiol (Lond) 295:63P

Harrow ID, Hue B, Pelhate M, Sattelle DB 1980 Cockroach giant interneurones stained by cobalt-backfilling of dissected axons. J Exp Biol 84:341-343

Heidmann T, Changeux J-P 1978 Structural and functional properties of the acetylcholine receptor protein in its purified and membrane-bound states. Annu Rev Biochem 47:317-357

Hildebrand JG 1980 Development of putative acetylcholine receptors in normal and deafferented antennal lobes during the metamorphosis of *Manduca sexta*. In: Sattelle DB et al (eds) Receptors for neurotransmitters, hormones and pheromones in insects. Elsevier/North-Holland Biomedical Press, Amsterdam, p 209-220

Jewess PJ, Clarke BS, Donnellan JF 1975 Isolation of a housefly head protein fraction that exhibits high affinity binding of cholinergic ligands. Croat Chem Acta 47:459-464

Mansour NA, Eldefrawi ME, Eldefrawi AT 1977 Isolation of putative acetylcholine receptor proteins from housefly brain. Biochemistry 16:4126-4132

Pitman R M 1971 Transmitter substances in insects: a review. Comp Gen Pharmacol 2:347-371

Prescott DJ, Hildebrand JG, Sanes JR, Jewett S 1977 Biochemical and developmental studies of acetylcholine metabolism in the central nervous system of the moth, *Manduca sexta*. Comp Biochem Physiol 56C:77-84

Rudloff E 1978 Acetylcholine receptors in the central nervous system of *Drosophila melanogaster*. Exp Cell Res 111:185-190

Sanes JR, Hildebrand JG 1976 Acetylcholine and its metabolic enzymes in developing antennae of the moth, *Manduca sexta*. Dev Biol 52:105-120

Sanes JR, Prescott DJ, Hildebrand JG 1977 Cholinergic neurochemical development of normal and deafferented antennal lobes during metamorphosis of the moth *Manduca sexta*. Brain Res 119:389-402

Sattelle DB 1977 Cholinergic synaptic transmission in invertebrate central nervous systems. Trans Biochem Soc 5:849-852

Sattelle DB 1978 The insect central nervous system as a site of action of neurotoxicants. In: Shankland DL et al (eds) Pesticide and venom neurotoxicity. Plenum Press, New York, p 7-26

Sattelle DB 1980 Acetylcholine receptors of insects. Adv Insect Physiol 15:215-315

Sattelle DB, Callec J-J 1977 Actions of isothiocyanates on the central nervous system of *Periplaneta americana*. Pestic Sci 8:735-747

Sattelle DB, McClay AS, Dowson RJ, Callec J-J 1976 The pharmacology of an insect ganglion: actions of carbamylcholine and acetylcholine. J Exp Biol 64:13-23

Sattelle DB, David JA, Harrow ID, Hue B 1980 Actions of α-bungarotoxin on identified insect central neurones. In: Sattelle DB et al (eds) Receptors for neurotransmitters, hormones and pheromones in insects. Elsevier/North-Holland Biomedical Press, Amsterdam, p 125-139

Schmidt-Nielsen BK, Gepner JI, Teng NNH, Hall LM 1977 Characterization of an α-bungarotoxin binding component from *Drosophila melanogaster*. J Neurochem 29:1013-1029

Triggle DJ, Triggle CR 1976 Chemical pharmacology of the synapse. Academic Press, London

Yamamura HI, Enna SJ, Kuhar MJ (eds) 1978 Neurotransmitter receptor binding. Raven Press, New York

DISCUSSION

Hildebrand: You mentioned that the binding sites for α-bungarotoxin hadn't yet been located ultrastructurally. Have the cellular locations of any putative neurotransmitter receptors, including these cholinergic receptors, been successfully examined by electron microscopy?

Sattelle: Using a biotin–α-bungarotoxin:avidin–ferritin labelling technique, Dr N. J. Lane has recently located α-bungarotoxin-binding sites at synapses in the neuropil of cockroach central ganglia. It has also been demonstrated that the biotin–α-bungarotoxin complex retains the ability to block physiological responses to α-BGTX (N. J. Lane et al, unpublished observations).

Dudai: Mecamylamine blocked the ACh response in the fast coxal depressor cell (D_f); yet mecamylamine has no effect on BGTX-binding sites in *Drosophila*, according to our studies (Dudai 1978). Does mecamylamine do anything in the dorsal unpaired median (DUM) cells that are not responsive to α-BGTX?

Sattelle: We have not tried that experiment, but we attempted to protect the extrasynaptic ACh receptors against the actions of α-bungarotoxin by pretreating the fast coxal depressor motor neuron (D_f) with reversible antagonists (J. A. David & D. B. Sattelle, unpublished). We first blocked the depolarization induced by ionophoretic application of ACh using the bath-application of either cobratoxin or mecamylamine. Then, still in the presence of the reversible antagonist, α-BGTX was bath-applied to the preparation for a period in excess of that required to completely block the ACh response. Both compounds were then washed off. We recover the ACh response in experiments involving cobratoxin but not when mecamylamine is used. So it sounds as if the physiological results are consistent with the binding results.

Olsen: Were you implying that the D_f cells differ pharmacologically from the giant interneurons? Was mecamylamine more potent on D_f and *d*-tubocurarine less potent?

Sattelle: Cell bodies of giant interneuron 2 (GI2) are more difficult to work on routinely, so most of the pharmacology has been established using D_f.

We haven't tried mecamylamine on the giant interneurons. *d*-Tubocurarine and atropine are about equally potent on the extrasynaptic receptors on D_f. Curare seems to be more potent at the cercal afferent, giant interneuron synapses, but possible presynaptic effects cannot be ruled out.

Olsen: But are you implying that a curare-sensitive, α-BGTX-insensitive receptor and an α-BGTX-sensitive, curare-insensitive receptor both exist?

Sattelle: It would be premature to make any such conclusion at this stage.

A. T. Eldefrawi: The correspondence between the T_{50} values of various cholinergic drugs that inhibit α-BGTX binding and the results from physiological experiments on the GI cells is much better with *Drosophila* than with the cockroach. Could this be related to the extrasynaptic receptors? When you look at binding in the cockroach you are really looking at a combination of synaptic and extrasynaptic components, whereas in *Drosophila* they may be mostly synaptic components with a high affinity.

Hall: In the saturation binding experiments done in my laboratory on the cockroach extracts there was a straight-line Scatchard plot, so there was no evidence for multiple classes of binding components. But since there is so much of the extrasynaptic receptor in the cockroach, the binding experiments might be looking at the extrasynaptic receptor and not seeing the relatively small amounts of the synaptic receptor.

Sattelle: I am not sure that the differences referred to by Professor Eldefrawi between *Drosophila* and the cockroach are really as significant as that.

Narahashi: There are nicotinic receptors which are not particularly sensitive to α-BGTX in vertebrate sympathetic ganglion cells. Can you say whether there are typical nicotinic and muscarinic receptors in the insect CNS preparations?

Sattelle: That is an interesting point. A lot of nerve tissue will bind α-BGTX and show an apparently specific, saturable binding component, but in only some cases is α-BGTX physiologically effective. In the cockroach, ACh responses are sensitive to α-BGTX in at least two of the cell types (D_f and GI2) but DUM neuron ACh responses are much less sensitive. The pharmacology of the cholinergic receptors on DUM neurons remains to be determined.

Narahashi: Did you test atropine?

Sattelle: Yes; it is not very effective, either in the binding studies or in physiological studies.

Usherwood: Since you cannot characterize the binding of α-BGTX at the ultrastructural level, how can you be sure that differences in accessibility of this toxin to binding sites are not responsible for the differences in sensitivity to α-BGTX between preparations?

Sattelle: One always worries about accessibility but we are less worried about it when we are applying drugs to the cell bodies in the periphery of desheathed ganglia.

Barnard: α-Bungarotoxin has, in fact, a reasonably fast rate constant of association (about 10^5 M^{-1} s^{-1} at 25 °C) with the ACh receptor of the vertebrate neuromuscular junction, either in homogenates or membrane preparations (Colquhoun & Rang 1976, Barnard et al 1977) or when solubilized (Dolly & Barnard 1975, Brockes & Hall 1975, Lo et al 1981); the observed rates would give a half-time for the reaction of ~1 min at 10^{-7} M-toxin, and so on in proportion. Where rate constants have been measured in the past for parallel systems involving neuronal ACh receptors from invertebrates, including insects (Shain et al 1974, Schmidt-Nielsen et al 1977, Dudai 1978), they are remarkably similar to the former. However, when the rate of blockade of electrophysiological response by α-BGTX is measured on an intact, superficially accessible end-plate of vertebrate muscle, this is 10–30 times slower, an effect attributed to local diffusion barriers (Barnard 1977, Dolly et al 1977). So I presume that this *is* likely to be the explanation for the long times to blockade which you mention.

As a general comment, your description of the behaviour with respect to α-BGTX of these particular synapses in the cockroach provides one of the best correlations yet made between physiological blockade by such a neurotoxin and its binding to receptors at a neuronal location. The only equivalent ones that I know are on the binding of α-BGTX to *Aplysia* ganglia (Shain et al 1974), and in the optic tectum in lower vertebrates, as I shall review later in the symposium (p 109–117). No such correlation is available as yet in the mammalian CNS. You have given convincing evidence for the actual physiological effect of the observed binding of the toxin. This parallel could be instructive for the neuropharmacology of such systems in the vertebrates.

Cull-Candy: How convinced are you of the similarity between the junctional and the extrajunctional ACh receptor? In vertebrate muscle there is strong evidence for a difference in the properties of the channels, and also differences in the pharmacology of the junctional and extrajunctional receptors.

Sattelle: We need more data to answer your question. The only antagonists we have investigated in detail are α-BGTX and quinuclidinyl benzilate (QNB). We get the same qualitative differences between those two when tested on synaptic and extrasynaptic membranes.

M. E. Eldefrawi: Nicotine is one of the major drugs to have a strong effect on α-BGTX binding. It is also one of the drugs to which insects are probably more sensitive than vertebrates. Recent studies (Abood et al 1980, Romano & Goldstein 1980) have argued that there is stereospecific binding of nicotine to brain that is sensitive to cholinergic nicotinic drugs such as *d*-tubocurarine, yet insensitive to the α-neurotoxins. These are non-cholinergic nicotine receptors. That adds to the complexity and warns us that we should be careful about expecting insect pharmacology to be analogous to vertebrate pharmacology.

Dudai: Dr Sattelle, with regard to your failure to detect muscarinic responses in your system, one should remember that the muscarinic ACh receptors, as defined by the binding of QNB, are found, at least in *Drosophila*, in low amounts which may make their detection difficult. We also found that when they are assayed in *Drosophila*, the muscarinic binding sites are modulated by GTP analogues (Dudai 1981). This may suggest that they are coupled to a nucleotide-mediated effector system. Such receptors may, for example, be presynaptic. Did you try to confirm the electrophysiological results mentioned by Meyer & Edwards (1980) on the muscarinic response in the cockroach? And in your system would you detect presynaptic muscarinic receptors?

Sattelle: Meyer & Edwards (1980) demonstrated a QNB-binding component in the terminal ganglion of the cricket. That was one reason why we tested QNB on the cercal afferent, giant interneuron system. As you saw, it was not very effective. The properties of putative presynaptic receptors remain to be investigated.

REFERENCES

Abood LG, Reynolds DT, Bidlack JM 1980 Stereospecific ^{3}H-nicotine binding to intact and solubilized rat brain membranes and evidence for its non-cholinergic nature. Life Sci 27:1307-1314

Barnard EA 1977 Properties and function of the acetylcholine receptor of vertebrate skeletal muscle. In: Roberts GCK (ed) Drug action at the molecular level. Macmillan, London

Barnard EA, Coates V, Dolly JO, Mallick B 1977 Binding of α-bungarotoxin and cholinergic ligands to acetylcholine receptors in the membrane of skeletal muscle. Cell Biol Int Rep 1:99-106

Brockes JP, Hall ZW 1975 Acetylcholine receptors in normal and denervated rat diaphragm muscle. II. Comparison of junctional and extrajunctional receptors. Biochemistry 14:2100-2106

Colquhoun D, Rang HP 1976 Effects of inhibitors on the binding of iodinated α-bungarotoxin to acetylcholine receptors in rat muscle. Mol Pharmacol 12:519-535

Dolly JO, Barnard EA 1975 Complete purification of the acetylcholine receptor protein from mammalian muscle. FEBS (Fed Eur Biochem Soc) Lett 57:267-271

Dolly JO, Albuquerque EX, Sarvey JM, Mallick B, Barnard EA 1977 Binding of perhydrohistrionicotoxin to the post-synaptic membrane of skeletal muscle in relation to its blockade of acetylcholine-induced depolarization. Mol Pharmacol 13:1-14

Dudai Y 1978 Properties of an α-bungarotoxin-binding cholinergic nicotinic receptor from *Drosophila melanogaster*. Biochim Biophys Acta 539:505-517

Dudai Y 1981 Modulation of a putative muscarinic receptor from *Drosophila melanogaster* by ions and a guanyl nucleotide. Comp Biochem Physiol C Comp Pharmacol, in press

Lo MMS, Dolly JO, Barnard EA 1981 Molecular forms of the acetylcholine receptor from vertebrate muscles and *Torpedo* electric organ. Interactions with specific ligands. Eur J Biochem 116:155-163

Meyer MR, Edwards JS 1980 Muscarinic cholinergic binding sites in an orthopteran central nervous system. J Neurobiol 11:215-219

Romano C, Goldstein A 1980 Stereospecific nicotine receptors on rat brain membranes. Science (Wash DC) 210:647-650

Schmidt-Nielsen BK, Gepner JI, Teng NNH, Hall LM 1977 Characterization of an α-bungarotoxin binding component from *Drosophila melanogaster*. J Neurochem 29:1013-1029

Shain W, Greene LA, Carpenter DO, Sytkowski AJ, Vogel Z 1974 *Aplysia* acetylcholine receptors: blockade by and binding of α-bungarotoxin. Brain Res 72:225-240

Properties of dopamine receptors at a neuroglandular synapse

C. R. HOUSE and B. L. GINSBORG*

*Department of Veterinary Physiology and *Department of Pharmacology, University of Edinburgh, Edinburgh, EH9 1QH, UK*

Abstract The salivary glands of the cockroach *Nauphoeta cinerea* (Olivier) receive a dual innervation from the suboesophageal ganglion and the stomatogastric nerve. Acinar cells are covered by a plexus of axons containing numerous varicosities. Radiochemical assay and the Falck–Hillarp technique indicate the presence of catecholamines. The neurotransmitter is probably released from the axon varicosities on the acinar surface and also from terminals deep in the extracellular spaces between acinar cells.

Receptor activation has been studied by recording responses to nerve stimulation and agonists applied ionophoretically or in the bathing fluid. The responses consist of changes in the rate of fluid secretion by the gland and alterations in the potential and conductance of acinar cells. The order of agonist potency, the failure of α- and β-adrenergic agonists to evoke electrical responses, and the effectiveness of certain dopamine-like agonists suggest that dopamine receptors mediate the electrical and secretory events caused by nerve stimulation.

The evidence for the idea that calcium ions act as second messengers during the activation of secretion is discussed.

We are concerned here with the problem of how one kind of secretory cell—a neuron—causes the transport of substances from another—an epithelial gland cell. The invasion of an action potential into the nerve endings induces the release of one or more transmitters which react with receptors in the gland cell and thus evoke the secretion of enzymes, ions and water. We have examined some aspects of this general problem by doing experiments on the innervated salivary gland of the cockroach, *Nauphoeta cinerea*.

This paper deals with two features of neuroglandular transmission, namely the pharmacological nature of the receptors and the physiological events following the reaction of the transmitter with its receptors.

1982 Neuropharmacology of insects. Pitman, London (Ciba Foundation symposium 88) p 32-47

Anatomy

The cockroach salivary gland is attractive to the experimentalist wishing to study the nervous control of secretion. Its gross structure is similar to that of mammalian salivary glands. However, the cockroach gland can be dissected with less trauma because of the absence of a blood supply. There is no encasing capsule, so the gland cells are accessible; moreover, the cells are not associated with smooth muscle or myoepithelial cells.

Gland cells

Each gland comprises a branching system of ducts terminated by acini of about 100 μm diameter (Whitehead 1971, Bland & House 1971). The acini contain peripheral and central cells (Fig. 1A). Peripheral cells have numerous mitochondria and extensive foldings of their apical membranes. Central cells contain rough endoplasmic reticulum, Golgi bodies and large granules of about 2 μm diameter. Probably these acinar cells produce the primary secretion of enzymes, ions and water in response to nervous activity (Bland & House 1971, Smith & House 1979, Maxwell 1981).

The duct cells near to the acini are also innervated and probably secrete a mucous component stored in large dense granules (Bland & House 1971). Further from the acini the duct cells are structurally different since they contain numerous mitochondria sandwiched between long folds of their basal membranes but no prepackaged material for release. Physiological evidence suggests that these duct cells modify the ionic composition of the fluid secreted by the acini (Smith & House 1979).

Salivary neurons

Each gland receives a dual innervation from the suboesophageal ganglion and the stomatogastric nerve (Whitehead 1971, Bowser-Riley 1978, Maxwell 1978). Most, if not all, of the acinar axons derive from the salivary duct nerve originating in the suboesophageal ganglion (Bowser-Riley 1978). Extracellular recording and stimulation methods show that the salivary duct nerve is functionally important (Whitehead 1973, Ginsborg & House 1976, Smith & House 1979). By contrast, the stomatogastric nerve does not participate directly in producing secretion or electrical activity in gland cells; it might function as a neuromodulator or control some other aspect of salivation not yet examined (Bowser-Riley 1978).

The acini are covered by a plexus of axons with numerous varicosities

(Whitehead 1971, Bowser-Riley 1978); within the spaces between acinar cells there are also nerve terminals (Maxwell 1978). Both varicosities and terminals contain a catecholamine, and histochemical methods suggest that these structures possess dopamine-filled vesicles (Bland et al 1973, Fry et al 1974, Maxwell 1980). However, the crucial evidence is lacking that dopamine is released by nerve stimulation in sufficient quantity to stimulate the gland cells.

The neuroglandular synapses in the acini are not distinguished by pre- or postsynaptic membrane thickenings. Transmitter released from varicosities on the acinar surface would have to travel about 1 μm to the nearest gland cell membrane, whereas the length of the diffusion path for the release sites on the nerve terminals is about 20 nm. The location of the receptors in the peripheral and central cells is uncertain but we assume that they occur in their basal membranes.

Physiology

Within an acinus there is probably a division of labour, with the peripheral cells transporting ions and water and the central cells secreting enzymes into the lumen (Bland & House 1971, Maxwell 1981). The saliva of cockroaches contains a number of enzymes including amylase, which is made and stored in acinar cells. The main ions in the saliva are Na^+, Cl^- and K^+ and it is likely that water transport into the saliva is driven by the active transport of Na^+ across the peripheral cells. The salivary secretion evoked by nerve stimulation or externally applied dopamine is identical both in maximal rates of production and in ionic composition (Smith & House 1979).

Transmitter release

When the salivary duct nerve is stimulated electrically two kinds of response can be recorded from the salivary gland. Acinar cells hyperpolarize (Fig. 1B) because of a rise in the potassium permeability of their basal membranes. A marked rise in the salivary secretion rate also occurs (Fig. 1C). The sluggish nature of the secretory response means that its time course cannot be a guide to the kinetics of transmitter release or the ensuing receptor activation. The electrical response, however, might be more useful in this respect.

We have taken the amplitude of the hyperpolarization evoked by a single nerve stimulus as a rough measure of transmitter release. As judged by this criterion, several factors can influence the transmitter output per impulse.

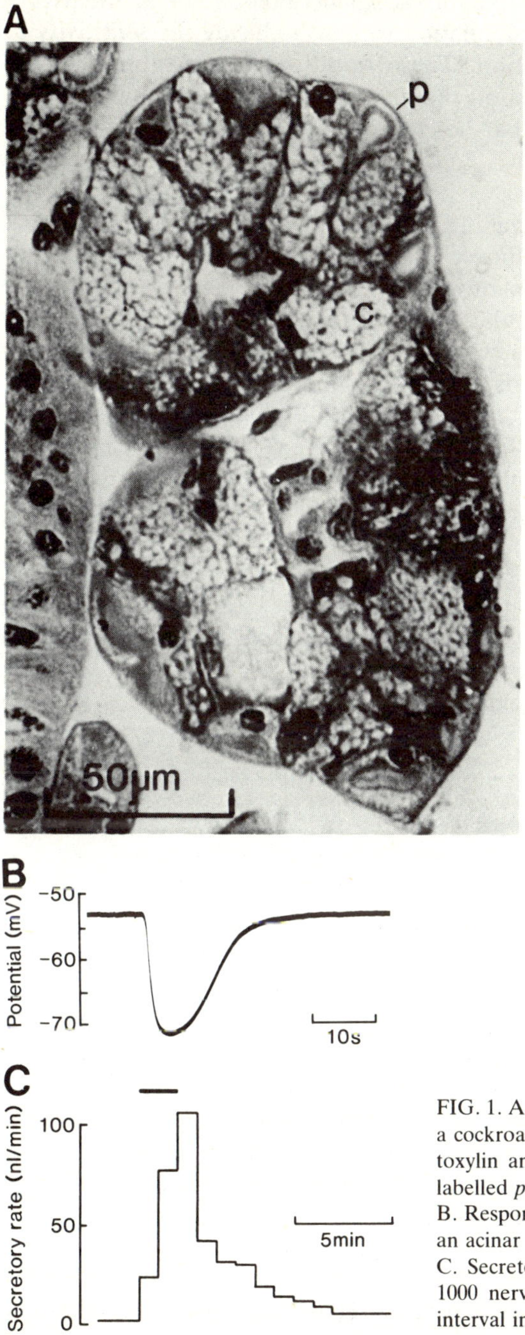

FIG. 1. A. Histological section of an acinus from a cockroach salivary gland stained with haematoxylin and eosin. Peripheral and central cells labelled *p* and *c* respectively.
B. Response of the basal membrane potential of an acinar cell to a single nerve stimulus.
C. Secretory response of a gland to a train of 1000 nerve stimuli given at 10 Hz during the interval indicated by the horizontal bar.

Silinsky (1974) found that bretylium and guanethidine reduced the hyperpolarization evoked by nerve stimulation without reducing the sensitivity to dopamine in the bathing solution. These results are compatible with the proposed actions of these drugs on the mammalian adrenergic system. The response to nerve stimulation but not to dopamine can also be reduced or abolished by bathing glands in solutions with low calcium concentrations (Ginsborg et al 1980a).

A number of substances, including acetylcholine and derivatives of dopamine, augment the electrical response to a single stimulus (House & Ginsborg 1979). Another possible potentiator of transmitter release is octopamine (Bowser-Riley & House 1976); this is of interest since octopamine probably has a neuromodulatory role elsewhere in the insect nervous system (Evans 1980). Finally, the presence of tetraethylammonium (TEA) ions in the external medium at concentrations above 1 mM also increases the size of the response to a nerve stimulus. TEA blocks voltage-sensitive K^+ channels in axonal membranes and might therefore prolong the duration of the action potential invading the salivary nerve terminals and hence augment transmitter release.

We conclude that the mechanism of dopamine release from the salivary nerves is rather similar to that of acetylcholine release at the vertebrate neuromuscular junction (Katz 1969) since both neuroglandular and neuromuscular transmissions are blocked by tetrodotoxin (B. L. Ginsborg & C. R. House, unpublished) or low extracellular calcium concentrations and are enhanced by TEA.

Receptor properties

A number of substances can mimic the transmitter by evoking electrical and secretory responses from gland cells (House & Ginsborg 1979). These include dopamine, noradrenaline, adrenaline, 5-hydroxytryptamine (5-HT, serotonin), apomorphine and the rigid analogue of dopamine, 2-amino-6,7-dihydroxy-1,2,3,4-tetrahydronaphthalene (ADTN). Dopamine is the most potent agonist so far tested on cockroach glands. With the exception of 5-HT and apomorphine all agonists can match the maximal responses to nerve stimulation. There is evidence that 5-HT reacts with a different set of receptors from those for the neurotransmitter and dopamine (House et al 1973, Bowser-Riley et al 1978).

It seems likely that apomorphine is a partial agonist for the dopamine receptors. When it reacts with these receptors, however, it produces a long-lasting hyperpolarization of the acinar cells (Ginsborg et al 1976) and

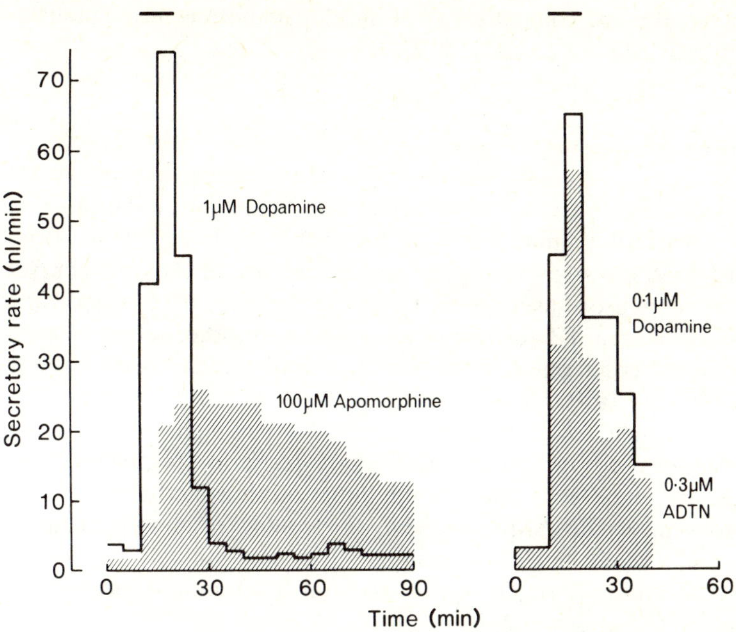

FIG. 2. Secretory responses of cockroach salivary gland cells to dopamine and other agonists. Left panel shows superimposed responses from a gland exposed to dopamine or apomorphine in the bathing solution for the period indicated by the horizontal bar. Right panel shows the corresponding comparison of responses of another gland to dopamine or 2-amino-6,7-dihydroxy-1,2,3,4-tetrahydronaphthalene (ADTN).

also prolonged fluid secretion (Fig. 2). A similar phenomenon has been found with apomorphine stimulation of tick salivary glands (Kaufman 1977).

The other dopamine-like agonist, ADTN, is almost equipotent with dopamine and can evoke normal acinar hyperpolarization (House & Ginsborg 1976) and secretory responses (Fig. 2).

The order of agonist potency (House & Ginsborg 1979), the failure of selective α- and β-adrenergic agonists to evoke responses (Ginsborg et al 1976) and the effectiveness of apomorphine and ADTN suggest that specific dopamine receptors are present in cockroach salivary glands. In keeping with this view is the observation that the neuroleptic drug, α-flupenthixol, an antagonist of responses to dopamine in mammals (e.g. Miller et al 1974, Docherty & McQueen 1978), suppresses both electrical and secretory responses to nerve stimulation or dopamine (House & Ginsborg 1976, Breward et al 1980).

We have suggested that there might be two classes of dopamine receptor in this gland (Ginsborg & House 1976, Bowser-Riley et al 1978). However, our results do not seem to be compatible with the dopamine receptor classification proposed by Kebabian & Calne (1979).

Second-messenger hypothesis

The latency of the electrical response of acinar cells to nerve stimulation or an ionophoretic pulse of dopamine at the acinar surface is about 500 ms. Both the long latency and the strong temperature dependence of the time courses of responses to the neurotransmitter and dopamine suggest the presence of a second-messenger system (Blackman et al 1979a,b). In other secretory cells attention has been focused on cyclic AMP and calcium ions as possible second messengers (Berridge 1979, Douglas 1978).

The presence of cyclic AMP or dibutyryl cyclic AMP in high concentrations (about 1 mM) does not produce electrical or secretory responses from cockroach salivary gland cells. Nor do inhibitors of phosphodiesterase (the enzyme hydrolysing cyclic AMP) increase or prolong responses to nerve stimulation or externally applied dopamine. It seems unlikely that cyclic AMP acts as a second messenger for processes involved in fluid secretion or the hyperpolarization of acinar cells.

We have also examined the influence of extracellular calcium ions on the acinar cell hyperpolarization evoked by nerve stimulation or the ionophoretic application of dopamine. Glands bathed for long periods (about 60 min) in 'calcium-free' solutions with EGTA still respond to dopamine whereas neuroglandular transmission fails within a few minutes (Ginsborg et al 1980a). However, during a sequence of dopamine pulses the responses progressively decline to zero amplitude. We have suggested that acinar cells contain an exhaustible store of calcium required for cellular activation and that this store is readily accessible to extracellular calcium ions, because the response to dopamine rapidly recovers when calcium is restored to the medium. Indeed, in addition to causing rapid recovery the readmission of calcium to the bathing solution also causes a hyperpolarization which is particularly large and prolonged if cobalt is present in the 'calcium-free' solution (Ginsborg et al 1980b). This suggests that 'calcium deprivation' is followed by a pronounced entry of calcium into acinar cells on its readmission to the bathing medium. Thus a rise in the cytosolic calcium concentration can mediate a rise in the potassium permeability of the basal membranes of acinar cells and thereby mimic one of the actions of the neurotransmitter. Moreover, the calcium ionophore A23187 causes a hyperpolarization of the acinar cells in cockroach salivary glands (Mitchell & Martin 1980).

We were interested to know whether changes in the extracellular calcium concentration had effects on fluid secretion. The rate of secretion evoked by

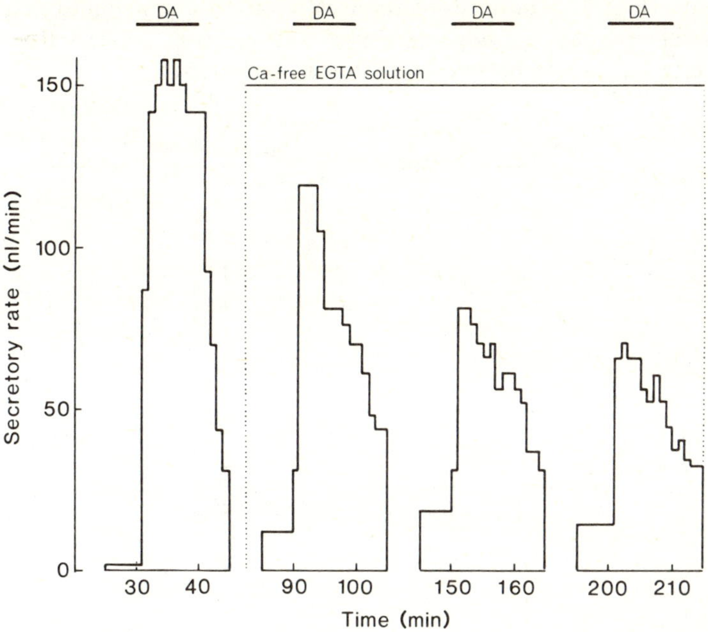

FIG. 3. Effect of calcium-free solution containing 0.5 mM-EGTA on secretory responses of salivary gland cells to dopamine. The dopamine (DA) was applied in the bathing solution at a concentration of 1 μM for the periods indicated by the horizontal bars.

dopamine is relatively insensitive to calcium deprivation (Fig. 3). However, during a sequence of dopamine applications the amplitudes of the responses fall and, moreover, during an individual response the peak rate is not so well maintained as it is in the presence of extracellular calcium. It might be expected that a long stimulation by dopamine in a 'calcium-free' solution would yield a response that would gradually decline almost to zero. Thus the exhaustible store of calcium might also be of significance for fluid secretion evoked by dopamine.

Could calcium ions released from a store into the cytosol bring about fluid secretion? One way of attempting to answer this question is to experimentally enhance the entry of calcium ions into the gland cells and monitor the rate of secretion. Calcium readmission to the bathing solution after a period of deprivation causes an increase in ion permeability, probably due to an influx of calcium into the acinar cells (Ginsborg et al 1980a). When calcium ions are restored there is also a transient increase in fluid secretion. This readmission response to calcium is quite small when magnesium is present in the 'calcium-free' solution but becomes substantial when cobalt is used as a calcium substitute (Fig. 4). These results indirectly raise the possibility that calcium ions might act as second messengers for neurally evoked fluid secretion. Calcium ions might be released into the cytosol from a store as a result of receptor activation.

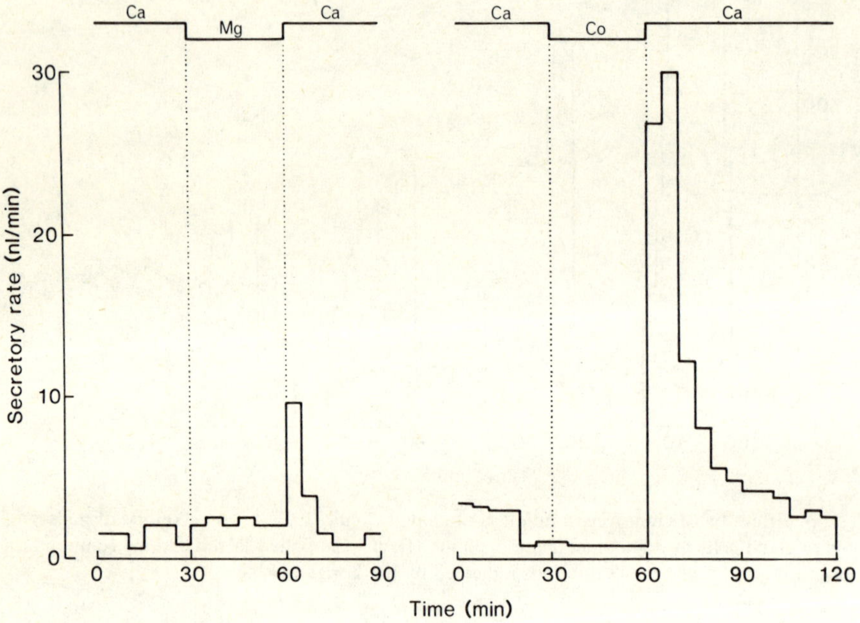

FIG. 4. Secretory responses to the readmission of calcium to the bathing solution. Left panel shows readmission response after a salivary gland was bathed in a solution in which calcium had been replaced by magnesium. Right panel shows the corresponding result when cobalt was used as a calcium replacement. In both experiments the normal calcium concentration was 5 mM.

Conclusion

The pharmacology of transmission between nerve and salivary acinar cells in *N. cinerea* suggests that dopamine receptors play an important part in activating glandular secretion. Since the different kinds of acinar cells are electrically coupled (Ginsborg et al 1974) the acinus might behave as a functional unit in which the central cells secrete amylase and the peripheral cells simultaneously transport Na^+, Cl^- and K^+ ions (Fig. 5).

The kinetics of the electrical response to nerve stimulation suggest that a second-messenger system operates after the interaction of the transmitter with its receptors. We propose that calcium ions act as second messengers in

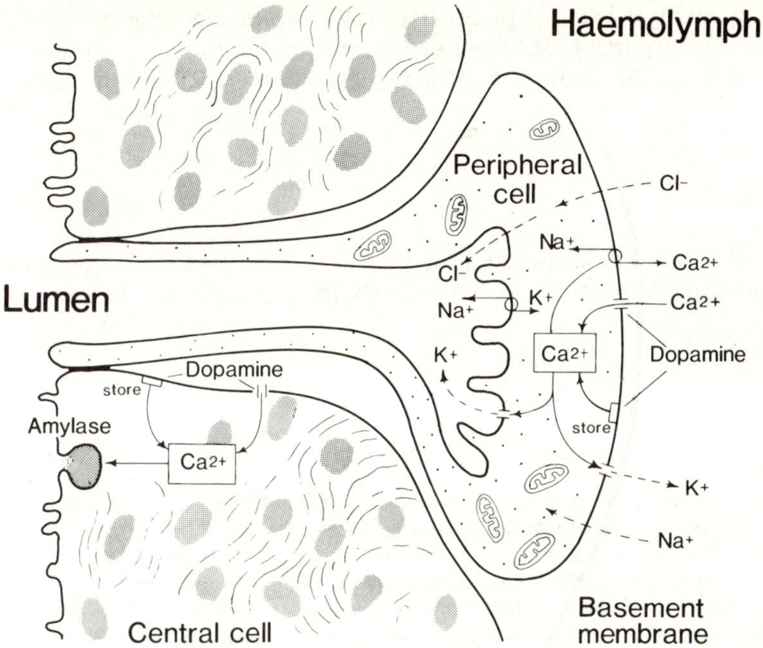

FIG. 5. Diagram illustrating how calcium could act as a second messenger in acinar cells. Exocytosis of granules containing amylase is probably triggered by a rise in the cytosolic calcium concentration near to the apical membrane of the central cell. A rise in the cytosolic calcium concentration in the peripheral cell could produce an influx of Na^+ ions across the basal membrane due to (a) the hyperpolarization of the membrane, (b) an increase in Na^+ permeability or (c) an Na^+:Ca^{2+} exchange pump. It seems possible that active transport of Na^+ occurs at the apical membrane of the peripheral cell. Although dopamine might cause the opening of calcium channels in the basal membranes this route of calcium entry into the cytosol is probably not essential for relatively brief secretory responses.

the acinar cell cytosol. It is possible that calcium ions are released into the cytosol of the central cell near to the apical membrane where exocytosis of enzyme occurs. Calcium ions released into the cytosol of the peripheral and central cells bring about a rise in potassium permeability of the basal membranes (and possibly of the apical too). The ensuing hyperpolarization of the basal membranes increases the electrochemical gradient for entry of Na^+ into the cells from the haemolymph; the Na^+ permeability of the basal membrane might also be increased after receptor activation. Another way in which calcium ions could increase sodium entry is through an Na^+:Ca^{2+} exchange pump which would help to maintain a low cytosolic calcium concentration. Our evidence is not in favour of an active step for Cl^- transport into the saliva; possibly Cl^- ions might accompany Na^+ in a coupled manner across the basal membrane. The Na^+ load in the peripheral cells might be cleared by an Na^+:K^+ pump in the folded apical membranes of these cells. The net transport of ions from the haemolymph into the lumen of the peripheral cell should create a sufficient local osmotic gradient to attract water.

Acknowledgements

We are indebted to the Science and Medical Research Councils for financial support and wish to thank Mr Colin Warwick for preparing the diagrams in this article.

REFERENCES

Berridge MJ 1979 Relationship between calcium and the cyclic nucleotides in ion secretion. In: Binder HJ (ed) Mechanisms of intestinal secretion. Alan R Liss, New York, p 65-81

Blackman JG, Ginsborg BL, House CR 1979a On the effect of ionophoretically applied dopamine on salivary gland cells of *Nauphoeta cinerea*. J Physiol (Lond) 287:67-80

Blackman JG, Ginsborg BL, House CR 1979b On the time course of the electrical response of salivary gland cells of *Nauphoeta cinerea* to ionophoretically applied dopamine. J Physiol (Lond) 287:81-92

Bland KP, House CR 1971 Function of the salivary glands of the cockroach, *Nauphoeta cinerea*. J Insect Physiol 17: 2069-2084

Bland KP, House CR, Ginsborg BL, Laszlo I 1973 Catecholamine transmitter for salivary secretion in the cockroach. Nat New Biol 244:26-27

Bowser-Riley F 1978 The salivary glands of the cockroach *Nauphoeta cinerea* (Olivier). A study of its innervation by light and scanning electron microscopy. Cell Tissue Res 187:525-534

Bowser-Riley F, House CR 1976 The actions of some putative neurotransmitters on the cockroach salivary gland. J Exp Biol 64:665-676

Bowser-Riley F, House CR, Smith RK 1978 Competitive antagonism by phentolamine of responses to biogenic amines and the transmitter at a neuroglandular junction. J Physiol (Lond) 279:473-489

Breward J, House CR, Smith RK 1980 α-Flupenthixol: an antagonist of dopamine-evoked fluid secretion by an insect salivary gland preparation. Br J Pharmacol 69:123-129

Docherty RJ, McQueen DS 1978 Inhibitory action of dopamine on cat carotid chemoreceptors. J Physiol (Lond) 279:425-436

Douglas WW 1978 Stimulus–secretion coupling: variations on the theme of calcium-activated exocytosis involving cellular and extracellular sources of calcium. In: Respiratory tract mucus. Elsevier/Excerpta Medica/North-Holland, Amsterdam (Ciba Found Symp 54) p 61-87

Evans PD 1980 Biogenic amines in the insect nervous system. Adv Insect Physiol 15:317-473

Fry JP, House CR, Sharman DF 1974 An analysis of the catecholamine content of the salivary gland of the cockroach. Br J Pharmacol 51:116P-117P

Ginsborg BL, House CR 1976 The responses to nerve stimulation of the salivary gland of *Nauphoeta cinerea* Olivier. J Physiol (Lond) 262:477-487

Ginsborg BL, House CR, Silinsky EM 1974 Conductance changes associated with the secretory potential in the cockroach salivary gland. J Physiol (Lond) 236:723-731

Ginsborg BL, House CR, Silinsky EM 1976 On the receptors which mediate the hyperpolarization of salivary gland cells of *Nauphoeta cinerea* Olivier. J Physiol (Lond) 262:489-500
Ginsborg BL, House CR, Mitchell MR 1980a On the role of calcium in the electrical responses of cockroach salivary gland cells to dopamine. J Physiol (Lond) 303:325-335
Ginsborg BL, House CR, Mitchell MR 1980b A calcium-readmission response recorded from *Nauphoeta* salivary gland acinar cells. J Physiol (Lond) 304:437-447
House CR, Ginsborg BL 1976 Actions of a dopamine analogue and a neuroleptic at a neuroglandular synapse. Nature (Lond) 261:332-333
House CR, Ginsborg BL 1979 Pharmacology of cockroach salivary secretion. Comp Biochem Physiol C Comp Pharmacol 63:1-6
House CR, Ginsborg BL, Silinsky EM 1973 Dopamine receptors in cockroach salivary gland cells. Nat New Biol 245:63
Katz B 1969 The release of neural transmitter substances. Liverpool University Press, Liverpool
Kaufman WR 1977 The influence of adrenergic agonists and their antagonists on isolated salivary glands of ixodid ticks. Eur J Pharmacol 45:61-68
Kebabian JW, Calne DB 1979 Multiple receptors for dopamine. Nature (Lond) 277:93-96
Maxwell DJ 1978 Fine structure of axons associated with the salivary apparatus of the cockroach, *Nauphoeta cinerea*. Tissue Cell 10:699-706
Maxwell DJ 1980 Histochemical properties of axons associated with the salivary apparatus of the cockroach, *Nauphoeta cinerea*. Tissue Cell 12:703-711
Maxwell DJ 1981 Morphological changes in gland cells and axons resulting from stimulation of the salivary nerves of the cockroach, *Nauphoeta cinerea*. Tissue Cell 13:141-151
Miller RJ, Horn AS, Iversen LL 1974 The action of neuroleptic drugs on dopamine-stimulated adenosine cyclic 3′,5′-monophosphate production in rat neostriatum and limbic forebrain. Mol Pharmacol 10:759-766
Mitchell MR, Martin AR 1980 An electrophysiological study of the effects of ionophore A23187 on *Nauphoeta* salivary glands. Q J Exp Physiol Cogn Med Sci 65:309-320
Silinsky EM 1974 The effects of bretylium and guanethidine on catecholaminergic transmission in an invertebrate. Br J Pharmacol 51:367-371
Smith RK, House CR 1979 Ion and water transport by isolated cockroach salivary glands. J Membr Biol 51:325-346
Whitehead AT 1971 The innervation of the salivary gland in the American cockroach: light and electron microscopic observations. J Morphol 135:483-505
Whitehead AT 1973 Innervation of the American cockroach salivary gland: neurophysiological and pharmacological investigations. J Insect Physiol 19:1961-1970

DISCUSSION

Usherwood: Does temperature differentially affect latency, rise time, decay time, and amplitude of the response of the salivary gland to dopamine?

House: We didn't look at that in detail in a large number of cells. In one experiment (Blackman et al 1979) we found that the time-to-peak (tp), the latency and the time of decay from peak to half-maximum shifted roughly by the same proportion with temperature. Whether the same similarities hold for all cells remains an open question.

Narahashi: The first response to nerve stimulation was potentiated by a high concentration (10^{-5} M) of octopamine but there was a decreased response to a second stimulus. Could this effect be due to stimulation of transmitter release by octopamine? If octopamine stimulates transmitter release the first response will be potentiated, so when the second stimulus is applied the transmitter has been temporarily depleted, causing a decrease in the response.

House: Octopamine appears to be a very weak dopamine agonist (Bowser-Riley & House 1976). It may be a partial agonist which acts as an inhibitor of the transmitter itself, because a partial agonist occupies a large fraction of the receptors. So octopamine has two opposing effects, I think. The first effect is a small change in the membrane potential. Then in addition to this direct effect there is a presynaptic effect, because octopamine appears to enhance the first response to nerve stimulation. So perhaps two or three times the amount of transmitter comes out on the first stimulus and thereafter octopamine successfully competes with the transmitter for its receptors.

Evans: Do you know anything about the cellular localization of octopamine in this preparation? Could the stomatogastric innervation be a modulatory octopamine input?

House: It could be, but I have no evidence on this point.

M. E. Eldefrawi: The picture emerges that calcium has an important physiological role here. A common tool that one would think of immediately is the calcium ionophore (A23187), which should imitate the effect of nerve stimulation or addition of dopamine, in this case. The other tool is a calcium antagonist, such as verapamil, which should inhibit the response. Have you used either approach to confirm the role of calcium?

House: Mitchell & Martin (1980) have shown that the ionophore A23187 produces a hyperpolarizing response of the acinar cells. We haven't been able to produce a secretory response with this ionophore and we haven't made any extensive study of calcium antagonists.

Cull-Candy: What would happen if you injected calcium into the cells?

House: I imagine it would cause a hyperpolarization of the acinar cells.

Penzlin: Why do you get such a late calcium effect? With a calcium influx I would expect the effect to be much earlier, but you seemed to have an hour's delay.

House: When we reduced the calcium in the bathing fluid the response to nerve stimulation was abolished almost instantaneously but the response to ionophoretic application of dopamine persisted. We thought it was perhaps a question of waiting long enough to make sure that the concentration of extracellular calcium had reached a very low value. We therefore waited for at least an hour and then stimulated. We still found a response, but then progressive stimulation showed the decline in subsequent responses.

Berridge: Muscarinic ACh receptors, which also gate calcium, have long latencies (Purves 1976). There is reason to believe that the long latency may result from an enzymic step, which is described in my paper (p 260–270).

You mentioned various agents that affect the dopamine receptor. Can you use these to distinguish between D_1 and D_2 receptors similar to those seen in vertebrates? With most agonists one can distinguish between receptors that affect cyclic AMP and those that work through calcium. With dopamine, the vertebrate D_1 receptor stimulates cyclic AMP production and the D_2 receptor causes an inhibition of adenylate cyclase and a reduction in cyclic AMP concentration. As yet, we don't know of a vertebrate dopamine receptor which works through calcium, which apparently is what you found. How does the pharmacology of the cockroach receptor compare with that of vertebrates?

House: I am worried about the dopamine receptor classification (Kebabian & Calne 1979) that is evolving in vertebrates. The classification depends, for example, on the use of apomorphine, injected in pulses into whole animals; usually the response to apomorphine or dopamine is some sort of inhibition. If apomorphine is interacting with dopamine receptors as it does here in the cockroach gland, it could be more effective, apparently, than dopamine. It might give only a small amount of persistent inhibition but over a long period it would produce a much larger effect than dopamine does over a short period.

Berridge: Have any specific dopamine agonists been used?

House: ADTN possibly comes closest to being a relatively specific agonist.

Berridge: Does that act on the D_1 or D_2 receptor?

House: I think ADTN interacts with both groups in vertebrates.

Dudai: We should be very careful about assigning to radioactive ligands used to label putative receptors in invertebrates the same role as they have in vertebrates. For example, dihydroergocryptine in vertebrates is used mainly to label α-adrenergic receptors (although it does interact with dopamine and serotonin receptors too). In *Drosophila* the same ligand seems, as revealed by *in vitro* experiments, to label mainly serotoninergic binding sites (Y. Dudai & S. Zvi, unpublished results 1981). Whenever we use labelled ligands in invertebrates we should assign them a role which is based on electrophysiological evidence in invertebrates. Most of these drugs have overlapping pharmacological profiles.

Usherwood: I agree that we must be careful not to assume that drugs will have the same effects on excitable systems of insects and of vertebrates, but we must also remember that information on the effects of pharmacological agents on vertebrate preparations is probably incomplete.

Berridge: When considering receptor specificity we should concentrate not only on receptor binding but also on the output signal from the receptor.

What the receptor tells us is less important than what the receptor tells the cell.

Piek: Your conclusion is that dopamine activates the potassium channel. How do you explain the increase in hyperpolarization with TEA, since this compound blocks potassium channels?

House: At the concentrations of TEA used in the experiments that I showed, TEA is acting presynaptically. TEA blocks voltage-sensitive potassium channels but the potassium channels we are talking about are voltage-sensitive only in the axons of the cells of the nerve terminals, not in the acinar cells. Thus at low concentrations TEA does not act directly on the acinar cell hyperpolarization.

Usherwood: Could you expand on the role of potassium in this system?

Mordue: And how are sodium and potassium interrelated in your system? Is there just a general cation effect?

House: You are talking about the transport of potassium from the haemolymph into the saliva. In the presence of a K^+-free bathing fluid we still get fluid secretion in response to dopamine. Secretion is however abolished in the absence of extracellular sodium. In the blowfly, for example, if the concentration of one ion is increased at the expense of the other the salivary gland still secretes normally. The cockroach preparation, however, is very sodium-dependent.

As for the role of potassium generally, I can't say much. In response to stimulation the same transient loss of K^+ occurs in the salivary glands of vertebrates. This may be an inevitable result of the increased calcium concentration in the cell but I don't know what function loss of cellular potassium serves. Certainly potassium loss across the basal membrane increases the electrochemical gradient for sodium entry and this is of some benefit for sodium transport from haemolymph to saliva.

Usherwood: Do the acinar cells always hyperpolarize on stimulation?

House: When we evoke the hyperpolarization we occasionally see a long-lasting small depolarization, but we can't get this component in isolation.

Donnellan: Is there any evidence for things like calmodulin, as a calcium-binding protein, being involved in this salivary gland response?

House: I haven't looked at that.

Berridge: Do you have any idea what the membrane calcium store is?

House: No.

Berridge: People working on pancreas have postulated that calcium is stored on the membrane and is rapidly released into the cytosol during the action of agonists (Matthews et al 1973). Much of the calcium bound to the inner part of the cell membrane is probably associated with polyphosphoinositides. It might be worth investigating these components as the possible calcium store because they are rapidly destroyed during stimulation. It is also known that vertebrate neural tissue contains a higher proportion of these polyphosphoinositides, which change very rapidly on excitation (Hawthorne & Pickard 1979).

M. E. Eldefrawi: Phosphoinositide is definitely one major factor that acts as a membrane buffer or calcium-buffering system, so to speak. Another calcium-buffering system that we have found is the cholinergic receptor of the motor end-plate (Rübsamen et al 1976). The receptor protein itself is a calcium-binding protein that releases three to six ions of calcium for every receptor site. When the receptor is activated by an agonist, calcium is immediately released. If an antagonist is applied to the receptor, calcium goes back on (Chang & Neumann 1976). There are undoubtedly several membrane-buffering systems for calcium and those two are probably good examples in vertebrates.

REFERENCES

Blackman JG, Ginsborg BL, House CR 1979 On the effect of ionophoretically applied dopamine on salivary gland cells of *Nauphoeta cinerea*. J Physiol (Lond) 287:67-80

Bowser-Riley F, House CR 1976 The actions of some putative neurotransmitters on the cockroach salivary gland. J Exp Biol 64:665-676

Chang HW, Neumann E 1976 Dynamic properties of isolated acetylcholine receptor proteins: release of calcium ions caused by acetylcholine binding. Proc Natl Acad Sci USA 73:3364-3368

Hawthorne JN, Pickard MR 1979 Phospholipids in synaptic function. J Neurochem 32:5-14

Kebabian JW, Calne DB 1979 Multiple receptors for dopamine. Nature (Lond) 277:93-96

Matthews EK, Petersen OH, Williams JA 1973 Pancreatic acinar cells: acetylcholine-induced membrane depolarization, calcium efflux and amylase release. J Physiol (Lond) 234:689-701

Mitchell MR, Martin AR 1980 An electrophysiological study of the effects of ionophore A23187 on *Nauphoeta* salivary glands. Q J Exp Physiol 65:309-320

Purves RD 1976 Function of muscarinic and nicotinic acetylcholine receptors. Nature (Lond) 261:149-150

Rübsamen H, Hess GP, Eldefrawi AT, Eldefrawi ME 1976 Interaction between calcium and ligand-binding sites of the purified acetylcholine receptor studied by use of a fluorescent lanthanide. Biochem Biophys Res Commun 68:56-63

Properties of modulatory octopamine receptors in the locust

PETER D. EVANS

ARC Unit of Invertebrate Chemistry and Physiology, Department of Zoology, University of Cambridge, Downing Street, Cambridge, CB2 3EJ, UK

Abstract An identified octopamine-containing neuron in the metathoracic ganglion of the locust, *Schistocerca americana gregaria*, modulates the actions of the slow excitatory motor neuron to the extensor tibiae muscle of the hindleg. The actions of the modulatory neuron, and of exogenously added octopamine, depend on the frequency of stimulation of the motor neuron. The effects of a wide range of agonists and antagonists of the octopamine receptors in the extensor tibiae muscle indicate that there are several classes of octopamine receptors, all of which respond maximally to phenolamines and are stereospecific for the D(-)isomer of octopamine. In addition the octopamine receptors are all selectively blocked by α-adrenergic blocking agents and activated by α-adrenergic agonists. The octopamine receptors are also activated by formamidine pesticides such as chlordimeform, suggesting that these receptors may serve as useful targets in the development of novel pesticides.

Neuromodulators can be distinguished from classical neurotransmitters on the basis of their release sites and their functional roles (Evans 1980a). Neuromodulators are released from neurosecretory terminals which, in contrast to the terminals of neurons releasing neurotransmitters, do not form discrete synaptic endings on their target cells. Neuromodulators act either to change the quality of information being transmitted through conventional synapses or to change the spontaneous activity of target cells. They thus act as local neurohormones and may have to diffuse considerable distances to their sites of action where they can influence whole populations of target cells.

The biogenic amine, octopamine (Fig. 1), modulates neuromuscular transmission and muscular contraction in skeletal muscle in insects. These effects have been intensively studied in the large extensor tibiae muscle of the hindleg of the locust, *Schistocerca americana gregaria* (Evans & O'Shea 1977, 1978, O'Shea & Evans 1979, Buchan & Evans 1980, Evans & Siegler 1982).

1982 Neuropharmacology of insects. Pitman, London (Ciba Foundation symposium 88) p 48-69

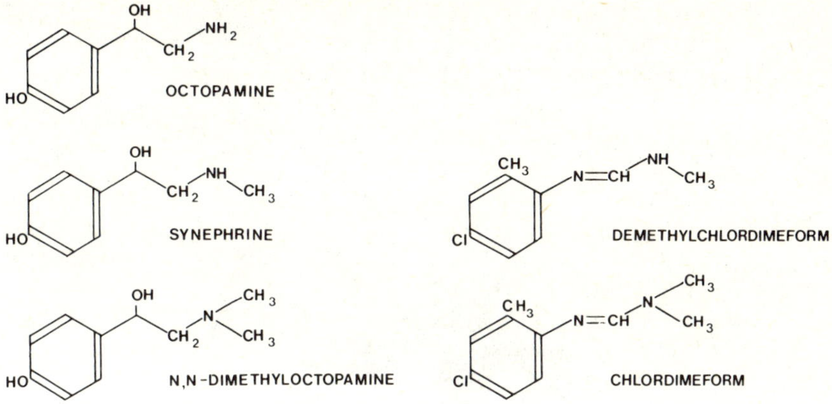

FIG. 1. Structures of phenolamines and formamidine pesticides.

This muscle is innervated by a single physiologically identified octopamine-containing neuron (Hoyle 1975, Hoyle & Barker 1975, Evans & O'Shea 1977, 1978) that has been designated DUMETi (dorsal unpaired median cell to extensor tibiae muscle) (Hoyle et al 1974). In the locust central nervous system (CNS) the axon of this cell runs anteriorly from the cell body before bifurcating to give bilaterally symmetrical branches. These branches innervate the extensor tibiae muscles in the left and right hindlegs (Fig. 2). The extensive dendritic branching of this neuron in the CNS presumably reflects the regions of synaptic input and output of the cell, together with regions of possible central modulatory outputs. In the periphery the terminals of DUMETi occur as 'blindly ending neurosecretory terminals' without making any specific synaptic endings (Hoyle et al 1974). This contrasts with the endings of the three motor neurons to this muscle (two excitatory and one inhibitory) which all form conventional synaptic structures on the muscle fibres.

The modulatory functions of the octopamine released from the DUMETi neuron in the locust extensor tibiae neuromuscular preparation will be outlined here, and the pharmacological similarities and differences between the various classes of octopamine receptors present will be described.

Effects of octopamine on extensor tibiae muscle

(a) Inhibition of a myogenic rhythm

The extensor tibiae muscle of the locust hindleg contains a specialized proximal bundle of muscle fibres that exhibit a myogenic rhythm. This bundle

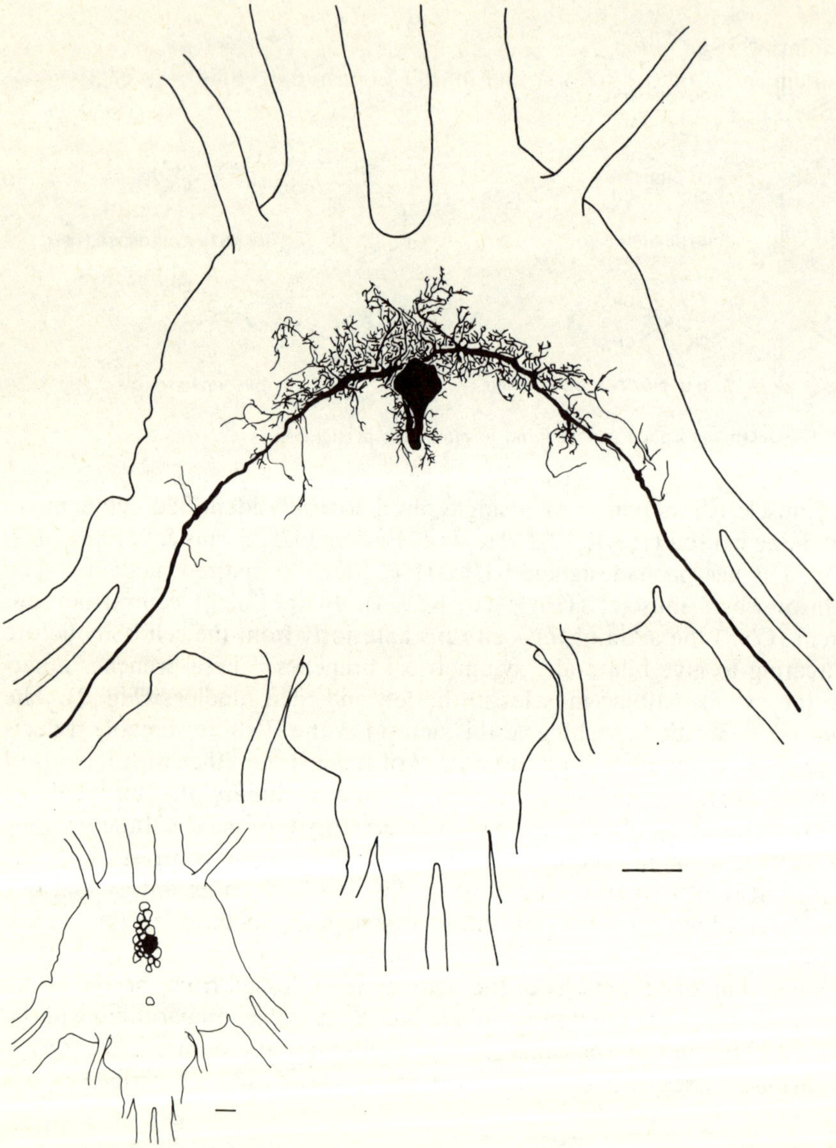

FIG. 2. Anatomy of the DUMETi neuron from the metathoracic ganglion of *Schistocerca americana gregaria*. The cell was filled intracellularly with cobaltous ions which were subsequently precipitated as the sulphide. The preparation was then subjected to whole-mount silver intensification using a modification of Timm's method (Bacon & Altman 1977). Inset shows position of DUMETi soma within the group of dorsal median neurons. Scale bars, 100 μm.

can be dissected out without inhibiting the rhythm (see Evans 1980a). Both stimulation of the DUMETi neuron and the application of exogenous octopamine to the muscle bundle inhibit this rhythm (Hoyle 1975, Evans & O'Shea 1978). The rhythmic muscle contractions probably serve to pump haemolymph along the leg and their inhibition prevents any interference with the responses of the muscle to specific patterns of neuronal input (see below). The pharmacological control of this myogenic bundle appears complex in that additional separate receptors for 5-hydroxytryptamine (serotonin) and the pentapeptide proctolin both accelerate the myogenic rhythm (see Evans 1980a).

(b) Modulation of neuromuscular transmission and muscle contraction

The widespread distribution of the DUMETi neuron in the extensor muscle suggests that inhibition of the myogenic rhythm cannot be its only function (O'Shea & Evans 1979). Nonetheless, direct stimulation of the neuron induces no contraction or electrical activity in the muscle. When the DUMETi neuron is stimulated at the same time as the motor neurons, however, this alters their actions on the muscle in the following ways.

If the slow excitatory motor neuron to the extensor (SETi) is stimulated at low frequencies (around 1 Hz) each impulse in the neuron elicits a single twitch in the muscle. Stimulation of the DUMETi neuron, or application of octopamine to the muscle, induces several changes in the twitch characteristics (O'Shea & Evans 1979, Buchan & Evans 1980). The amplitude of the twitch and its rate of contraction are increased (maximum increase in amplitude 40%) and there is a much larger increase (maximum 150%) in the rate of relaxation of twitch tension (Evans 1981) (Fig. 3i). There is a concomitant increase in the size of the excitatory junctional potentials in the muscle fibres. All these effects are of long duration, far outlasting the period of DUMETi stimulation or octopamine application. When the same experiment is repeated on twitch tension generated by stimulation of the fast excitatory motor neuron (FETi), a large increase in the rate of relaxation of twitch tension is observed but only very small increases in the amplitude and rate of contraction are seen (Fig. 3ii).

The simplest explanation of these results is that the relaxation responses originate postsynaptically from the activation of octopamine receptors on the muscle fibres, while the increases in amplitude and contraction rate, which are specific for SETi-generated tension, are mediated presynaptically by the activation of octopamine receptors on the SETi terminals. The existence of such presynaptic receptors is indicated by the fact that octopamine can increase the frequency but not the amplitude of spontaneous miniature

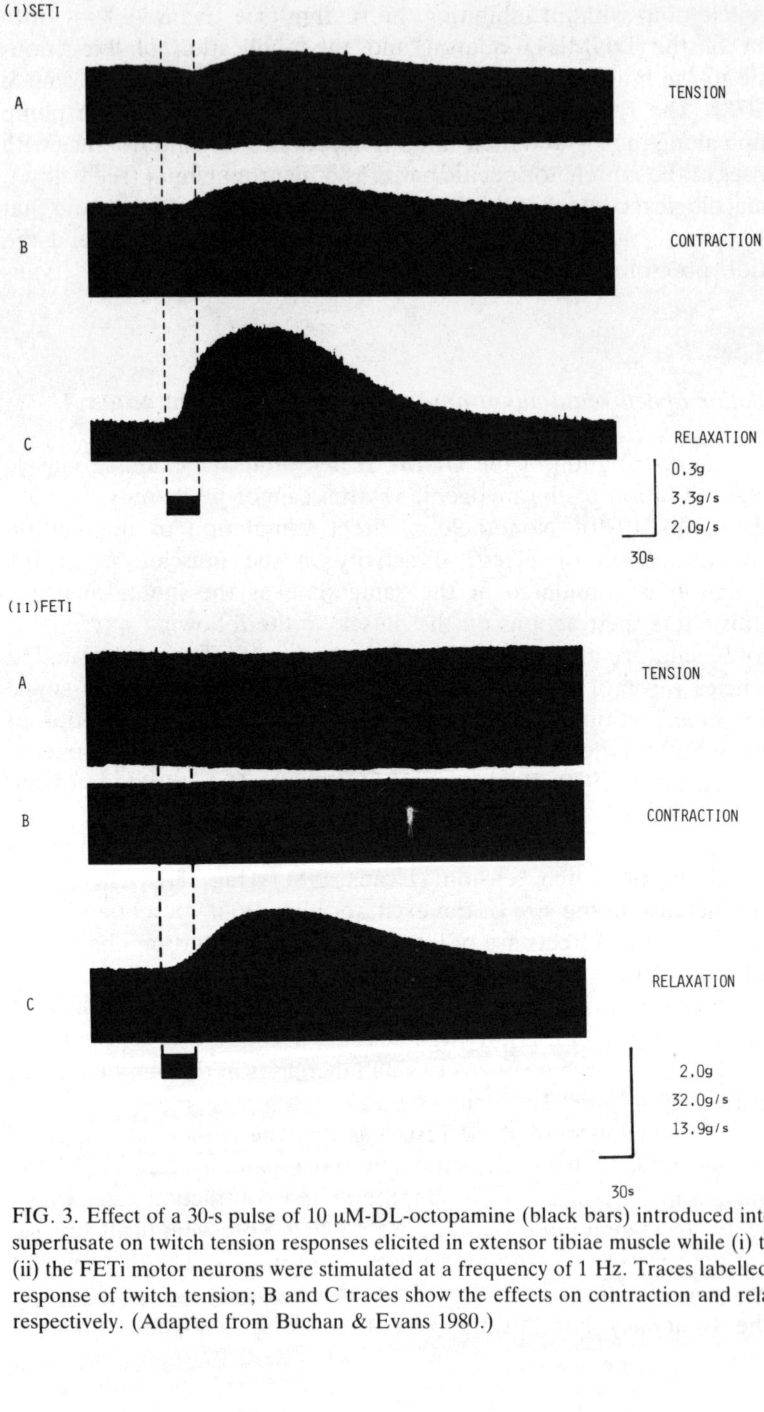

FIG. 3. Effect of a 30-s pulse of 10 μM-DL-octopamine (black bars) introduced into the muscle superfusate on twitch tension responses elicited in extensor tibiae muscle while (i) the SETi and (ii) the FETi motor neurons were stimulated at a frequency of 1 Hz. Traces labelled A show the response of twitch tension; B and C traces show the effects on contraction and relaxation rates respectively. (Adapted from Buchan & Evans 1980.)

end-plate potentials (MEPPs) from the SETi terminals (O'Shea & Evans 1979). In similar studies on muscle fibres innervated only by FETi, no change was found in MEPP frequency (Evans 1981). The above evidence, while suggesting that the octopamine-mediated increase in SETi twitch amplitude and contraction rate is produced at a presynaptic site, does not of course rule out the possibility of an additional postsynaptic component in these responses.

Since the locust rarely fires its motor neurons continuously at 1 Hz, we investigated the effects of octopamine application and DUMETi stimulation at several higher firing frequencies of SETi that produce tetanic tension in the muscle (Evans & Siegler 1982). Both effects depend on the frequency at which the motor neuron is stimulated. Fig. 4 shows that when SETi is

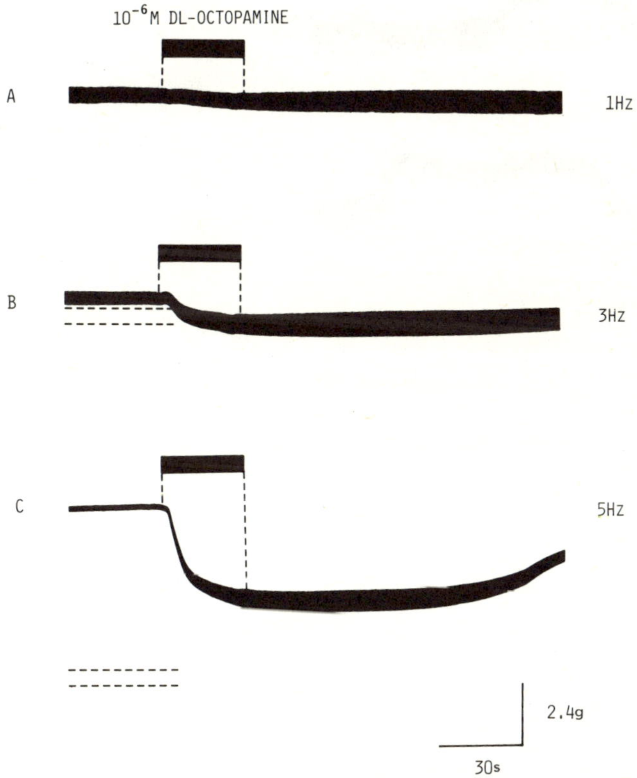

FIG. 4. Effect of a 30-s pulse of 10^{-6} M-DL-octopamine (black bars) introduced into the muscle superfusate on SETi-induced tension at various stimulation frequencies in the extensor tibiae muscle of a locust hindleg. Horizontal dashed lines indicate level of tension responses at 1 Hz for traces B and C. Increasing SETi stimulation frequency from 1–3 Hz and from 3–5 Hz increases the maintained tension in the muscle which is relaxed by octopamine application.

stimulated at 3 or 5 Hz, where there is an incomplete fusion of twitches, the predominant modulatory effect of octopamine is a prolonged relaxation of maintained tension. This effect reaches a maximum at stimulation frequencies between 5 and 7 Hz and decreases at higher frequencies (Evans & Siegler 1982). It is interesting to note that proctolin increases basal and maintained tension in this muscle (May et al 1979), again in a frequency-dependent manner over a prolonged time course (Fig. 5) (P. D. Evans, unpublished results). Thus, maintained tension in the locust extensor muscle is modulated antagonistically by the actions of a biogenic amine and a peptide. Whether

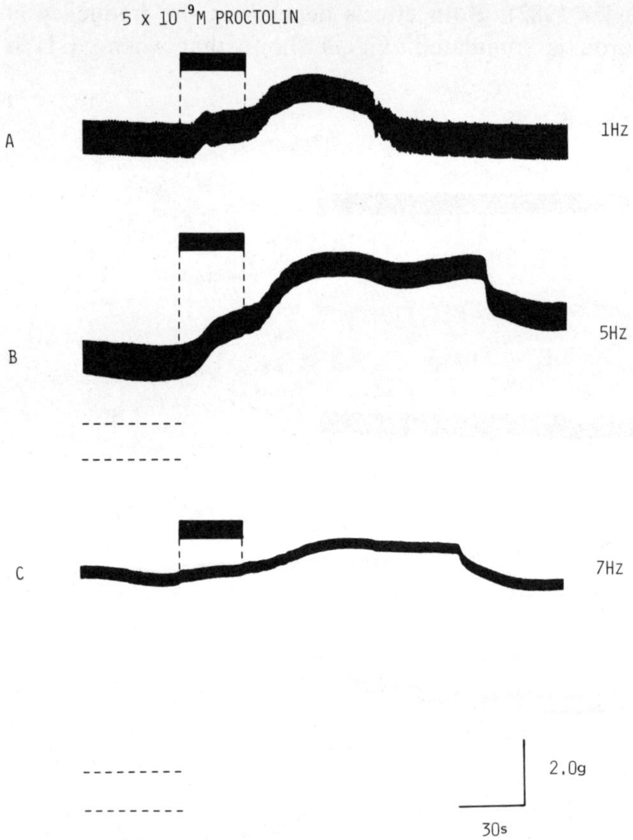

FIG. 5. Effect of a 30-s pulse of 5×10^{-9} M-proctolin (black bars) introduced into the muscle superfusate on SETi-induced tension at various stimulation frequencies in the extensor tibiae muscle of a locust hindleg. Horizontal dashed lines indicate level of tension responses at 1 Hz for traces B and C. Proctolin increases level of maintained tension in muscle. The effect is frequency-dependent and in the example shown is maximal at 5 Hz. Note abrupt relaxation of muscle during recovery phase.

these effects of proctolin have a physiological significance awaits a demonstration that the muscle is actually exposed to such concentrations of proctolin in the intact animal.

The locust extensor tibiae muscle also exhibits 'catch' tension (Wilson & Larimer 1968). In this condition a given frequency of motor neuron stimulation can maintain more tension after a brief train of higher frequency stimulation than before it. This effect would obviously be an advantage to the animal for maintaining posture but would be a considerable hindrance during locomotion. Octopamine application or DUMETi stimulation selectively reduces catch tension in this muscle (Evans & Siegler 1982). Fig. 6A shows the action of octopamine in reducing catch tension elicited by stimulating SETi continuously at 15 Hz with an interposed 0.5 s pulse at 100 Hz. Catch tension is also shown by the muscle if SETi is stimulated with a pattern based on SETi activity during walking (Burns & Usherwood 1979). Under these stimulation conditions both octopamine (Fig. 6B) and DUMETi stimulation reduce the amount of tension maintained between bursts of SETi activity as well as increasing the rate of relaxation of tension (Evans & Siegler 1982).

(c) Function of DUMETi in relation to behaviour

The above evidence suggests that the DUMETi neuron changes the extensor tibiae muscle in such a way that instead of favouring maintenance of posture it favours rapid movements, such as occur during walking. Stimulating DUMETi reduces interference from the myogenic rhythm and removes any residual catch tension from the muscle, thus ensuring that its past history of excitation will not interfere with the response to a given stimulus pattern. The muscle's responses will then be more closely matched to its pattern of neuronal input. This suggestion receives further support from the observations that DUMETi fires before any movement involving brisk tibial extension and that its firing frequency is increased when the animal is aroused (Hoyle & Dagan 1978). Therefore, DUMETi firing is timed appropriately to reduce the catch property of the extensor muscle when locomotion begins; because of the long-lasting effects of DUMETi, its action would continue for some time during locomotion. These functions are consistent with the general idea put forward by several authors that octopamine and other biogenic amines contribute to arousal and increased behavioural responsiveness (see Evans 1980a).

Pharmacology of octopamine receptors

Octopamine receptors can be assigned to different classes on the basis of their interactions with various agonists and antagonists (Evans 1981). All the

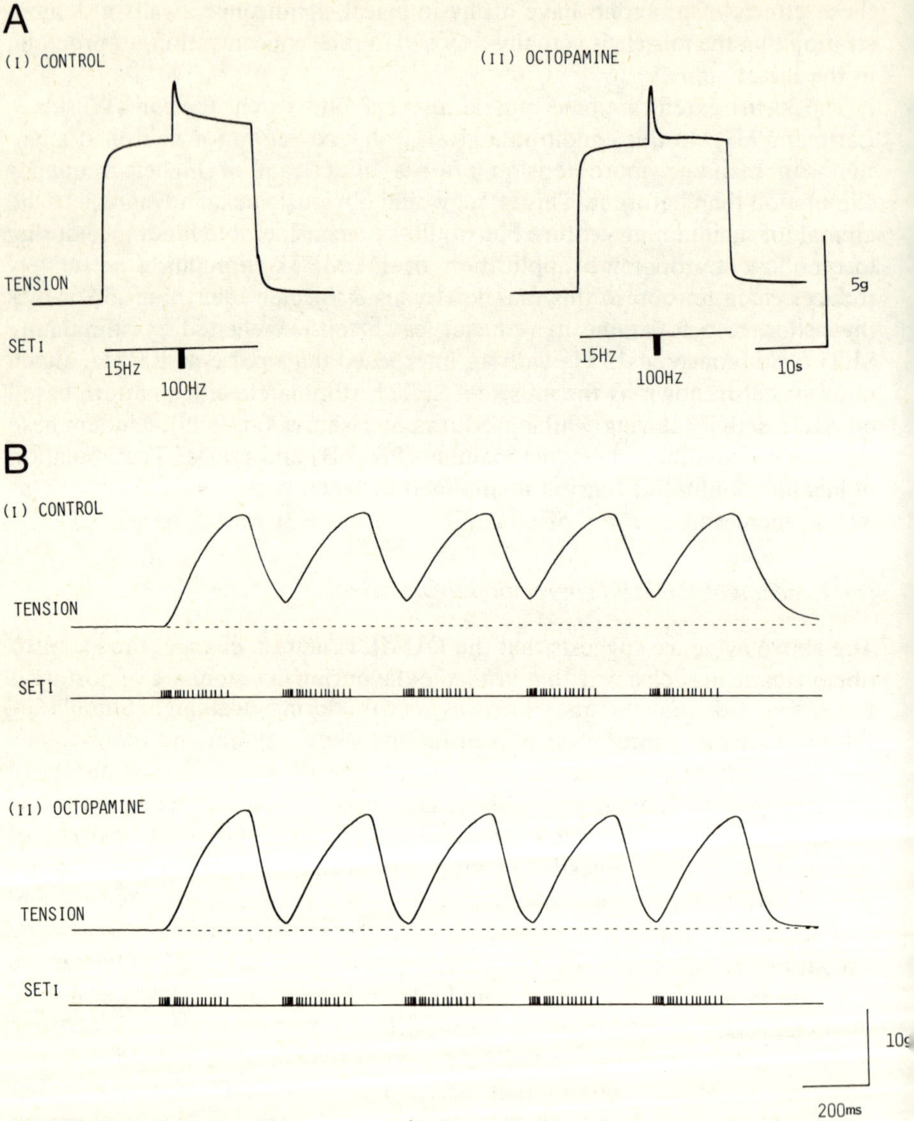

FIG. 6. Relaxation of catch tension by 10^{-6} M-DL-octopamine.
A. Tension profile produced by stimulating SETi continuously at 15 Hz except for an interposed period of 0.5 s at 100 Hz (black bar), (i) before and (ii) during the application of octopamine to the muscle.
B. Tension profile in response to five cycles of SETi stimulation in a pattern derived from a walking animal, (i) without and (ii) with octopamine in the muscle superfusate. There is a more rapid rise and fall of tension with octopamine and a reduction in the tension maintained between steps. (Adapted from Evans & Siegler 1982.)

classes, however, have some features in common that separate octopamine receptors from those specific for other neurotransmitters (Evans 1980b, 1981).

(a) General properties of octopamine receptors

Characteristically, octopamine receptors respond maximally to biogenic amines that have a single hydroxyl group (in the *para* position) on the aromatic ring (phenolamines) rather than to amines lacking ring hydroxyl groups (phenylamines) or those having two ring hydroxyl groups (catecholamines). This feature is exhibited by all the octopamine receptors in the locust extensor tibiae muscle (Evans 1981) and by other insect octopamine receptors, including those modulating light production in fireflies (Carlson 1968) and those stimulating adenylate cyclase activity in cockroach brain homogenates (Harmer & Horn 1977). Synephrine, the *N*-methylated analogue of octopamine (see Fig. 1 for structure), consistently appears to be more potent than octopamine itself in intact cellular preparations such as the firefly light organ and the locust extensor muscle. This contrasts with studies on the stimulation of adenylate cyclase activity in broken cell preparations of firefly light organ (Nathanson 1979) and cockroach brain (Harmer & Horn 1977), where octopamine is slightly more potent than synephrine. This raises the problem encountered in all pharmacological studies of intact preparations, namely that different drugs may have a differential access to receptor sites or may be inactivated at different rates. It is known from studies on biogenic amine uptake in cockroach nerve cord that *N*-methylated amines are taken up less readily than their non-methylated precursors by a high affinity uptake system (Evans 1978). However, the responses of the locust extensor tibiae muscle to octopamine and synephrine are not changed by exposure to desimipramine, a potent inhibitor of the high affinity uptake system for octopamine (Evans 1978, 1981).

All the responses described so far for octopamine have also been shown to be stereospecific for the D(−) isomer (see Evans 1980a,b, 1981). This isomer is the naturally occurring form in insects (Goosey & Candy 1980, Starratt & Bodnaryk 1981).

Another characteristic of all octopamine receptors studied in insects, crustaceans and molluscs (see Evans 1980a, 1981) is that they are specifically blocked by α-adrenergic blocking agents such as phentolamine, in preference to β-adrenergic blocking agents such as propranolol. In addition, in the locust extensor muscle all the octopamine receptors studied are selectively activated by α-adrenergic agonists such as clonidine and tolazoline in preference to β-adrenergic agonists such as isoprenaline (Evans 1981).

(b) Evidence for several classes of octopamine receptor

In the locust extensor tibiae neuromuscular preparation the octopamine receptors present can be differentiated into three classes on the basis of their interactions with a range of agonists and antagonists (Evans 1981). It has been proposed that the receptors on the myogenic bundle be designated $octopamine_1$ class receptors to distinguish them from the $octopamine_2$ class receptors mediating the modulation of neuromuscular transmission and muscle contraction in the rest of the extensor muscle. It has further been proposed that the presynaptic octopamine receptors on the SETi nerve terminals which mediate increases in the amplitude of twitch tension should be designated $octopamine_{2A}$ class receptors to distinguish them from the $octopamine_{2B}$ class receptors mediating the relaxation of twitch and tetanic tension in this muscle (Evans 1981).

$Octopamine_1$ receptors can be distinguished from $octopamine_2$ receptors by the fact that chlorpromazine and yohimbine are much better blockers of the former than is metoclopramide, whereas the converse is true for the latter receptors (Table 1). This distinction is supported by our evidence that clonidine is a more potent agonist than naphazoline at $octopamine_1$ receptors while the converse is true for $octopamine_2$ receptors (Table 2).

$Octopamine_{2A}$ can be distinguished from $octopamine_{2B}$ receptors, since blocking agents such as cyproheptadine, mianserin and metoclopramide are better blockers of 2A than 2B receptors, while the converse is true for chlorpromazine (Table 1). This distinction is again supported by agonist studies. Naphazoline is a much better agonist than tolazoline at 2A receptors and tolazoline is much better than clonidine at 2B receptors (Table 2).

(c) Relationship between α-adrenoreceptors and octopamine receptors

The similarities between the pharmacological profiles of octopamine receptors and those of vertebrate α-adrenoreceptors are striking and suggest that the specificity of drugs currently available for characterizing aminergic receptors is very poor. In general, however, α-receptors are thought to mediate their actions via changes in calcium permeability, activation of guanylate cyclase activity and inhibition of adenylate cyclase activity (see Berridge 1980), while octopamine receptors are reported to increase adenylate cyclase activity (see Evans 1980a). A possibility that must be considered is that each of the classes of octopamine receptor may mediate its actions through different mechanisms, as has been found for many forms of other aminergic receptors (Berridge 1980).

Evidence is increasing for the existence of specific octopamine receptors in

TABLE 1 Action of antagonists on octopamine receptors

Drug	*Receptor class* *Octopamine$_1$* *(EC_{50})(M)*	*Octopamine$_{2A}$* *(EC_{50})(M)*	*Octopamine$_{2B}$* *(EC_{50})(M)*
Metoclopramide	—	1.0×10^{-6}	9.5×10^{-6}
Yohimbine	2.8×10^{-7}	—	—
Chlorpromazine	2.6×10^{-8}	1.6×10^{-4}	7.0×10^{-5}
Cyproheptadine	3.7×10^{-8}	2.2×10^{-6}	5.1×10^{-5}
Mianserin	4.5×10^{-6}	1.2×10^{-6}	2.0×10^{-5}

EC_{50} for octopamine$_1$ receptors is the concentration of drug required to reduce the response of myogenic rhythm to a 5-min pulse of 10^{-7} M-DL-octopamine by 50%. EC_{50} for octopamine$_{2A}$ and octopamine$_{2B}$ receptors is the concentration of drug required to reduce the response of SETi-induced twitch amplitude and relaxation rate, respectively, to a 30-s pulse of 10^{-6} M-DL-octopamine by 50% (SETi fired at 1 Hz). For further details of experiments and additional drugs see Evans (1981).

TABLE 2 Action of agonists on octopamine receptors

Drug	*Receptor class* *Octopamine$_1$* *(EC_{50})(M)*	*Octopamine$_{2A}$* *(EC_{50})(M)*	*Octopamine$_{2B}$* *(EC_{50})(M)*
Clonidine	6.8×10^{-10}	6.4×10^{-6}	2.0×10^{-5}
Naphazoline	1.2×10^{-8}	1.3×10^{-8}	2.2×10^{-7}
Tolazoline	1.5×10^{-9}	3.2×10^{-6}	6.0×10^{-7}

EC_{50} for octopamine$_1$ receptors is the concentration of a 5-min pulse of the drug required to reduce the frequency of the myogenic rhythm by 50%. EC_{50} for octopamine$_{2A}$ and octopamine$_{2B}$ receptors is concentration of a 30-s pulse of drug required to produce a 50% of maximal response to octopamine in SETi twitch amplitude and relaxation rate, respectively (SETi fired at 1 Hz). For further details of experiments and additional drugs see Evans (1981).

the vertebrate nervous system (see Evans 1981). In addition an increasing diversity of α-adrenergic receptor subclasses is being reported, based on physiological and biochemical studies using many drugs of only limited specificity. Thus in future studies of the isolation and pharmacology of presumed α-adrenoreceptors in vertebrates it will be necessary to establish that a single population of receptors is being studied and not a mixture of octopamine receptors and α-adrenoreceptors. Further studies are obviously required to see whether the classification of octopamine receptors outlined above can be extended to other invertebrate and vertebrate nervous systems. It is apparent, however, that any general classification scheme for aminergic receptors should reflect the close relationship between octopamine receptors and α-adrenoreceptors.

(d) Actions of formamidine pesticides on octopamine receptors

Recent information from work on firefly light organs (Hollingworth & Murdock 1980) and on the locust extensor tibiae muscle (Evans & Gee 1980) indicates that the formamidine pesticides such as chlordimeform produce at least some of their effects by interacting with octopamine receptors. The structural relationship between the phenolamines (octopamine and synephrine) and chlordimeform and its more potent demethylated derivative is shown in Fig. 1. In the locust the actions of the formamidines were qualitatively similar to the actions of octopamine but had a much longer time course. This has been suggested to be due to the increased tissue solubility of the formamidines, together with an increased half-time of receptor occupancy. The actions of the formamidines could be blocked by phentolamine but not by propranolol (Evans & Gee 1980).

Sublethal doses of formamidine pesticides produce many of their protective effects on crops by modifying insect behaviours such as feeding or mating, presumably by an interaction with receptor sites in the CNS. This is again consistent with the activation of a modulatory system such as that described for octopamine in the locust. It seems likely, therefore, that octopamine receptors can provide novel target sites for the development of a new range of pesticides.

Other roles for octopamine

The effects of octopamine in insects have many parallels with the actions of noradrenaline released from the sympathetic nervous system and the actions of adrenaline released from the adrenal glands of vertebrates (see Evans 1980a, 1981, Evans & Siegler 1982). Thus, as for noradrenaline and adrenaline in vertebrates, octopamine appears to have several roles to play in insects and its specific role in any particular part of the insect nervous system needs to be studied.

Apart from functioning as a neuromodulator in the locust, as described above, octopamine might also act as a true neurotransmitter. It is released from classical synaptic structures in the firefly light organ (see Evans 1980a) and in the glandular lobe of the locust corpus cardiacum, where it mediates the release of hyperlipaemic hormone (Orchard & Loughton 1981). Octopamine also affects carbohydrate metabolism in insect nerve cord, muscle and fat body (see Evans 1980a); it is thought to mediate these effects by being released into the haemolymph as a circulating neurohormone.

In the future it is likely that further roles will be identified for octopamine as its functions are examined in other areas of the insect nervous system.

Studies on the insect CNS will undoubtedly provide additional evidence for a true neurotransmitter role for octopamine, while studies on neurosecretory release structures will provide further information on the role of octopamine in the release of hormones. In functional terms it is likely that further examples of the antagonistic modulation of octopamine effects by 5-hydroxytryptamine and peptide hormones will be described. Most of the available evidence suggests that octopamine plays a central role in coordinating the behavioural responsiveness of insects, especially under stressful conditions.

REFERENCES

Bacon JP, Altman JS 1977 A silver intensification method for cobalt-filled neurones in wholemount preparations. Brain Res 138:359-363

Berridge MJ 1980 Receptor coupling to adenylate cyclase and calcium gates. In: Sattelle DB et al (eds) Receptors for neurotransmitters, hormones and pheromones in insects. Elsevier/North-Holland Biomedical Press, Amsterdam, p 27-38

Buchan PB, Evans PD 1980 Use of an operational amplifier signal differentiator reveals that octopamine increases the rate of development of neurally evoked tension in insect muscle. J Exp Biol 85:349-352

Burns MD, Usherwood PNR 1979 The control of walking in Orthoptera. II. Motor neuron activity in normal free-walking animals. J Exp Biol 79:69-98

Carlson AD 1968 Effects of drugs on luminescence in larval fireflies. J Exp Biol 49:195-199

Evans PD 1978 Octopamine: a high-affinity uptake mechanism in the nervous system of the cockroach. J Neurochem 30:1015-1022

Evans PD 1980a Biogenic amines in the insect nervous system. Adv Insect Physiol 15:317-473

Evans PD 1980b Octopamine receptors in insects. In: Sattelle DB et al (eds) Receptors for neurotransmitters, hormones and pheromones in insects. Elsevier/North-Holland Biomedical Press, Amsterdam, p 245-258

Evans PD 1981 Multiple receptor types for octopamine in the locust. J Physiol (Lond) 318:99-122

Evans PD, Gee JD 1980 Action of formamidine pesticides on octopamine receptors. Nature (Lond) 287:60-62

Evans PD, O'Shea M 1977 The identification of an octopaminergic neurone which modulates neuromuscular transmission in the locust. Nature (Lond) 270:257-259

Evans PD, O'Shea M 1978 The identification of an octopaminergic neurone and the modulation of a myogenic rhythm in the locust. J Exp Biol 73:235-260

Evans PD, Siegler MVS 1982 Octopamine mediated relaxation of maintained and catch tension in locust skeletal muscle. J Physiol (Lond) 324:in press

Goosey MW, Candy DJ 1980 The D-octopamine content of the haemolymph of the locust *Schistocerca americana gregaria* and its elevation during flight. Insect Biochem 10:393-397

Harmer AJ, Horn AS 1977 Octopamine-sensitive adenylate cyclase in cockroach brain: effects of agonists, antagonists and guanylyl nucleotides. Mol Pharmacol 13:512-520

Hollingworth RM, Murdock LL 1980 Formamidine pesticides: octopamine-like actions in a firefly. Science (Wash DC) 208:74-76

Hoyle G 1975 Evidence that insect dorsal unpaired median (DUM) neurones are octopaminergic. J Exp Zool 193:425-431

Hoyle G, Barker DL 1975 Synthesis of octopamine by insect dorsal median unpaired neurones. J Exp Zool 193:433-439

Hoyle G, Dagan D 1978 Physiological characteristics and reflex activation of DUM (octopaminergic) neurones of locust metathoracic ganglion. J Neurobiol 9:59-79

Hoyle G, Dagan D, Moberly B, Colquhoun W 1974 Dorsal unpaired median insect neurons make neurosecretory endings on skeletal muscle. J Exp Zool 187:159-165

May TE, Brown BE, Clements AN 1979 Experimental studies upon a bundle of tonic fibres in the locust extensor tibialis muscle. J Insect Physiol 25:169-181

Nathanson JA 1979 Octopamine receptors, adenosine 3′,5′-monophosphate and neural control of firefly flashing. Science (Wash DC) 203:65-68

Orchard I, Loughton BG 1981 Is octopamine a transmitter mediating hormone release in insects? J Neurobiol 12:143-153

O'Shea M, Evans PD 1979 Potentiation of neuromuscular transmission by an octopaminergic neurone in the locust. J Exp Biol 79:169-190

Starratt AN, Bodnaryk RP 1981 Stereoisomeric identity of octopamine in the central nervous system of invertebrates. Insect Biochem 11:645-648

Wilson DM, Larimer JL 1968 The catch property of ordinary muscle. Proc Natl Acad Sci USA 61:909-916

DISCUSSION

Piek: Can you explain Graham Hoyle's finding that β-adrenergic antagonists are active against the octopamine receptor, but not α-antagonists?

Evans: If Hoyle would publish what drugs he used and their concentrations, I could attempt to explain it. In our experiments β-adrenergic antagonists, such as propranolol, can block octopamine receptors but only at concentrations that are several orders of magnitude higher than those required for α-adrenergic antagonists.

Neumann: Can you classify chlordimeform and the demethylated analogue in relation to the three octopamine receptors?

Evans: I have no information on this at the moment.

Donnellan: What is the removal mechanism for octopamine? Does the longevity of the response suggest that diffusion is responsible for its removal?

Evans: The cockroach CNS contains a specific high affinity uptake mechanism for octopamine. An effective blocking agent for this mechanism is desimipramine. We tried to potentiate responses to octopamine in the periphery in the locust, by putting desimipramine in, but we got no significant effects. In fact at high concentrations, desimipramine started to antagonize the octopamine responses, so we couldn't increase the concentration of the drug enough to block an uptake system, if it exists.

Donnellan: Is any particular enzyme involved in the overall removal of octopamine?

Evans: I have no evidence for or against that. *N*-Acetyltransferase activity against a wide range of biogenic amines is present in almost every tissue of the locust.

Mordue: There is evidence from Candy's group that octopamine is released during flight in the locust; there are surprisingly large amounts in the circulation at this time (see Goosey & Candy 1980). What happens in the whole animal? Is the extensor tibiae muscle affected by octopamine during flight or is there some mechanism that prevents blood-borne octopamine from getting at the sites?

Evans: The raised concentrations of octopamine reported during locust flight by Goosey & Candy (1980) would certainly be above threshold for the activation of the octopamine receptors on the extensor tibiae muscle, if it were ever exposed to them. However, it is not clear whether the muscles of the legs would be exposed to the same concentration changes during flight as those measured for the body cavity. Further, it seems likely that the concentration changes experienced by any tissue will differ depending on its distance from the release site. Thus, if octopamine is released during flight from the corpora cardiaca, to sample blood from the neck, as Goosey & Candy (1980) did, may give one a very high localized concentration close to the release site and not the general circulating concentration of the haemolymph, which might be considerably reduced by both dilution and tissue metabolism.

Mordue: We need to know not only what happens to locally released octopamine but also what happens to it when it is circulating in the haemolymph.

Evans: If labelled octopamine is injected into the haemolymph it is metabolized in a few minutes into *N*-acetylated products (P. D. Evans, unpublished). There is a widespread inactivation mechanism in the whole insect that would bring down the general circulating levels.

Berridge: In the mammalian heart, adrenaline modulates contraction in almost exactly the same way as octopamine seems to regulate locust leg muscle. There is a positive inotropic effect and an increased rate of relaxation, but in the mammalian heart these are both postsynaptic events. Could you review the evidence that the increased contraction in response to octopamine is presynaptic?

Evans: The relaxation response is obviously a postsynaptic response. The other responses have a very different time course to that. Our evidence for presynaptic octopamine receptors on the terminals of the slow motor neurons is that octopamine increases the frequency of the spontaneous miniature end-plate potentials (MEPPs) but doesn't change their amplitude (p 53). We could find no effect on MEPPs from the fast motor neuron. We think this suggests a presynaptic site of action for octopamine on the terminals of the slow motor neuron. We also find that the pharmacological profile of the receptor mediating the change in MEPPs is very similar to the pharmacological profile of the octopamine$_{2A}$ class of receptor (p 59). This doesn't, however, rule out a contribution from the postsynaptic mechanism to the increase in contraction size.

Kelly: The peaks in Fig. 3 (p 52) all came at the same times and they returned to normal at the same times. It was only the half-time after the peak, the falling phase, that was different. Could that be a property of the elasticity of the muscle? I don't think it is evidence that the receptors are involved for a different length of time.

Evans: A close examination of the peaks in Fig. 3 reveals that the peak responses of the relaxation rate occur some 10–20 s before the peak responses in tension amplitude and contraction rate. We have not examined the elastic properties of the muscle.

Usherwood: Since octopamine increases the rate of relaxation of the extensor tibiae muscle, the octopaminergic input to this muscle could assist in promoting movements of the tibia during walking. Does the inhibitory (GABA) input to the extensor muscle also play a role in walking, as suggested by Runion & Usherwood (1968), Usherwood & Runion (1970) and Burns & Usherwood (1978, 1979), and how are the inhibitory and octopaminergic systems related functionally?

Evans: The catch tension in the muscle is also reduced by firing the common inhibitor (the third motor fibre to this muscle). We looked at the effect on the preparation of firing the common inhibitor and of firing DUMETi and found very different time courses. The slow motor neuron (SETi) is maximally sensitive to octopamine at a frequency of 7 Hz but we can repeat the same thing at different frequencies. When we fire DUMETi at 10 Hz we get a dramatic reduction in tension over a long time course. If we fire the common inhibitor in the same preparation for an equivalent period of time at 50 Hz there is an almost immediate reduction in tension. As soon as we stop firing the inhibitor the tension returns. So they are of very different magnitudes and different time courses. I think the common inhibitor is just reducing tension and competing against the excitation produced by the slow excitatory motor neuron, whereas DUMETi acts through a very different mechanism.

Usherwood: How do you envisage octopamine acting at the postsynaptic level?

Evans: There are three types of octopamine receptor, so when we apply octopamine we presumably stimulate all three kinds of octopamine receptor in the muscle. Measurements on whole muscle don't tell us what is going on at any specific type of receptor. The long time course suggests that the modulatory phenomenon may be affecting adenylate cyclase or calcium, or both. Octopamine is said to stimulate adenylate cyclase but I don't believe that all classes of octopamine receptor necessarily stimulate this cyclase. I haven't got the evidence to divide them into different functional categories.

Sattelle: Do the various physiological effects you describe in response to octopamine differ in their sensitivity to octopamine?

Evans: Yes. The myogenic rhythm is the most sensitive. It is affected at concentrations of 10^{-10} or 10^{-11} M. The modulatory phenomena are sensitive at 10^{-9} or 10^{-10} M.

Mordue: Locusts have different types of walking behaviour. One might expect to find differences in octopamine-modulated actions in the juvenile hopper stages, compared with adult walking.

Cull-Candy: Why are there no octopamine receptors on the fast motor neuron terminals?

Evans: The fast motor neuron (FETi) is used in kicking and jumping, which are explosive responses, and modulating those by a certain percentage either way might not make much difference functionally to the system. There will just be an all-or-none explosive response, whereas a more graded response may be needed in controlling posture, or the speed of walking. We have no evidence for that but, as Professor Mordue implies, this would be worth looking at. Octopamine doesn't change the MEPP frequency of muscle fibres innervated by the fast motor neuron.

Cull-Candy: That is a good control for showing that octopamine works on the slow motor neurons.

Evans: Yes, it is restricted to the SETi terminals, but we can't rule out that it is a property of the muscle fibres that are innervated solely by the slow motor neuron.

Kelly: Each new group working on neurotransmitters develops a new system of classification. In the mammalian adrenergic system they use α_1, β_1 and β_2. Why did you use 1, 2A and 2B rather than alpha and beta, with 1 and 2 signifying pre- and postsynaptic?

Evans: The adrenergic system has become totally confused and I didn't want to draw any rigid parallels between these octopamine receptors and α- and β-adrenergic receptors. The octopamine receptors seem to fall into two main groups, which is why I called them octopamine$_1$ and octopamine$_2$. The distinctions between 2A and 2B are more hazy than the distinctions between 1 and 2 and I felt it would be safer not to refer to a class 3. If someone later shows that A and B are the same but that there is a problem of accessibility to the receptor or something like that, the structure of the classification will not have to be changed. I didn't want to say that octopamine$_1$ and octopamine$_2$ are pre- and postsynaptic because the picture in vertebrates is now very confused, with α_1 and α_2 receptors being found both presynaptically and postsynaptically. The plethora of alpha receptors coming out from biochemical studies is confusing the matter. I have had long discussions about this with Mike Berridge, who would prefer a functional basis for classifying receptors, but we lack the information for doing this.

Kelly: I would suggest that work on invertebrate systems has not yet told us anything new about vertebrate nervous systems. Ascher & Kehoe's (1975) studies of ACh receptors in gastropods, for example, haven't contributed anything new to our understanding of the vertebrate nervous system. Strychnine blocks some ACh effects in the gastropod nervous system, but I don't think this can be seen in the vertebrate CNS. The glutamate system in insects has also been disappointing in terms of finding specific antagonists. The GABA system in invertebrates is not blocked by bicuculline, though it is blocked in vertebrates. I would have thought that the same applies with clonidine and phentolamine; the locust isn't going to tell us anything new about the vertebrate nervous system.

Usherwood: I cannot agree that invertebrate preparations have contributed nothing to basic neuropharmacology. For example, extrajunctional glutamate receptors were first described in invertebrate preparations and presynaptic receptors, including autoreceptors, which are currently the subject of intensive study in vertebrate systems, were discovered in insect preparations in the 1960s (Usherwood & Machili 1966, 1968).

Kelly: I was talking specifically in terms of antagonists, not agonists. In many ways the GABA story was led by the invertebrate workers and it has been very helpful to have people breaking the road in front of us. That is true of all the agonists, but with antagonists the receptors in invertebrates look distinctly different from those in the vertebrate nervous system.

Cull-Candy: I'm not sure that is correct for GABA receptors. What about picrotoxin? This may simply act on chloride channels in both vertebrates and invertebrates.

Kelly: People have had a hard job proving bicuculline to be a GABA antagonist in invertebrates (Feltz 1971, and see review by Kelly & Peart 1975).

Cull-Candy: Didn't Takeuchi & Onodera (1972) show that bicuculline reduces the GABA response at the crayfish nerve-muscle junction? Takeuchi & Takeuchi (1969) and more recently Constanti (1978) have suggested that picrotoxin may block the chloride channel at the GABA receptor.

Kelly: McLennan (1973) would dispute those results. In this particular case, invertebrate work appears to be no easier to interpret than vertebrate work.

Usherwood: Much effort has been devoted to the discovery of antagonists for glutamate receptors on vertebrate neurons but the complexity of the preparations used to test possible antagonists and the methodology employed complicates interpretations of the sites and modes of action of test compounds. When compounds claimed to be antagonists of glutamate receptors on vertebrate neurons have been tested on invertebrate glutamatergic systems the results have been disappointing, most of them exhibiting weak agonist properties. It is true that we have so far failed to discover an efficient antagonist for glutamate receptors on insect muscle, but I am confident that antagonists for insect glutamate receptors will show some activity on vertebrate glutamatergic systems.

Kelly: I hope you are right. You say that invertebrate workers were the first to describe extrajunctional receptors, but Curtis has always looked on glutamate as an extrajunctional receptor, though he didn't use that word (Curtis et al 1960). He probably didn't think in those terms but he certainly showed that the reversal potential was different from that for the junctional potential.

Cull-Candy: I don't think Curtis's observations are explained in terms of an extrajunctional receptor. They are more readily explained if recording and current-injecting electrodes were far away from the synapse—so the membrane potential at the synapse was not actually being observed during current injection.

M. E. Eldefrawi: Work on *Aplysia* showed that tubocurarine inhibited sodium and chloride conductances induced by activation of different neurotransmitters, thus acting as a channel blocker (Carpenter et al 1977). Recently, Katz & Miledi (1978) and Colquhoun et al (1979) have shown that even in vertebrates, curare acts as a channel blocker in addition to its action as a competitive antagonist. Here, work on invertebrates led to a reinvestigation of the vertebrate pharmacology which we had thought was clearly understood.

Kelly: Blackman was the first to publish on the channel-blocking effect of hexamethonium in the frog sympathetic ganglia (see Blackman 1970).

Neumann: How representative of invertebrates are locusts? It may just be that invertebrates are more diverse and that several solutions are possible, one of which might fit the vertebrate nervous system.

Usherwood: From a pharmacological viewpoint the terms vertebrate and invertebrate introduce the false impression of a discontinuity. There is obviously a diversity of structure and function in the nervous systems of both vertebrates and invertebrates and it is important to appreciate the dangers of extrapolation even between related species. However, at the cellular level it is possible to identify basic principles in pharmacology which apply across the animal kingdom without regard for the presence or absence of backbones!

Dudai: Vertebrate α-adrenergic receptors from various sources show marked differences in their pharmacological profile; so what is a representative vertebrate species? There may not be one.

Kelly: The reasons why Dr Evans finds octopamine unsatisfactory in an invertebrate system are not necessarily going to help us with other unsatisfactory systems in the vertebrate CNS. It would be much more interesting to get a chemist to look more closely at the octopamine molecule. It is a simpler molecule than adrenaline or noradrenaline in some respects. There is more potency when a CH_3 group is lost and one starts to see a variety of agonists which haven't been seen in other systems.

Evans: I find the locust system useful because I can identify drugs that block one of the receptors and then look at the actions of the other receptors. Vertebrate workers might get something from this if they looked at their octopamine receptors to see whether the same classes of drugs are active.

Kelly: There are many things that you can do that we can't do in vertebrates. Denervating the locust extensor tibiae system must be relatively easy. You could remove the slow motor neuron and your preparation would survive long enough to get effective degeneration. The locust survives for days rather than hours after denervation and you could get rid of presynaptic receptors in this way.

Narahashi: People tend to regard whatever material is used first as the prototype. In many cases a particular material was used because it was suitable for a particular experiment. The squid axon has been regarded as a prototype because it is so convenient for the detailed analysis of membrane ionic conductances. It is important to do an experiment using the most convenient preparation, whether vertebrate or invertebrate, and then apply that technique and knowledge to material relevant to the question.

REFERENCES

Ascher P, Kehoe J 1975 Amine and amino acid receptors in gastropod neurons. In: Iversen LL et al (eds) Handbook of psychopharmacology, vol 4: Amino acid neurotransmitters. Plenum, New York, p 265-310

Blackman JG 1970 Dependence on membrane potential of the blocking action of hexamethonium at a sympathetic synapse. Proc Univ Otago Med Sch 48:4-5

Burns MD, Usherwood PNR 1978 Mechanical properties of locust extensor tibiae muscles. Comp Biochem Physiol 61A:85-95

Burns MD, Usherwood PNR 1979 The control of walking in Orthoptera. II: motor neurone activity in normal free-walking animals. J Exp Biol 79:69-98

Carpenter DO, Swann JW, Yarowsky PJ 1977 Effect of curare on responses to different putative neurotransmitters in *Aplysia* neurons. J Neurobiol 8:119-132

Colquhoun D, Dreyer F, Sheridan RE 1979 The actions of tubocurarine at the frog neuromuscular junction. J Physiol (Lond) 293:247-284

Constanti A 1978 The 'mixed' effect of picrotoxin on the GABA dose/conductance relation recorded from lobster muscle. Neuropharmacology 17:159-167

Curtis DR, Phillis JW, Watkins JC 1960 The chemical excitation of spinal neurones by certain acidic amino acids. J Physiol (Lond) 150:656-682

Feltz A 1971 Competitive interaction of β-guanidino propionic acid and γ-aminobutyric acid on the muscle fibre of the crayfish. J Physiol (Lond) 216:391-401

Goosey MW, Candy DJ 1980 The D-octopamine content of the haemolymph of the locust, *Schistocerca americana gregaria* and its elevation during flight. Insect Biochem 10:393-397

Katz B, Miledi R 1978 A re-examination of curare action at the motor endplate. Proc R Soc Lond B Biol Sci 203:119-133

Kelly JS, Peart PM 1975 Amino acid receptors in CNS. II. GABA in supraspinal regions. In: Iversen LL et al (eds) Handbook of psychopharmacology, vol 4: Amino acid neurotransmitters. Plenum, New York, p 129-209

McLennan H 1973 γ-Aminobutyric acid antagonists in crustacea. Can J Physiol Pharmacol 51:774-775

Runion HI, Usherwood PNR 1968 Tarsal receptors and leg reflexes in the locust and grasshopper. J Exp Biol 49:421-436

Takeuchi A, Onodera K 1972 Effect of bicuculline on the GABA receptor of the crayfish neuromuscular junction. Nature (Lond) 236:55-56

Takeuchi A, Takeuchi N 1969 A study of the action of picrotoxin on the inhibitory neuromuscular junction of the crayfish. J Physiol (Lond) 205:377-391

Usherwood PNR, Machili P 1966 Chemical transmission at the insect excitatory neuromuscular synapse. Nature (Lond) 210:634-636

Usherwood PNR, Machili P 1968 Pharmacological properties of excitatory neuromuscular synapses in the locust. J Exp Biol 49:341-361

Usherwood PNR, Runion HI 1970 Analysis of the mechanical responses of metathoracic extensor tibiae muscles of free-walking locusts. J Exp Biol 52:39-58

Properties of postsynaptic channels activated by glutamate and GABA in locust muscle fibres

S. G. CULL-CANDY

Department of Biophysics, University College London, Gower Street, London WC1E 6BT, UK

Abstract Analysis of voltage-clamped noise has been used to investigate the life-time and conductance of the single excitatory synaptic channels that are opened during ionophoretic application of glutamate to locust muscle fibres. Similar information may be obtained with the patch clamp technique, which has been used to study extrajunctional glutamate receptors. In addition, properties of channels activated by neurally released transmitter have been obtained by analysing the spontaneous miniature excitatory junctional currents that result from the action of transmitter on the muscle membrane. These techniques have been used to look at the channel kinetics of glutamate receptors under various conditions. A number of factors influence the way channels operate. These include the membrane potential of the muscle fibre, the agonist which opens the channel, the temperature of the preparation and the species of cation in the bathing medium.

The conductance and life-time of inhibitory channels activated by γ-aminobutyric acid in these muscle fibres have also been obtained by analysing current noise. Inhibitory and excitatory receptors show marked differences in their channel kinetics.

Neurotransmitters such as glutamate, γ-aminobutyric acid (GABA) and acetylcholine (ACh) usually act on synaptic receptors to increase the postsynaptic permeability to small cations and anions. Transmitter–receptor interaction is thought to cause the brief and transient opening of ionic channels. A number of methods have been developed to give information about the operation of ionic channels associated with synaptic receptors. These include analysis of the membrane current noise that occurs in the prolonged presence of transmitter or agonist as a result of fluctuations in the number of open channels (Katz & Miledi 1972, Anderson & Stevens 1973). A

1982 Neuropharmacology of insects. Pitman, London (Ciba Foundation symposium 88) p 70-87

second method is the patch clamp technique for measuring the currents passing through individual channels (Neher et al 1978). A third approach has been the analysis of the relaxation that occurs after a perturbation, for example after a jump in agonist concentration when a transmitter packet is released from a nerve terminal (Katz & Miledi 1972, Anderson & Stevens 1973). This paper will consider properties of excitatory glutamate- and inhibitory GABA-activated channels in locust muscle (*Schistocerca gregaria*), revealed using these approaches.

Junctional and extrajunctional glutamate receptors

Locust muscle fibres possess excitatory junctional glutamate receptors (Usherwood & Machili 1968, Beránek & Miller 1968) and at least two distinct types of extrajunctional receptors. The extrajunctional D (depolarizing) and H (hyperpolarizing) receptors differ from each other in their pharmacological properties and in the kinetics and selectivity of their associated ionic channels (Cull-Candy 1976). So far, analysis of channel kinetics of glutamate receptors has mainly been restricted to excitatory junctional receptors and extrajunctional D-receptors.

Fig. 1 illustrates typical extrajunctional responses to glutamate and the glutamate analogue, ibotenate, applied to a normal muscle fibre from adjacent barrels of an ionophoretic micropipette. Both D- and H-receptors are activated by glutamate, while only H-receptors are activated by ibotenate. The marked difference in the time course of D- and H-responses suggests differences in the kinetic properties of the channels associated with the two types of extrajunctional receptors. To avoid complications which would result from simultaneous activation of both membrane channels we abolish the H-response by removing Cl^- from the medium in the experiments on glutamate current noise described below.

In several respects the distribution of glutamate sensitivity in locust muscle shows interesting similarities to the distribution of ACh sensitivity in normal and denervated vertebrate muscle (see Axelsson & Thesleff 1959, Katz & Miledi 1964). Junctional glutamate sensitivity is relatively localized whereas extrajunctional sensitivity is present along the entire length of muscle fibres. After denervation (Usherwood 1969, Cull-Candy 1975, 1978, Clark et al 1979) or local damage (Cull-Candy 1975) extrajunctional sensitivity increases approximately 100 times above the control levels. Furthermore, the extrajunctional sensitivity which is normally present in innervated fibres is 20–30 times higher in the tendon region (Cull-Candy 1978). The increased glutamate sensitivity that is found under these conditions is caused by an increase in D-receptors; H-receptors appear unaffected.

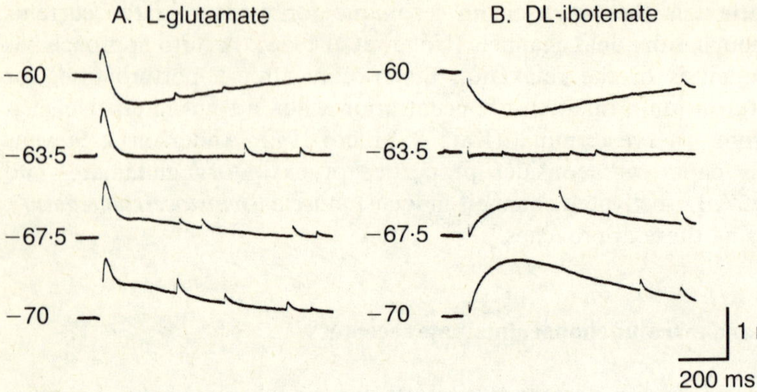

FIG. 1. Extrajunctional responses to L-glutamate (A) and D,L-ibotenate (B) applied ionophoretically from a twin-barrel micropipette. The membrane potential is held with a second intracellular microelectrode. Glutamate produces a DH-response at −60 mV. The H-response is annulled at −63.5 mV and a D-response alone is present. Further hyperpolarization of the membrane potential reverses the direction of the H-response. Ibotenate activates only H-receptors, producing an H-response at −60 mV. No response is seen at −63.5 mV, the equilibrium potential for the H-response. Further hyperpolarization of the membrane reverses the H-response. Miniature junction potentials are recorded at all sites in these multiple innervated fibres. Calibration, 1 mV and 200 ms. (From Cull-Candy 1976 by permission of *The Journal of Physiology*.)

Glutamate current noise

Ionophoretic application of glutamate to an excitatory junctional region of a voltage-clamped locust muscle fibre increases the membrane current to a new level during the period of application. The increased membrane current is accompanied by an increase in the membrane noise level (Anderson et al 1976).

Analysis of the membrane current noise into its frequency components using Fourier analysis (see Anderson et al 1978 for further details) yields a plot of the variance per hertz (A^2 s) against frequency (Hz), known as a power density spectrum. For an exponentially decaying process (such as is expected for a population of open synaptic channels) the spectrum has a characteristic Lorentzian form represented by the continuous curves in Fig. 2. The mean life-time of the glutamate-activated channel is obtained from the spectrum as the time constant, τ_{noise}, of the exponential decay according to $\tau_{noise} = 1/(2\pi f_c)$, where f_c is the frequency where the variance is reduced to half the plateau level. For glutamate-operated channels the mean life-time estimated from noise spectra is τ_{noise} (−80 mV) = 2.5 ± 0.13 ms at 20 °C (Anderson et al 1976, 1978).

The single-channel conductance, γ, was determined both from the variance of the noise and from the zero frequency asymptote of the Lorentzian curve fitted to the noise spectrum (see Anderson et al 1978). The conductance of the single channel opened by glutamate is γ(glutamate) = 122 ± 0.4 pS (mean ± SEM, 34 estimates in five cells) at 23 °C.

Patch clamp of extrajunctional glutamate-activated channels

The patch clamp technique has been used to examine channels opened by extrajunctional D-receptors in locust muscle (Patlak et al 1979, Cull-Candy et al 1981). The extrajunctional channel has a conductance and life-time which is similar to the values estimated by means of noise analysis for junctional channels. Thus from our measurements the extrajunctional channel has a conductance of 150 pS and a mean life-time of 2 ms at 23 °C.

In recent experiments it has been possible to examine the concentration dependence of individual receptor–channel complexes by using internally perfused patch clamp electrodes (see Cull-Candy et al 1981). We have found that the reciprocal of the channel closed time increases with about the second power of the glutamate concentration, suggesting that at least two glutamate molecules cooperate in activating a single channel.

Factors influencing synaptic channels

By analysing membrane noise and miniature excitatory junctional currents (MEJCs) we have shown that a number of factors affect the characteristics of excitatory synaptic channels in locust muscle fibres (Anderson et al 1976, 1978, Cull-Candy & Miledi 1980).

Membrane potential sensitivity

The decay time of the excitatory junctional current (EJC) and the time constant underlying glutamate noise (Fig. 2) both become briefer with membrane hyperpolarization, indicating that the channel life-time is reduced (see also Dudel 1974, Onodera & Takeuchi 1978). Fig. 3 shows the relationship between τ_{noise} and membrane potential, V_m, obtained in two muscle fibres at different temperatures. In most fibres the mean life-time of the channel, which depended exponentially on V_m, showed an e-fold change for about a 185 mV change in V_m at 20 °C.

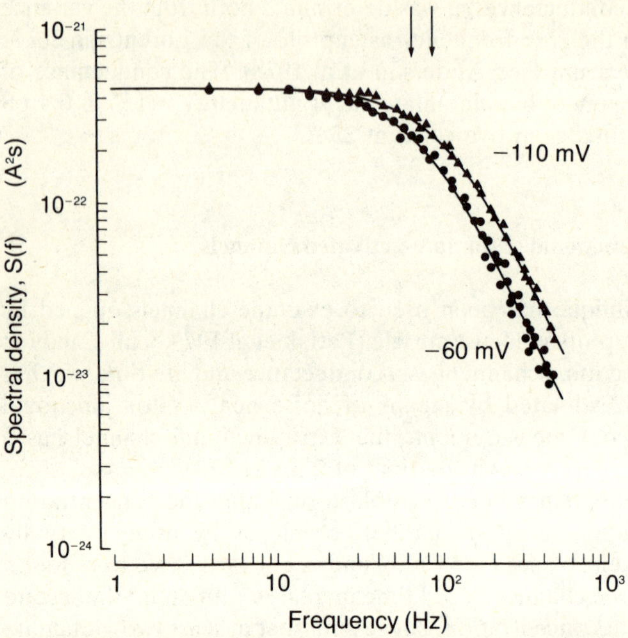

FIG. 2. Effects of clamp potential on power spectrum of current noise produced by application of glutamate to a voltage-clamped nerve–muscle junction. Spectra were obtained at the same site at $V_m = -60$ mV (circles) and $V_m = -110$ mV (triangles), T = 23 °C. The half-power frequency (indicated by arrow) is $f_c = 64$ Hz for the −60 mV spectrum, hence $\tau_{noise} = 2.49$ ms. Mean membrane current = 14 nA and the single-channel conductance calculated from the Lorentzian curve is $\gamma = 103$ pS. The half-power frequency for the −110 mV spectrum is $f_c = 91$ Hz, hence $\tau_{noise} = 1.75$ ms. Mean membrane current = 20 nA and the single-channel conductance is $\gamma = 100$ pS. Vertical axis applies to the −60 mV curve; multiplication by 0.508 gives spectral density values for the −110 mV curve. (From Anderson et al 1978 by permission of *The Journal of Physiology*.)

Agonist-activated channels

Membrane current noise produced by glutamate and the glutamate agonist, quisqualate, has been examined when these drugs were applied to the same junctional region from adjacent barrels of a multi-barrel micropipette (Anderson et al 1976, 1978). The quisqualate noise spectrum was shifted towards lower frequencies compared with the glutamate spectrum, indicating that quisqualate-activated channels have a longer mean life-time. As shown in Fig. 4, quisqualate opens channels with a mean life-time of 3.57 ms compared with 1.64 ms for glutamate at a clamp potential of −110 mV. However, the voltage sensitivity of the channel life-time and the conductance of a single

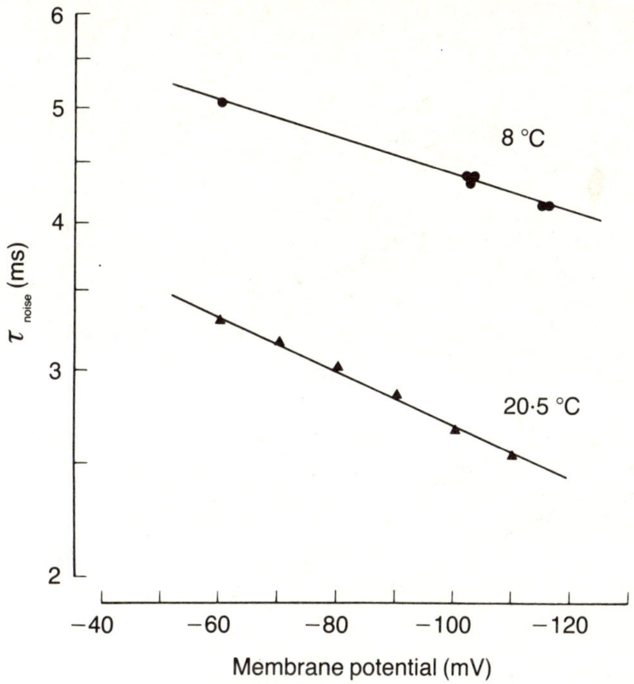

FIG. 3. Membrane potential dependence of channel life-time (τ_{noise}) plotted for two cells. Channel life-time was calculated from the half-power frequency of the noise spectrum ($\tau_{noise} = 1/(2\pi f_c)$) obtained at various clamp potentials and at two temperatures. The lines fitted to the points represent the relationship $\tau(V_m) = \tau(0) \exp(V_m/H)$ where $\tau(V_m)$ is the time constant at V_m and $\tau(0)$ is the time constant at 0 mV. H is a constant characterizing the voltage sensitivity of channel life-time. H increases from 188.7 mV at 20.5 °C to 270.3 mV at 8 °C, reflecting a reduced voltage sensitivity. At −110 mV channel life-time changes from 2.5 ms at 20.5 °C to 4.3 ms at 8 °C. (Replotted from data of Anderson et al 1978.)

channel appear identical whether the channel is opened by glutamate or quisqualate. The single channel conductance for quisqualate is: γ(quisqualate) = 120 ± 3.9 pS (mean ± SEM, 10 estimates) (Anderson et al 1978). The dependence of channel life-time on the agonist that opens the channels has been observed for both ACh receptors and glutamate receptors (Katz & Miledi 1973, Colquhoun et al 1975, Anderson et al 1976, Crawford & McBurney 1976, Clark et al 1980).

Since different drugs open channels with different life-times, agreement between the time constants underlying MEJCs and agonist-induced noise provides an additional test for the identity of a natural transmitter. Thus the observation that glutamate and the natural transmitter produce channels with

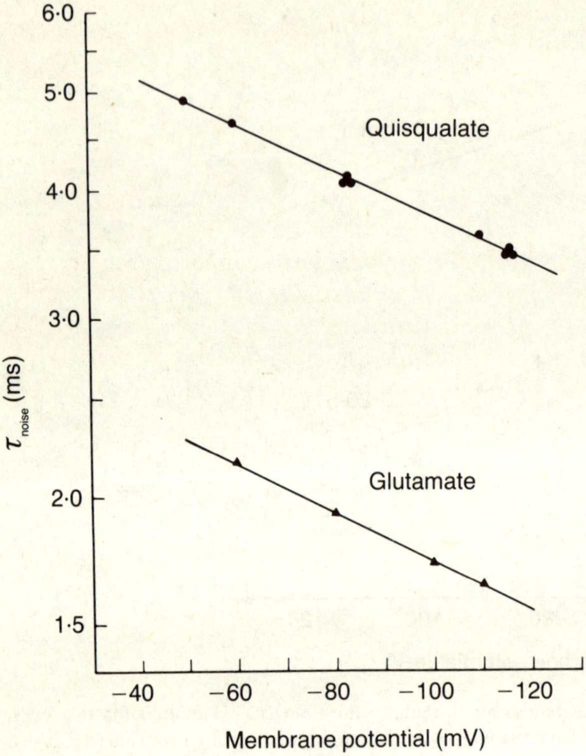

FIG. 4. Comparison of mean life-time (τ_{noise}) of channels opened by glutamate and quisqualate, as a function of clamp potential. Channel life-time was calculated from the half-power frequency of glutamate or quisqualate noise spectra obtained from the same junctional site in a muscle fibre. The mean life-time of channels opened by quisqualate is 2.2 longer than that of channels opened by glutamate. The potential sensitivity of the mean channel life-time is similar for channels opened by glutamate and quisqualate. (Replotted from data of Anderson et al 1978.)

similar life-times in locust (Anderson et al 1976, Cull-Candy & Miledi 1980) and crab nerve–muscle junctions (Crawford & McBurney 1976) reinforces the possibility that glutamate may be the transmitter at these synapses.

Temperature dependence

Lowering the temperature decreases the conductance of the glutamate-activated channel and prolongs its life-time. Over the temperature range 25–6 °C the temperature dependence is such that a 1 °C drop in temperature increases channel life-time by about 0.2 ms and decreases the single-channel conductance by 3.3 pS (Anderson et al 1978, Cull-Candy & Miledi 1980). In this temperature range the activation energy of the rate constant for channel closing ($1/\tau_{noise}$ or $1/\tau_{EJC}$) is 11.4 kcal/mol (~48 kJ/mol) at $V_m = -100$ mV, hence Q_{10} is 1.8. The activation energy for the single-channel conductance is 6.1 kcal/mol (~25.5 kJ/mol), hence Q_{10} is 1.2. As shown in Fig. 3, the voltage sensitivity of channel life-time is also temperature-dependent. The voltage sensitivity is decreased as temperature is lowered, which is opposite to the situation described for ACh receptors (Anderson & Stevens 1973). The temperature dependence of the rate constant for channel closing and the single-channel conductance fits an Arrhenius plot. However, both channel parameters undergo a sharp change in their temperature dependence at about 6 °C. This change in temperature dependence could possibly be caused by a change in the fluidity of the membrane lipid associated with the receptor or a conformational change in the protein molecules of the receptor (Anderson et al 1978, Cull-Candy & Miledi 1980).

Effect of cations

The properties of transmitter-activated synaptic channels have been examined in the presence of isotonic Ca^{2+} and the absence of external Na^+. In this bathing medium inward synaptic current is carried mainly by Ca^{2+}. MEJCs can be recorded for many hours in isotonic Ca^{2+} (Cull-Candy & Miledi 1980), as previously described for ACh-activated channels at vertebrate end-plates (Katz & Miledi 1969, Bregestovski et al 1979). τ_{MEJC} is markedly reduced in isotonic Ca^{2+}, indicating that the duration of the open state of the channel is decreased when the cation carrying the current is Ca^{2+} rather than Na^+. Furthermore, the MEJC amplitude is markedly reduced, possibly suggesting a decrease in single-channel conductance. Similar changes in channel properties are found at vertebrate end-plates in isotonic Ca^{2+} (Bregestovski et al 1979).

Junctional and extrajunctional GABA receptors

Ionophoretic application of GABA to voltage-clamped tonic muscle fibres has indicated the presence of extrajunctional as well as junctional GABA sensitivity in this muscle. Furthermore, extrajunctional GABA receptors are also present in extensor tibiae fibres (Cull-Candy & Miledi 1981), although only a small percentage of these fibres receive inhibitory innervation (Usherwood & Grundfest 1965). Typical extrajunctional responses to GABA obtained from two voltage-clamped fibres are illustrated in Fig. 5. Extrajunc-

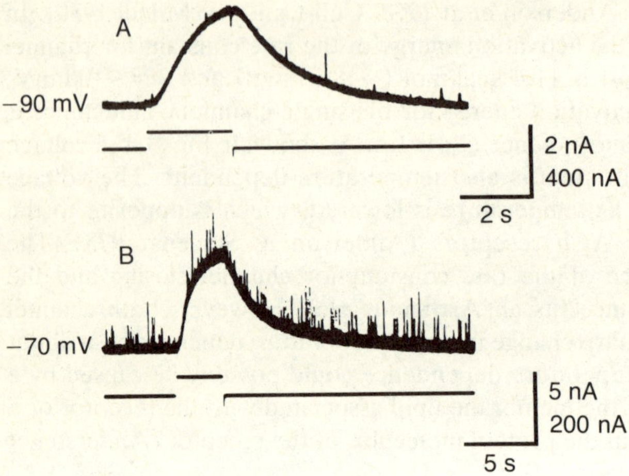

FIG. 5. Membrane current change produced by ionophoretic application of GABA to extrajunctional regions of two voltage-clamped muscle fibres. Inward membrane currents occur in response to GABA at the clamp potentials used. A, clamp potential −90 mV; B, clamp potential −70 mV. Calibrations: A, 2 nA, 400 nA, 2s; B, 5 nA, 200 nA, 5s. (From Cull-Candy & Miledi 1981 by permission of The Royal Society.)

tional GABA responses can still be detected when junctional and extrajunctional glutamate receptors have been completely desensitized by 10^{-2}M-glutamate. It therefore seems that the extrajunctional GABA responses do not result from activation of glutamate H-receptors, although they both mediate an increase in conductance to Cl^- ions.

GABA current noise

When GABA is applied ionophoretically to a junctional or an extrajunctional region of a voltage-clamped muscle fibre the membrane current changes to a new steady level and the membrane noise level increases. As expected, the direction and amplitude of the current change depends on the relative levels of clamp potential and GABA equilibrium potential. It is apparent, even at a qualitative level, that GABA current noise is markedly smaller than glutamate current noise when membrane currents of similar size are compared (with a similar driving force on the ions).

Power density spectra of GABA noise can usually be fitted with a single Lorentzian component (Fig. 6), although some spectra showed evidence of a second, higher frequency component (see also Dudel et al 1980). For spectra which are reasonably described by a single Lorentzian component the mean

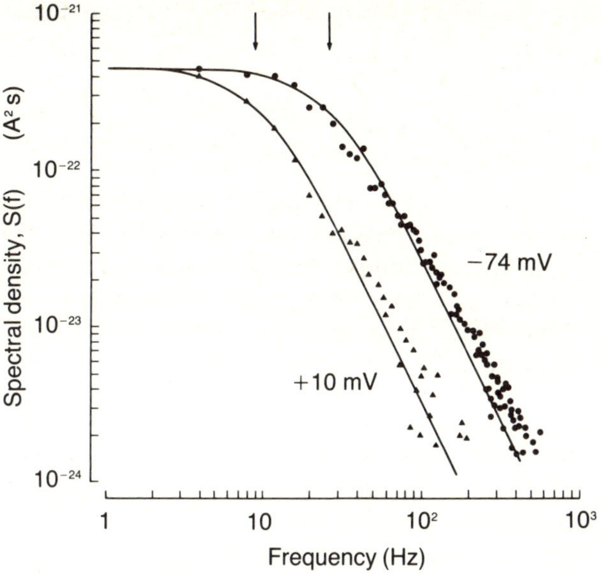

FIG. 6. Effect of clamp potential on power spectrum of current noise produced by application of GABA to a voltage-clamped junctional region. Spectra were obtained at the same site at V_m = −74 mV (circles) and V_m = +10 mV (triangles) at T = 21 °C. The half-power frequency (indicated by arrow) for −74 mV spectrum is f_c = 25 Hz, hence τ_{noise} = 6.37 ms. Mean membrane current = 31.6 nA and the single-channel conductance calculated from the Lorentzian curve γ = 17.1 pS (GABA equilibrium potential −10 mV in this cell). The half-power frequency for +10 mV spectrum is f_c = 9 Hz, hence τ_{noise} = 17.68 ms. Mean membrane current = 8 nA and the single-channel conductance is γ = 21.1 pS. Vertical axis applies to −74 mV curve; multiplication by 0.27 gives spectral density values for +10 mV curve. (From Cull-Candy & Miledi 1981 by permission of The Royal Society.)

life-time of junctional GABA-activated channels is, τ_{noise} (−80 mV) = 4.08 ± 1.08 ms at 21 °C. The conductance of the single junctional channel is, γ(GABA) = 22.62 ± 2.29 pS (± SEM, 21 estimates in five cells). Similar values of open time and conductance were obtained for extrajunctional GABA-activated channels (Cull-Candy & Miledi 1981).

Voltage sensitivity of channel life-time

Fig. 6 illustrates the effect of membrane potential on power spectra of GABA current noise obtained from a junctional site voltage-clamped at −74 mV and +10 mV. The mean life-time of GABA-activated channels is decreased by hyperpolarization, being exponentially dependent on clamp potential. An e-fold change in channel life-time is normally produced by a change of about 180 mV in potential.

Discussion

Various factors affect the channel kinetics of glutamate receptors in locust muscle. These include the membrane potential of the muscle fibre, the agonist which opens the channel, the temperature at which channels are operating and the species of cation in the bathing medium (Anderson et al 1976, 1978, Cull-Candy & Miledi 1980). Channel kinetics of GABA receptors in locust muscle are also affected by the membrane potential, However, the potential sensitivity of both glutamate- and GABA-activated channels differs markedly from that of ACh-activated channels at the vertebrate end-plate. Channels opened by ACh are prolonged by membrane hyperpolarization (Kordaš 1969, and see Table 1), whereas the life-time of channels opened by glutamate and GABA is reduced.

From our noise measurements the single-channel conductance of the excitatory junctional channel in locust muscle is 122 pS compared with only 23 pS for the inhibitory junctional channel (see Anderson et al 1976, 1978, Cull-Candy & Miledi 1981). Similar values have been obtained for the single-channel conductance of extrajunctional D-receptors measured by the patch clamp technique (Patlak et al 1979, Cull-Candy et al 1981). It is interesting to compare properties of junctional channels in these fibres with excitatory and inhibitory channels opened by glutamate, GABA and ACh in the muscle membranes of other species. Some findings from various laboratories are summarized in Table 1. Postsynaptic channels opened by GABA and ACh have conductances that lie in the range of approximately 10–25 pS and similar values have been found for channels opened by glutamate in crayfish muscle (Stettmeier et al 1979). The conductance of the excitatory

TABLE 1 Glutamate-, GABA- and acetylcholine-activated junctional channels in muscle fibres

Synapse	*Life-time (ms)*	*Conductance (pS)*	*Reference*
Crab junction (glutamate)	1.4	—	Crawford & McBurney 1976
Locust junction (glutamate)	2.5	122	Anderson et al 1976
Crayfish junction (glutamate)	1.6	22	Stettmeier et al 1979
Crayfish junction (GABA)	5.0	9	Dudel et al 1980
Locust junction (GABA)	4.0	23	Cull-Candy & Miledi 1981
Frog end-plate (ACh)	1.0	25	Katz & Miledi 1972, Anderson & Stevens 1973

In most cases T = 20–23 °C and membrane potential is −80 mV.

junctional channel opened by glutamate in locust muscle is therefore unusually large while the conductance of GABA-activated channels is similar to the values reported for other inhibitory and excitatory nerve–muscle junctions.

Acknowledgement

This work was supported by the Medical Research Council.

REFERENCES

Anderson CR, Cull-Candy SG, Miledi R 1976 Glutamate and quisqualate noise in voltage-clamped locust muscle fibres. Nature (Lond) 261:151-153

Anderson CR, Cull-Candy SG, Miledi R 1978 Glutamate current noise: post-synaptic channel kinetics investigated under voltage clamp. J Physiol (Lond) 282:219-242

Anderson CR, Stevens CF 1973 Voltage clamp analysis of acetylcholine produced end-plate current fluctuations at frog neuromuscular junction. J Physiol (Lond) 235:655-692

Axelsson J, Thesleff S 1959 A study of supersensitivity in denervated mammalian skeletal muscle. J Physiol (Lond) 147:178-193

Beránek R, Miller PL 1968 The action of iontophoretically applied glutamate on insect muscle fibres. J Exp Biol 49:83-93

Bregestovski PD, Miledi R, Parker I 1979 Calcium conductance of acetylcholine-induced endplate channels. Nature (Lond) 279:638-639

Clark RB, Gration KAF, Usherwood PNR 1980 Ion channels opened by glutamate agonists on locust muscle membrane. J Physiol (Lond) 301:60P-61P

Clark RB, Gration KAF, Usherwood PNR 1979 Desensitization of glutamate receptors in innervated and denervated locust muscle fibres. J Physiol (Lond) 290:551-568

Colquhoun D, Dionne VE, Steinbach JH, Stevens CF 1975 Conductance of channels opened by acetylcholine-like drugs in muscle end-plate. Nature (Lond) 253:204-206

Crawford AC, McBurney RN 1976 The post-synaptic action of some putative excitatory transmitter substances. Proc R Soc Lond B Biol Sci 192:481-489

Cull-Candy SG 1975 Effect of denervation and local damage on extrajunctional L-glutamate receptors in locust muscle. Nature (Lond) 258:530-531

Cull-Candy SG 1976 Two types of extrajunctional L-glutamate receptors in locust muscle. J Physiol (Lond) 255:449-464

Cull-Candy SG 1978 Glutamate sensitivity and distribution of receptors along normal and denervated locust muscle fibres. J Physiol (Lond) 276:165–181

Cull-Candy SG, Miledi R 1980 Factors affecting the channel kinetics of glutamate receptors in locust muscle fibres. In: Sattelle DB et al (eds) Receptors for neurotransmitters, hormones and pheromones in insects. Elsevier/North-Holland Biomedical Press, Amsterdam, p 162-173

Cull-Candy SG, Miledi R 1981 Junctional and extrajunctional membrane channels activated by GABA in locust muscle fibres. Proc R Soc Lond B Biol Sci 211:527-535

Cull-Candy SG, Miledi R, Parker I 1981 Internal perfusion of patch-clamp electrodes for analysis of synaptic channels. J Physiol (Lond) 317:3P-4P

Dudel J 1974 Nonlinear voltage dependence of excitatory synaptic current in crayfish muscle. Pflügers Arch Eur J Physiol 352:227-241

Dudel J, Finger W, Stettmeier H 1980 Inhibitory synaptic channels activated by γ-aminobutyric acid (GABA) in crayfish muscle. Pflügers Arch Eur J Physiol 387:143-151

Katz, B, Miledi R 1964 Further observations on the distribution of acetylcholine-reactive sites in skeletal muscle. J Physiol (Lond) 170:379-388

Katz B, Miledi R 1969 Spontaneous and evoked activity of motor nerve endings in calcium Ringer. J Physiol (Lond) 203:689-706

Katz B, Miledi R 1972 The statistical nature of the acetylcholine potential and its molecular components. J Physiol (Lond) 244:665-700

Katz B, Miledi R 1973 The characteristics of 'end-plate noise' produced by different polarizing drugs. J Physiol (Lond) 230:707-717

Kordaš M 1969 The effect of membrane polarization on the time course of the end-plate current in frog sartorius muscle. J Physiol (Lond) 204:493-502

Neher E, Sakmann B, Steinbach JH 1978 The extracellular patch clamp: a method for resolving currents through individual open channels in biological membranes. Pflügers Arch Eur J Physiol 375:219-228

Onodera K, Takeuchi A 1978 Effects of membrane potential and temperature on the excitatory post-synaptic current in the crayfish muscle. J Physiol (Lond) 276:183-192

Patlak JB, Gration KAF, Usherwood PNR 1979 Single glutamate-activated channels in locust muscle. Nature (Lond) 278:643-645

Stettmeier H, Dudel J, Finger W 1979 Activation of synaptic channels by glutamate in crayfish muscle. Pflügers Arch Eur J Physiol 382:R34

Usherwood PNR 1969 Glutamate sensitivity of denervated insect muscle fibres. Nature (Lond) 223:411-413

Usherwood PNR, Grundfest H 1965 Peripheral inhibition in skeletal muscle of insects. J Neurophysiol 28:497-518

Usherwood PNR, Machili P 1968 Pharmacological properties of excitatory neuromuscular synapses in the locust. J Exp Biol 49:341-361

DISCUSSION

Usherwood: What are the implications of the voltage dependence of the glutamate channel life-time?

Cull-Candy: That is a contentious problem at the moment. For ACh receptors at the vertebrate end-plate the life-time of the channel is prolonged by membrane hyperpolarization, while for amino acid receptors in the locust muscle it becomes briefer. We were originally attracted by the idea that the life-time of the channel reflected the behaviour of a gating molecule, which was a dipole, in the cell membrane (Anderson et al 1976) and that one could explain the difference in membrane voltage sensitivity in terms of the dipole model, by hyperpolarization altering the electric field across the dipole. If so, the orientation of the dipole molecule associated with the ACh receptor may be the reverse of that for the amino acid receptor. This is only one of several possibilities, but the other present ideas are more complicated! The voltage sensitivity is thought to involve one of the reaction steps in the binding or the conformational change in the postsynaptic receptor.

Usherwood: What are the limitations of the noise analysis technique for understanding single-channel kinetics?

Cull-Candy: One limitation is that we are looking at a population of channels and of receptors, so we are obtaining a mean response. We are seeing the average channel, if you like, which is useful in some respects. Patch clamping is a more elegant way of looking at the response of an individual receptor or an individual channel.

Usherwood: Does the limited frequency response of the recording system introduce additional problems of interpretation of noise data?

Cull-Candy: The frequency response of the system has up to now been generally less good for noise analysis than it could be. People have usually filtered their data during analysis, or simply recorded over a limited bandwidth. It is possible to study high frequency events with noise analysis, though only a few people interested in rapid kinetics (faster than 1 ms) of the receptor have used the noise analysis method on synaptic receptors.

Barnard: It is interesting that voltage dependence of the channel life-time seems to be widespread in the channels opened directly by neurotransmitters. Yet it is also important to note that there are certain vertebrate nicotinic ACh receptors with, exceptionally, no voltage dependence (Sheridan 1976, MacDermott et al 1978), so it doesn't seem to be an absolute requirement for the operation of such a channel.

Cull-Candy: That is true. There are quite a few reports of receptors at vertebrate and invertebrate synapses which don't show voltage dependence.

Barnard: That seems to indicate that it is a secondary phenomenon. Yet it is curious that voltage dependence is maintained to the same quantitative degree in glutamate receptors in insects as in ACh receptors in vertebrates.

Cull-Candy: Voltage dependence is interesting because it is very convenient to be able to alter the kinetics of the channel almost instantaneously, and electrophysiologists can alter voltage rather well. The kinetics can't be changed so rapidly by altering temperature or other factors which influence channel kinetics.

Barnard: One of the current ideas about the origin of the channel voltage dependence at the structural level is that it is determined by how far above or below the surface of the membrane the ACh molecule is bound on a projecting receptor molecule, or where the ion-selectivity filter is located (Stevens 1978), because of interaction with the intense electrical field of the membrane itself. A fairly small difference in this receptor geometry between species might explain the differences in voltage dependence.

Cull-Candy: There are several types of explanation for the varying degrees of voltage sensitivity found in different synapses.

Barnard: Could you say something further about the dependence of the channel life-time on calcium? Where chloride is the permeant ion, can you find such changes, for example?

Cull-Candy: We haven't looked at the influence of the permeating anions on the inhibitory channel. Onodera & Takeuchi (1979) have found that if the iodide or bromide is exchanged for chloride ions there is a prolongation of the channel life-time.

Barnard: At the molecular level, one would like to know how the ion that is responsible for the conductance increase through the channel determines the channel life-time—since the latter is assumed to be given by the rate of the conformational change in the receptor protein that is induced by its binding of the agonist. At first sight one wouldn't expect a large and selective ionic effect on the latter process.

Cull-Candy: It is probably quite a complicated effect.

Usherwood: Have you studied the effect of calcium at the single-channel level using patch clamp analysis?

Cull-Candy: We haven't tried that yet.

Narahashi: The conductance of the glutamate-activated channel is almost five times that of the ACh-activated channels. The vertebrate ACh-activated channel has a diameter of 0.5 or 0.6 nm and a conductance of 25 pS, so the conductance of 120 pS means that the open glutamate channel is almost twice as wide as the ACh-activated channel. Have you looked at the ionic selectivity of the glutamate-activated channel?

Cull-Candy: We haven't really looked very closely at the selectivity of this channel, though we have been thinking along the same lines as you suggest.

M. E. Eldefrawi: Have you looked at the effect of the so-called channel blockers? We and others have done a great deal of work on the ACh receptor channel (Albuquerque et al 1980). If we assume that the glutamate channel is a larger one and probably has less selectivity than the ACh receptor–channel complex, would some of these same drugs affect the conductance of the glutamate channel?

Cull-Candy: We have some preliminary experiments on drugs that appear to block the channels. I suppose a model system for drugs which block synaptic channels has been the ACh channel at the vertebrate end-plate, and we have certainly looked at some drugs that block at the vertebrate end-plate. Some of these drugs affect the glutamate channel. However, there are also some which block ACh channels but have no effect on glutamate-activated channels.

Gration: Streptomycin is the only drug with channel-blocking properties that we have studied in detail (see Gration & Usherwood 1980). Its action is strongly voltage-dependent; for example, at -60 mV (the muscle resting potential) there is no significant difference between mean channel life-times ($\sim$2.5 ms at 5×10^{-6} M-glutamate) in control (streptomycin-free) preparations and those exposed to streptomycin (10^{-5} M), but when the membrane of streptomycin-treated muscles is hyperpolarized to −120 mV, channel life-times are shortened to 0.5–0.8 ms. Streptomycin also affects the frequency of channel opening in a voltage-dependent manner: when the membrane is hyperpolarized from −60 mV to −120 mV the frequency of opening decreases by 10–20-fold. This reduction presumably arises because streptomycin is a slowly dissociating channel blocker.

M. E. Eldefrawi: In these experiments you have eliminated the presynaptic effect of streptomycin.

Gration: That's right. We are just looking here at the extrajunctional (non-synaptic) glutamate channels.

Kelly: Does the quisqualate-activated receptor have the same temperature phase shift as the glutamate-activated one?

Cull-Candy: We haven't looked at that.

Hall: Do you see a shift in the temperature (6 °C) at which you get a break if you change the growth temperature of your locusts?

Cull-Candy: We have never tried that. The break (i.e. the change in temperature dependence of channel parameters) itself is membrane potential-dependent, which we assume is because the conformation of the membrane lipids might be altered by altering the potential across it. We have never tried acclimatizing the animals to low temperature, which might alter their membrane composition.

Hall: Could the lipid composition of the membrane be changed by diet or by the temperature at which the animals are kept?

Cull-Candy: I couldn't say, although as I mentioned we had thought of acclimatizing the animals.

Breer: The change in temperature dependence of channel parameters at around 6 °C is very low compared to the break points of, for example, membrane-bound enzymes, which depend on the lipid composition of the membrane. Do you know what membrane components are involved in your system?

Cull-Candy: No. We have previously suggested that there may be a change in the receptor-channel complex itself or there may be an indirect effect on channel properties, for example via a lipid phase transition (Anderson et al 1977).

Lunt: We need to know the fatty acid composition of the phospholipids. That information is not available, yet this is such a simple thing for a lipid biochemist to do.

Usherwood: Information is available on the phospholipid composition of the muscles and nervous system of the housefly, *Musca domestica* (Bridges & Price 1970), but not for the locust.

Berridge: I have been studying temperature effects on various membrane events in the salivary glands of *Calliphora* and find a similar break in the Arrhenius plot, but at a higher temperature (12–15 °C). This break point is not affected by varying the temperature at which the insects are reared. We are trying to alter the fatty acid composition by dietary manipulation, which we hope will shift the transition point.

Hall: If the growth temperature is changed in fish, the lipid composition of the membranes changes (Lagerspetz 1974). Is that true for insects too?

Berridge: As yet, we haven't determined the lipid composition.

Barnard: Can you compare temperature dependence in the junctional and the extrajunctional receptors?

Cull-Candy: We haven't compared it with noise analysis, but it could be done. All our noise experiments were done on junctional receptors.

Barnard: But you showed that the channel life-time was the same, at ambient temperature, in the junctional and the extrajunctional receptors, didn't you?

Cull-Candy: It is similar, about 2 ms.

Barnard: It would be interesting to know whether there is a different temperature dependence in the two cases, over the lower temperature range. There may well be, as a result of different local membrane compositions or a greater mobility in the membrane of the extrajunctional receptors.

Olsen: The kind of quantitative information we have heard here is exactly what biochemists need for determining the functional consequences of ligand binding to receptor proteins. We should be looking for the same kinds of dependence on temperature, concentration, and perhaps even voltage (if an intact cell membrane can be used), for the kinetics of neurotransmitter binding to receptors. Information on binding proteins corresponding to this kind of information obtained with electrodes will finally lead us somewhere. However, the biochemist needs a reasonable amount of material, and that is a problem with the insect muscle.

Berridge: What is the relationship between the locust muscle that Dr Cull-Candy has used and the one that Dr Evans used in his octopamine studies?

Cull-Candy: In our locust preparation the muscle is probably the same one as Peter Evans uses.

Berridge: Excitation of the fast extensor tibiae (FETi) induces the release of glutamate which acts through your channel system. When Peter Evans modulates this response with octopamine, is he modulating the same muscle system you have studied?

Cull-Candy: Yes.

Berridge: Is there any evidence that octopamine affects the channel properties?

Cull-Candy: I couldn't say, although it would be interesting to find out.

REFERENCES

Albuquerque EX, Tsai M-C, Aronstam RS, Witkop B, Eldefrawi AT, Eldefrawi ME 1980 Phencyclidine interactions with the ionic channel of the acetylcholine receptor and electrogenic membrane. Proc Natl Acad Sci USA 77:1224-1228

Anderson CR, Cull-Candy SG, Miledi R 1976 Glutamate and quisqualate noise in voltage-clamped locust muscle fibres. Nature (Lond) 261:151-153

Anderson CR, Cull-Candy SG, Miledi R 1977 Potential-dependent transition temperature of ionic channels induced by glutamate in locust muscle. Nature (Lond) 268:663-665

Bridges RG, Price GM 1970 Phospholipid composition of various organs from larvae of the housefly *Musca domestica*, fed on diets deficient in choline. Int J Biochem 1:483-490

Gration KAF, Usherwood PNR 1980 Interactions of glutamate with amino acid receptors on locust muscle. Verh Dtsch Zool Ges 122-132

Lagerspetz KYH 1974 Temperature acclimation and the nervous system. Biol Rev 49:477-514

MacDermott A, Dionne VE, Parsons RL 1978 Decay of bullfrog sympathetic ganglion fast excitatory postsynaptic current is voltage insensitive. Biophys J 21:52a

Onodera K, Takeuchi A 1979 An analysis of the inhibitory post-synaptic current in the voltage-clamped crayfish muscle. J Physiol (Lond) 286:265-282

Sheridan RE 1976 Voltage-dependent acetylcholine receptors in the skate electroplaque. Biophys J 16:212a

Stevens CF 1978 Interactions between intrinsic membrane protein and electric field. An approach to studying nerve excitability. Biophys J 22:295-306

Neurosecretory peptides and biogenic amines

W. MORDUE

Department of Zoology, Tillydrone Avenue, Aberdeen, AB9 2TN, UK

Abstract Neurosecretory peptides are distributed throughout the insect central nervous system and are of prime importance as regulators of physiological, biochemical, behavioural and developmental events. Peptides act as hormones, transmitters or neurohormonal modulators. Many target tissues respond to peptidergic and aminergic stimulation but the role of aminergic stimulation *in vivo* is unclear. An exception is octopamine, which has a clear role in flight metabolism in locusts. Comparative studies suggest that peptidergic control mechanisms are common in insects and control similar processes in all of them, such as cuticular tanning, plasticization, eclosion and diuresis. Peptides related to adipokinetic hormone control intermediary metabolism in a variety of ways and may regulate lipid, carbohydrate or amino acid metabolism.

In its infancy insect endocrinology was preoccupied with the control of development and morphogenesis. This has meant that our knowledge of neurohormones and their control over physiological and metabolic events has progressed significantly only in the last 10–15 years. The expansion of knowledge in this field is such that we must drastically revise our ideas of the extent and variety of neurohormones in insects. Directly acting neurosecretions are used by the organism to regulate intermittent but essential physiological processes or elicit key switches in metabolism to such an extent that we might forgive an over-zealous student for thinking that all biochemical processes are controlled hormonally. With such a wide variety of hormonal factors regulating a bewildering array of processes, I shall deal only with some of the more important of these.

1982 Neuropharmacology of insects. Pitman, London (Ciba Foundation symposium 88) p 88-108

(1) Myotropic neurohormones

(a) Malpighian tubules

In a few orders of insects, such as Orthoptera and Coleoptera, a spiral muscle which is not innervated is associated with each Malpighian (excretory) tubule, and contraction causes pronounced spiralling movements. Many tubules adhere so closely to the gut, fat body, and other tissues that such movements are very limited and the function of these muscles is not clear. Corpus cardiacum factors induce contractions in tubules from *Locusta migratoria, Schistocerca americana gregaria, Periplaneta americana* (cockroach) and *Carausius morosus* (stick insect) (see Goldsworthy & Mordue 1974). In the latter two species, 5-hydroxytryptamine (5-HT; serotonin) at concentrations of about 10^{-9} M also increases the rate of contractions. Exogenous 5-HT has no effect on Malpighian tubules in the locust (Mordue 1972) but adrenaline is stimulatory, albeit at rather high concentrations (10^{-4} M) (Mordue 1972, Rafaeli & Mordue 1981). Octopamine has no effect on locust tubules (A. J. Mordue, unpublished observations). In locusts, adrenergic stimulation induces increased fluid secretion as well as increased muscle contractions. Diuretic hormone (a peptide neurohormone; see section 4a) induces marked diuresis but scarcely affects muscle contraction (Rafaeli & Mordue 1981). Muscle contraction in *Locusta* tubules in response to adrenergic stimulation is blocked by phentolamine, but phentolamine does not block diuresis induced by the locust diuretic hormone. Not enough work has been done on serotoninergic tubules for any differential effects of the amine and neurohormone (diuretic hormone) on muscle contractions to be distinguished.

(b) Heart

There is little clear-cut evidence for the hormonal control of heart beat in insects. Heart-accelerating factors are present in the brain–corpora cardiaca complexes of a number of insects (see Goldsworthy & Mordue 1974, Miller 1979, for reviews) but their actions are discernible only in semi-isolated preparations. Since these so-called cardioaccelerators are without effect on heart beat in the intact insect their actions are probably pharmacologically rather than physiologically significant. The first heart accelerator described was neurohormone D, and an amino acid composition for this peptide, extracted from the cephalic endocrine complex of *Periplaneta americana*, has been proposed recently (Baumann & Gersch 1981). Neurohormone D increases both the rate and amplitude of heart beat, whereas neurohormone C, also from *Periplaneta*, increases the rate but decreases the amplitude. In locusts the corpus cardiacum exists as two more or less distinct storage and glandular lobes; extracts of storage lobes are similar in action to neurohormone D and those of glandular lobes to neurohormone C (see Goldsworthy & Mordue 1974). It is now clear that such neurohormone C activity in the glandular lobes is related to a pharmacological action of adipokinetic hormone (see section 5a and Stone & Mordue 1980). More recently the candidate neurotransmitter proctolin (see below) has been shown to increase both the rate and the amplitude of contraction of the cockroach semi-isolated heart preparation; the threshold concentration is about 10^{-9} M (Miller 1979). In spite of this potent physiological action it is still unclear whether proctolin is a transmitter in the insect heart. It remains to be resolved whether any peptide has a particular physiological role in visceral muscles. Nevertheless, the pharmacological responses of visceral muscles to a variety of peptides are of particular interest and provide sensitive bioassays as well as challenging the endocrinologist and neuromuscular physiologist to resolve the complexities of interaction between peptides and biogenic amines and their respective receptor sites.

(c) Proctolin

Proctolin is a pentapeptide, Arg-Tyr-Leu-Pro-Thr, that has been isolated from *Periplaneta* and causes slow graded contractions of the longitudinal muscles of the proctodeum. Such contractions can also be invoked by repetitive nerve stimulation and it has been suggested that proctolin is either an excitatory neurotransmitter or acts as a neurohormone to modulate muscle excitability (see Starratt & Steele 1980). Proctolin acts on preparations other than the hind-gut. Its effects on the heart have just been described. At concentrations as low as 10^{-10} to 10^{-9} M it induces myogenic contractions in the locust extensor tibialis. An increased rate of contraction of the hyperneural muscle in *Periplaneta* was elicited in response to proctolin when the innervation of the hyperneural muscle from the ventral nerve cord was intact. When the ventral nerve cord was removed, proctolin was unable to induce muscle contractions. Thus the site of action of proctolin in this preparation is probably distinct from the neuromuscular junctions of the hyperneural muscle (see Miller 1979).

Other peptides in the corpora cardiaca of insects also stimulate the hind-gut but are not identical with proctolin; whether these corpus cardiacum peptides have a physiological role is unclear. Many of these other peptides show the same wide distribution as proctolin and proctolin-like peptides in the nervous system; for example, cardioaccelerators are found in the brain and corpora cardiaca and also in the terminal abdominal ganglion (see Mordue & Stone 1979).

The wide distribution of proctolin and its potency of action on a number of neuromuscular preparations suggest that it is an important candidate peptidergic transmitter. When it acts on the hind-gut, proctolin is probably released from motor nerve endings. When it acts on heart and hyperneural and leg muscles, the sites of release are uncertain. The recent identification of the cell bodies responsible for the synthesis of proctolin and the immunocytochemical mapping of their axon pathways promise to provide much important information. We may consider proctolin as the first confirmed example of a peptide transmitter in insects. Peptidergic neurons with peptidergic transmitters are of course widely known in vertebrates.

(2) Peptide hormones and the cuticle

Ecdysteroids are key hormones in the determination of moulting and the detailed action of these steroids in eliciting tanning in the blowfly puparium has been investigated by Karlson & Sekeris (1966). However, it is likely that the all-important role of ecdysone in controlling tanning, which has been promulgated for some 30 years, is restricted to one particular stage in the development of flies (see Mordue & Stone 1979). In most insects, including flies, the initiation of hardening and darkening of the cuticle is regulated by a protein neurosecretory hormone—bursicon. Like the many peptide hormones found in insects, bursicon is distributed throughout the nervous system and is present in the brain neurosecretory cells and corpora cardiaca in very large amounts. The principal sites of release are from the ganglia in the thorax and abdomen in the wide variety of species that have been shown to possess bursicon (see Mordue & Stone 1979, Seligman 1980). Bursicon is released into the haemolymph before and during the moulting process but in blowflies bursicon is released only when the fly emerges onto a clear substrate and stops digging. Thus a variety of proprioceptive stimuli other than the shedding of the old cuticle are important in eliciting the release of this hormone. Once released, bursicon stimulates the hydroxylation of tyrosine to dihydroxyphenylalanine (dopa). Without bursicon, tyrosine levels remain high and dopa and its intermediates do not accumulate. It now seems that the rate-limiting step in cuticular tanning is hydroxylation; it is this step that limits the production of the tanning substrate *N*-acetyldopamine. Karlson's hypothesis (Karlson & Sekeris 1966) that ecdysone regulates tanning via its control over the synthesis of dopa decarboxylase (aromatic-L-amino acid decarboxylase, EC 4.1.1.28) probably does not hold true for tanning in most insects.

Bursicon acts via stimulation of cyclic AMP production, but there is some disagreement about the mechanisms after this step: bursicon may increase the permeability of the haemocytes and thus facilitate hydroxylation of tyrosine, but this is disputed (see Seligman 1980). Bursicon controls a variety of other post-eclosion events during the consolidation of the cuticle after the moult, including deposition of endocuticle, wing expansion, programmed cell death between the lamina of the wings, and general plasticization. In most insects the cuticle is untanned at ecdysis and it is possible that the expansion of the body and wings at ecdysis is facilitated instead by plasticization (see Reynolds 1980). However, only *Calliphora erythrocephala* (blowfly), the moth *Manduca sexta* and locusts have been investigated.

In *Rhodnius prolixus* (assassin bug), cuticular plasticization is essential to accommodate the very large blood meals but the mechanism is under direct nervous control (see Reynolds 1980). In *Rhodnius* the epidermal cells are supplied by axons from neurosecretory cells which exert a local control. Little is known about the properties of bursicon; it is not identical with but has some similarities to diuretic hormone of *Rhodnius* (Reynolds 1980). The actions of both hormones are mimicked by 5-HT but the pharmacological properties of the Malpighian tubule receptors are very different from those of the epidermal receptors. The 5-HT/plasticizing factor receptor is similar to that of other insect tissues and to that found in mammals (Reynolds 1980).

(3) Eclosion hormones

In Lepidoptera a neurosecretory peptide, eclosion hormone, triggers the emergence of the adult (Truman 1978). The hormone originates in the median neurosecretory cells and is released from the corpora cardiaca just before eclosion, i.e. in a 20-min burst some 2.5–3 h before emergence in *Manduca sexta* (see Reynolds & Truman 1980). The signal for release is keyed into a circadian clock situated in the brain; this clock, which is entrained by environmental variables, especially the photoperiod, controls the release of eclosion hormone and thus the time of emergence of the moth. In *Hyalophora cecropia* (silk moth), for example, the release of eclosion hormone initiates specific patterns of behaviour which are preprogrammed in the central nervous system (CNS). The execution of this programme extricates the moth from its pupal case. Eclosion hormone elicits this now well-quantified behaviour pattern in isolated abdomens and even from the completely isolated CNS maintained *in vitro* (Truman 1978). In the latter preparation the behaviour is detected as a pattern of action potentials in particular motor neurons. Injections of eclosion hormone into pharate moths still within their pupal skin causes precocious activation of the CNS and resulting precocious behaviour patterns associated with eclosion.

Reynolds & Truman (1980) recently summarized the chemical characteristics of eclosion hormone. It is a peptide with a relative molecular mass (M_r) of some 8500 (about 70 amino acid residues). In *Manduca* the half-life of eclosion hormone is about 45 min, with a threshold concentration for eliciting eclosion behaviour of about 15 nM (Reynolds & Truman 1980). Eclosion hormone releases a long and complex behavioural sequence even from the isolated CNS; moreover, the CNS has to be exposed to the hormone for only a few minutes before the pre-eclosion and eclosion behaviour patterns, which last for many hours, are triggered.

Eclosion hormone has other effects on the physiology and behaviour of insects. In addition to activating adult behaviour it releases pre-eclosion behaviour and initiates the 'programmed cell death' of some abdominal intersegmental muscles. These are longitudinal muscle bands in the 4th, 5th and 6th segments which are active as part of the eclosion mechanism but have no role once the adult has emerged. The eclosion hormone in switching off pupal behaviour turns off the motor neurons which supply these abdominal intersegmental muscles; without a motor supply they degenerate (Truman 1973). Eclosion hormone also promotes death of particular neurons in abdominal ganglia, with both motor and interneurons being lost. It is likely therefore that the turning off of pupal behaviour is accomplished by the selective destruction of neural circuits (see Truman 1973).

Truman et al (1981) have presented evidence that eclosion hormone(s) control all post-embryonic ecdyses in a number of moths. For all moults other than to the adult stage, eclosion hormone is released from the ventral ganglia; for the adult moult, release is from the corpora cardiaca. Preliminary evidence indicates that the hormones from the two sources are very similar. Moreover, eclosion hormone has been demonstrated in a number of non-lepidopteran species. This may indicate that this hormone triggers ecdysis in a wide variety of insects (Truman et al 1981).

(4) Hormones controlling water balance

There is abundant if sometimes contradictory evidence that the insect excretory system—the Malpighian tubules and rectum—is regulated by neurosecretory hormones; both diuretic and antidiuretic hormones have been proposed. This account will be restricted to two well-documented types of hormone: diuretic hormones which act on the Malpighian tubules and chloride transport-stimulating hormone (CTSH) which acts on the rectum.

(a) Diuretic hormones

As their biological design incorporates an impressive array of systems to reduce water loss to a minimum, insects could be faced with problems in disposing of excess dietary or metabolic water. To overcome this, many possess diuretic hormones of neurosecretory origin which stimulate fluid secretion by the Malpighian tubules. The consequent increased flow of primary urine into the hind-gut is greater than the absorptive capacity of the rectal epithelium and in consequence the insect loses water from its body. The control and timing of the release of diuretic hormone vary markedly with the life-style of the insect. Infrequent feeders such as *Rhodnius*, which takes very large blood meals, face an urgent requirement to excrete a large quantity of watery urine rapidly; insects which feed more frequently, such as locusts, need to be able to promote rapid loss of water on occasion, depending on the water content of the diet. In locusts, *Anisotarsus cupriensis, Pieris brassica, Carausius morosus, Dysdercus fasciatus* and *Periplaneta* diuretic hormones are products of the brain neurosecretory cells and are released from the corpora cardiaca. In *Rhodnius* and *Glossina morsitans* diuretic hormones are produced by cell bodies in the ventral nerve cord.

Diuretic hormones in general seem to be peptides of low M_r (about 2000); a form with a high M_r (>60 000) has also been described in *Rhodnius*, though it is presumed that the low M_r form is the physiologically active one (see Stone & Mordue 1980). Diuretic hormones are released in response to feeding; the stimulation of urine flow occurs rapidly after about 1 min in *Rhodnius* and diuresis is maintained by constant release of hormone. The response of the Malpighian tubule cells to diuretic hormones involves cyclic AMP as an intracellular messenger, and tubules from a variety of insects show increased urine flow in response to exogenous cyclic AMP. In addition to peptidergic stimulation and cyclic AMP application, the tubules respond to biogenic amines. In *Rhodnius* and *Carausius* 5-HT is a potent stimulant of increased urine flow; in *Locusta* and *Schistocerca* urine flow is insensitive to 5-HT but is stimulated by adrenaline. In *Rhodnius* it is thought that both hormone and amine act at the same receptor; however, in locusts we have demonstrated two distinct receptor sites (Rafaeli & Mordue 1981). The catecholamine receptor and the response to adrenaline can be blocked with phentolamine; this treatment does not prevent the normal increase in urine flow elicited by the peptidergic diuretic hormone. Treatment with canavanine, a non-protein amino acid, markedly reduces urine production by *in vitro* preparations of locust tubules. However, in subsequent treatment with either cyclic AMP or diuretic hormone a 700% increase in production rate has been seen, compared with the normal 62% increase in response to cyclic AMP or diuretic hormone (Rafaeli & Applebaum 1980). The mechanism of this potentiation is not understood.

Comparative studies (Rafaeli & Mordue 1981) have also revealed interesting cross-reactivity. It is known that diuretic hormone from *Carausius* stimulates tubules of *Rhodnius* but the reverse combination is without effect. Locust tubules are responsive to hormones from *Glossina* and *Periplaneta* and only poorly so to *Rhodnius* hormone. *Glossina* hormone stimulates both muscle contraction and diuresis in *Locusta* and both actions can be blocked with phentolamine. Thus there are certain similarities between these hormones from different insects but also marked differences.

In most insects urine flow is generated by the active transport of potassium and the primary urine is rich in potassium but contains little sodium; however, in *Glossina* and *Rhodnius* sodium is, or can be, the prime mover. It is unclear what events take place in response to treatment with 5-HT, adrenaline, cyclic AMP or diuretic hormone, or whether all these agents produce similar responses. In *Locusta* fluid secretion is not stimulated if the sodium concentration in the bathing fluid is less than 140 mM, even if potassium, the prime mover, is present (Morgan & Mordue 1981). Recently, we have measured transmural potentials and the potentials across the basal and apical membranes of the Malpighian tubules. The values obtained from a direct impalement of the tubule cells were −39.4 mV (basal membrane) and +4 mV (transmural) relative to the bathing medium; diuretic hormone or cyclic AMP increase fluid secretion without any associated change in these potentials. This finding contrasts with other studies which used a 'paraffin-liquid gap' technique to measure potential differences. For example, in *Rhodnius* (Maddrell 1971) the transmural potential in unstimulated tubules is negative; on addition of diuretic hormone a drastic 'spike' potential was observed. The transient changes in potential from negative to positive and return to a larger negative value are ascribed to the switching on of cation and anion pumps. The role of sodium in locusts and other insects is also unclear (see Morgan & Mordue 1981). It is not involved in a stoichiometric exchange with potassium, and if Na^+/K^+ exchange pumps exist they are both ouabain-insensitive and not essential for maintenance of fluid flow. Most workers have shown that ouabain does not inhibit stimulated secretion. Amiloride and ethacrynic acid, inhibitors of sodium transport, inhibit both basal and stimulated secretion in locusts. In spite of intensive investigations the relationships between transport of inorganic solutes and fluid flow are unresolved. Further, using double-barrelled microelectrodes that are selectively sensitive to K^+ or Cl^- or Na^+, we can explain potential differences across the apical and basal membranes of the cells in terms of intracellular concentrations of these ions; these studies, however, show no clear-cut relationships between solute transport and fluid secretion.

(b) Chloride transport-stimulating hormone

CTSH is present in the corpora cardiaca of locusts; its specific action is to stimulate electrogenic transport of Cl^- from the rectal lumen into the haemolymph. Isolated rectal tissue transports Cl^- at a high rate and short-circuit current is accounted for totally by Cl^- movement. Short-circuit current and Cl^- transport are affected markedly by CTSH, a neurosecretory peptide released from the storage lobes of the corpora cardiaca; as little as 0.005 of a gland is effective. This electrogenic transport of chloride across rectal tissue is not influenced by ouabain, furosemide, acetazolamide, vasoactive intestinal peptide, dogfish rectal gland extract or the Ca^{2+} ionophore, A23187. The stimulation of Cl^- transport is affected if the K^+ concentration falls below 10 mM; and CTSH stimulates absorption of K^+ and Cl^- but not Na^+. Cyclic AMP or GMP applied on the blood side causes the same increase in Cl^- transport as does 0.1 of a pair of corpora cardiaca, and homogenates of corpora cardiaca elicit a threefold increase in rectal tissue cyclic AMP. Agents such as theophylline and prostaglandins, which increase intracellular levels of cyclic AMP, also stimulate the short-circuit current. CTSH has an M_r of about 8000 and its chemical characteristics and responses in a variety of bioassays indicate that it is different from either diuretic hormone or adipokinetic hormone and related peptides. CTSH is released in response to feeding, as is diuretic hormone, and it is likely that the two hormones act in concert to recycle KCl through the Malpighian tubules and rectum (see Phillips et al 1980).

(5) Hormones and metabolism

(a) Lipid metabolism

Adipokinetic (hyperlipaemic) peptides are present in the corpora cardiaca of locusts, cockroaches, beetles and butterflies. The role of adipokinetic hormone (AKH) in the regulation of lipid mobilization and utilization during flight in locusts is well documented (see Goldsworthy & Mordue 1974, Goldsworthy & Cheesman 1978). AKH from locusts is the only hyperlipaemic peptide characterized fully (Stone et al 1976); both N- and C-terminals are blocked and it is uncharged. AKH has the following amino acid sequence:

$$\text{PCA-Leu-Asn-Phe-Thr-Pro-Asn-Trp-Gly-Thr-NH}_2$$

1 2 3 4 5 6 7 8 9 10

AKH is released from the glandular lobes of the corpora cardiaca at the onset of flight and acts on the fat body to release diacylglycerides which attach to acceptor lipoproteins; a new lipoprotein complex forms in the blood and this is the major carrier of glycerides from the fat body to the flight muscles. AKH also increases the oxidation of diacylglycerols in the flight muscle and simultaneously depresses glycolytic flux in the flight muscles (see Goldsworthy & Cheesman 1978). Recently octopamine in the haemolymph has been shown to increase during the first 15 min of flight in locusts. This burst of octopamine may be important in sustaining glucose oxidation by the flight muscles when this is prone to inhibition by the increasing levels of circulating diacylglycerol caused by release of AKH. The origin of this octopamine is uncertain; nor is it clear whether it is acting as a true hormone or as a neuromodulator of flight metabolism (Goosey & Candy 1980).

In structure–activity studies of AKH we have shown that only larger peptides, with a minimum of the first eight residues, have significant lipid-mobilizing activity. So far, all active compounds tested are uncharged. Substitutions in or deletions from the parent decapeptide drastically reduce biological activity and there is a precise requirement for 10 residues—however, it is of interest that nonapeptides have little detectable activity (see Mordue & Stone 1978). AKH is related structurally to the red pigment-concentrating hormone, a neurosecretory chromatophorin present in certain crustaceans. Both peptides have significant activity when tested in reciprocal assays (Mordue & Stone 1979). Comparative studies among various species of insects have also indicated some similarities between different corpus cardiacum hormones. For instance, locust AKH is chemically distinct from hyperglycaemic and adipokinetic factors from cockroaches but locust and cockroach peptides are active in each other's test systems. The Colorado beetle (*Leptinotarsa decemlineata*) uses proline as an energy substrate during flight and it is intriguing that either AKH or a closely related peptide regulates the synthesis of proline at the expense of stored triacylglycerols (Weeda et al 1980). From a comparative standpoint it is of interest that a number of insects utilize peptides to regulate intermediary metabolism (see below), especially flight metabolism, and that these peptides are similar to AKH.

(b) Peptides and carbohydrate metabolism

The disaccharide trehalose is the chief sugar found in insect haemolymph. There is much contradictory and confusing information on the possible roles of hyper- and hypotrehalosaemic peptides in insects (see Goldsworthy & Mordue 1974, Mordue & Stone 1979). Only in blowflies is there convincing evidence that a hypertrehalosaemic peptide is released from the corpus cardiacum during flight to maintain an adequate supply of trehalose. If release of this hormone is prevented, blood trehalose levels fall and the insect is unable to fly. The existence of a hypertrehalosaemic peptide has been known for some time in the cockroach but its structure and physiological role have not been resolved. The amino acid compositions obtained by two groups for this peptide (see Stone & Mordue 1980) show some agreement and are similar to the composition of a peptide extracted from corpora cardiaca of stick insects (G. Gade & W. Mordue, unpublished observations; see Mordue 1981). AKH at high doses has a hypertrehalosaemic effect but it is not clear whether this action is physiological or pharmacological.

Interest in the possible involvement of peptides in the control of carbohydrate metabolism is heightened by evidence that many insects appear to possess insulin-like and glucagon-like immunoreactive compounds. Blowflies contain a hypoglycaemic peptide in their brain neurosecretory cells and this peptide is similar to bovine insulin (Duve et al 1979). An extract of whole heads cross-reacted with antibodies raised against bovine insulin, increased glucose incorporation into lipid in rat adipocytes, and displaced ^{125}I-labelled insulin from preparations of liver plasma membrane. However, the peptide is thought not to be identical with bovine insulin. In *Manduca*, peptides from the corpora cardiaca react immunochemically with bovine glucagon and insulin antibodies (see Kramer 1980). The hyperglycaemic peptide, although immunochemically related to bovine glucagon, is chemically distinct. From the effects of trypsin digestion, which does not alter its glycogenolytic activity, it lacks lysine and arginine residues; moreover, high doses of glucagon are only weakly glycogenolytic in *Manduca*. Similarly, the insulin-like immunoreactive peptide is effective in lowering the blood sugar level but bovine insulin is not. The amino acid composition of AKH and the hyperglycaemic peptide is known (in part only for the latter), and neither resembles insulin or glucagon. The structure of the *Manduca* and *Calliphora* peptides affecting carbohydrate metabolism is awaited eagerly. At present, most identification is based on immunological reactivity only, and it is worth mentioning the findings of Julliard et al (1980) on the possible false identification of human placental proteins with β-endorphin and ACTH activities. This work is of particular importance if we consider the explosion in immunohistochemical identification of vertebrate peptides within the insect CNS (see for example El-Salhy et al 1980).

(c) Release of peptides

Neurohormones are released from axon terminals in response to nerve impulses arriving at these axon endings. An increase in the potassium concentration of the medium bathing isolated neurohaemal organs mimics the depolarization induced by such impulses and in the presence of calcium releases neurosecretory hormones into the medium. This manipulation has been used to elicit the release of diuretic, hyperglycaemic and hyperlipaemic hormones; it has proved unsuccessful in eliciting the release of CTSH. For most neurohaemal organs the amount of hormone released in response to potassium-induced depolarization is only a small part of the total stored; such treatment may release only the readily releasable hormone and much of the chemically extractable material may not be released under physiological conditions.

A whole variety of stimuli, such as feeding, flight and eclosion, act as cues for the release of neurohormones. Electrical stimulation of the afferent nerves to the corpora cardiaca (NCC I and NCC II) has demonstrated mechanisms for the selective release of hormones (see Goldsworthy & Mordue 1974). More recently Orchard & Loughton (1981) have suggested that octopamine functions as the neurotransmitter in the synapses between the NCC II axons and the glandular lobe cells of the corpora cardiaca, and that it is used to elicit release of AKH. However, I must stress that at present we know very little about the biosynthesis, transport, storage and release of insect neurohormones.

Conclusions

In this short résumé of the more important actions of neuropeptides I have tried to indicate their overall importance to the neuroendocrine biology of insects. It is clear that peptides have a variety of hormonal, transmitter and modulatory actions. Much of the present confusion as to whether effects are physiological or pharmacological stems from a lack of knowledge of chemical structures and in consequence a lack of information about the precise dosages used. However, rigorous experimental criteria have to be applied, and potency of action does not in itself indicate a physiological role. Progress is needed not only in studies of the modes of action of peptides but also in comparative studies which will show us how universal particular mechanisms are. The manipulation of the functioning of peptidergic systems by xenobiotics is still a totally unexplored field.

Acknowledgements

I am grateful to the Science and Engineering Research Council and to the Royal Society for research support.

REFERENCES

Baumann E, Gersch M 1981 Purification and identification of neurohormone D, a heart accelerating peptide from the corpora cardiaca of the cockroach *Periplaneta americana*. Insect Biochem, in press

El-Salhy M, Abou-el-ela R, Falkmer S, Grimelius L, Wilander E 1980 Immunohistochemical evidence of gastro-entero-pancreatic neurohormonal peptides of vertebrate type in the nervous system of the larva of a dipteran insect, the hoverfly, *Eristalis aeneus*. Regul Peptides 1:187-204

Duve H, Thorpe A, Lazarus NR 1979 Isolation of material displaying insulin-like immunological and biological activity from the brain of the blowfly *Calliphora vomitoria*. Biochem J 184:221-227

Goldsworthy GJ, Cheesman P 1978 Comparative aspects of the endocrine control of energy metabolism. In: Gaillard PJ, Boer HH (eds) Comparative endocrinology. Elsevier, Amsterdam, p 423-436

Goldsworthy GJ, Mordue W 1974 Neurosecretory hormones in insects. J Endocrinol 60:529-558

Goosey MW, Candy DJ 1980 The D-octopamine content of the haemolymph of the locust, *Schistocerca americana gregaria* and its elevation during flight. Insect Biochem 10:393-397

Julliard JH, Shibasaki T, Ling N, Guillemin R 1980 High molecular-weight immunoreactive β-endorphin in extracts of human placenta is a fragment of immunoglobulin G. Science (Wash DC) 208:183-185

Karlson P, Sekeris C 1966 Ecdysone, an insect steroid hormone, and its mode of action. Recent Prog Horm Res 22:473-502

Kramer KJ 1980 Insulin-like and glucagon-like hormones in insects. In: Miller TA (ed) Neurohormonal techniques in insects. Springer-Verlag, New York, p 116-136

Maddrell SHP 1971 The mechanisms of insect excretory systems. Adv Insect Physiol 8:199-331

Miller TA 1979 Nervous versus neurohormonal regulation of insect heartbeat. Am Zool 19:77-86

Mordue W 1972 Hormones and excretion in locusts. Gen Comp Endocrinol suppl 3:289-298

Mordue W 1981 Adipokinetic hormone and related peptides. Proc 8th Int Symp Neurosecretion: in press

Mordue W, Stone JV 1978 Structure and metabolism of adipokinetic hormone. In: Gaillard PJ, Boer HH (eds) Comparative endocrinology. Elsevier, Amsterdam, p 487-490

Mordue W, Stone JV 1979 Insect hormones. In: Barrington EJW (ed) Hormones and evolution. Academic Press, New York, p 215-271

Morgan PJ, Mordue W 1981 Stimulated fluid secretion is sodium dependent in the Malpighian tubules of *Locusta migratoria*. J Insect Physiol 27:271-279

Orchard I, Loughton BG 1981 Is octopamine a transmitter mediating hormone release in insects? J Neurobiol 12:143-153

Phillips JE, Mordue W, Meredith J, Spring J 1980 Purification and characteristics of the chloride transport stimulating factor from locust corpora cardiaca: a new peptide. Can J Zool 58:1851-1860

Rafaeli A, Applebaum SW 1980 Canavanine potentiates the response of insect Malpighian tubules to diuretic hormone and cyclic AMP. Nature (Lond) 283:872-873

Rafaeli A, Mordue W 1981 The responses of the Malpighian tubules of *Locusta* to hormones and other stimulants. Gen Comp Endocrinol, in press

Reynolds SE 1980 Cuticle plasticizing factors. In: Miller TA (ed) Neurohormonal techniques in insects. Springer-Verlag, New York, p 179-195

Reynolds SE, Truman JW 1980 Eclosion hormones. In: Miller TA (ed) Neurohormonal techniques in insects. Springer-Verlag, New York, p 196-215

Seligman IM 1980 Bursicon. In: Miller TA (ed) Neurohormonal techniques in insects. Springer-Verlag, New York, p 137-153

Starratt AN, Steele RW 1980 Proctolin: bioassay, isolation and structure. In: Miller TA (ed) Neurohormonal techniques in insects. Springer-Verlag, New York, p 1-30

Stone JV, Mordue W 1980 Isolation of insect neuropeptides. Insect Biochem 10:229-239

Stone JV, Mordue W, Batley K, Morris HR 1976 Structure of locust adipokinetic hormone, a neurohormone that regulates lipid mobilization during flight. Nature (Lond) 263:207-211

Truman JW 1973 How moths 'turn on': a study of the action of hormones on the nervous system. Am Sci 61:700-706

Truman JW 1978 Hormonal release of stereotyped motor programmes from the isolated nervous system of the cecropia silkmoth. J Exp Biol 74:151-174

Truman JW, Taghert PH, Coperhaver PF, Tublitz NJ, Schwartz LM 1981 Eclosion hormone may control all ecdyses in insects. Nature (Lond) 291:70-71

Weeda E, Koopmanschap AB, de Kort CAD, Beenakkers AMT 1980 Proline synthesis in the fat body of *Leptinotarsa decemlineata*. Insect Biochem 10:631-636

DISCUSSION

Evans: What is the best amine for stimulating the Malpighian tubule? Have you compared synephrine and adrenaline? And is there any adrenaline in the locust?

Mordue: Ada Rafaeli and I have recently reported on this (see Rafaeli & Mordue 1981). In more recent, as yet unpublished studies (A. J. Mordue & W. Mordue), octopamine is inhibitory, if anything, in its effect on the Malpighian tubules. I am not sure whether this is an antagonistic action or even an agonist effect which doesn't stimulate the tissue enough to give a fluid response. We haven't tried synephrine yet. I don't know of any evidence for a significant amount of adrenaline in the locust. It requires about 10^{-4} M-adrenaline to get consistent stimulation of the Malpighian tubule, though we see some effect at 5×10^{-5} M.

Evans: Maddrell et al (1971) showed that 5-HT may mimic the actions of the peptide diuretic hormone on insect Malpighian tubules. Could adrenaline be mimicking the action of an unknown peptide hormone in your system?

Mordue: In that paper, where Maddrell and colleagues (1971) tried to characterize the aminergic response and the peptidergic response, I am not sure they proved that the two receptors are the same. They may have a similar system to ours, with an aminergic and a peptidergic receptor.

Penzlin: Can you say anything about antidiuretic hormones in insects, especially in locusts? You didn't mention them.

Mordue: Some of my own work suggested the existence of antidiuretic hormones in *Locusta* and *Schistocerca*. I excluded it here because, if rigorous criteria are applied, ADH is an antidiuretic *factor* and has not been shown to be a true *hormone*. I still maintain that in the control of the functioning of the Malpighian tubules and the regulation of water balance it is not necessary to have antidiuresis. All that is needed is a lack of diuretic hormone. The tubules then switch down to a basal rate of secretion which puts 10 or 12 μl fluid per hour into the hind-gut. This rate matches the rate of rectal reabsorption. At such slow rates of Malpighian tubule secretion the animal can keep itself in water balance without having to switch off the Malpighian tubules.

We must look again to see whether the antidiuretic hormone acting on the rectum could be the chloride transport-stimulating hormone. The chloride transport is very significant and could be dragging water with it. With the rectal preparations set up as in frog skin experiments, at present it is not possible to measure water movement, but it should be possible to do that with isolated rectal sacs.

Berridge: You said you thought adipokinetic hormone was released in a burst at the onset of flight. When a locust takes off on a migratory flight, does that burst of hormone last all the time of the flight or does it have to release another burst? If it stops flying, how rapidly does AKH activity disappear from the blood?

Mordue: That is one of our difficulties. We need to improve our detection and estimation techniques. We are now using high performance liquid chromatography to detect the level of AKH in blood.

Berridge: How rapidly does the increased lipid return to normal?

Mordue: The lipid elevation goes on for some time after an insect is prevented from further flight. I think this continued increase is a function of delay in the target tissue. Extracting the blood for subsequent chemical determination of AKH is tedious, but if we do that we have the impression the hormone has been removed from the blood very quickly. It takes some hours for the lipids in the haemolymph to return to the resting level.

Berridge: Most neurohaemal release sites show inactivation. Once you have released a burst of hormone, can you release more AKH from that particular gland with high potassium concentrations?

Mordue: High potassium levels release a small amount of stored AKH but we don't know whether that is the readily releasable hormone or whether the rest of what we see there is sequestered in such a way that it is rarely released in physiological conditions, as happens in vertebrates.

Olsen: What is the volume in which picomoles of hormone are effective?

Mordue: The blood volume of an adult male is about 300 μl and that of a female locust, 500 μl.

Usherwood: Presumably the complex structure of the hormones contributes to high signal-to-noise ratios at their target sites. Have you any evidence that the hormones serve additional roles which befit their structural complexity, apart from activating cell membrane receptors?

Mordue: This is something we need to know. We are just beginning to get a better understanding of the circulatory levels. We infer how much AKH has been released from the raised levels of diglyceride during flight and work back to dose-response curves. More sophisticated monitoring techniques might show us that AKH is released at different times and in different amounts and is doing other things. For example, in work done with John Moorhouse (unpublished) we showed that if AKH is put onto motor neurons supplying the flight muscles the rate of firing increases greatly. We can do that with many other peptides too, so is this significant? Robinson & Goldsworthy (1977) would maintain that AKH has a very specific effect on the mitochondria, where it switches off carbohydrate utilization. AKH may have a whole series of effects which are hidden by its major effect of mobilizing so much diglyceride.

Kelly: Does having say 12 or 30 amino acids in sequence give better receptor specificity than a shorter chain?

Mordue: I don't know. We must characterize some of the other peptide hormones and see what minor substitutions or deletions are required to affect the biological activity radically. For AKH and the crustacean hormones the deletions or substitutions do confer very particular specificity. Eight residues might be a minimal requirement to give enough specificity in the system.

Kelly: In the mammalian system this has caused a lot of argument. Substances like cholecystokinin and gastrin have 33 amino acid sequences, yet sequences as short as four or eight contain the biological activity (see Dodd & Kelly 1981). A hormone like vasoactive intestinal peptide has 29 amino acids but activity is lost after quite small changes in length. In the mammalian system either argument can be sustained: there may be large molecules of which only very small fragments are active, or there may be large molecules which can't be altered. Your system is a bit like somatostatin, in which all the known changes in structure change the activity only five- or six-fold. The changes in activity that you see are not at all dramatic.

Mordue: Except that if we change the residue at position 5 in naturally occurring AKH, activity is destroyed. Moreover, if we delete leucine from AKH the resulting nonapeptide has no activity at all. In the interactions between AKH and the locust fat body receptors, 10 residues seem to be critical. There are also a whole set of structure-activity relationships worked out for the red pigment concentrating hormone system of crustacea (see Christensen et al 1979). Again there is a very particular requirement for the sequence of the residues and the type of residues. Tyrosine inserted into position 4 gives dramatically improved pharmacological activity. We have found no agonist or antagonist among any of our analogues.

Barnard: The reason for the structural complexity may lie in the same sort of phenomena as are now known in vertebrates. There, most or all of the gastrointestinal peptides and similar circulating hormones (e.g. insulin) also appear to act as neurotransmitters or modulators in the CNS (Havrankova et al 1978, Dockray & Gregory 1980). In the insect some circulating peptide hormones may have roles in the nervous system that have yet to be uncovered. This phenomenon seems to be related to the mode of biosynthesis of these peptides: they have to come from protein precursors, because no animal system is known that synthesizes directly a peptide of more than a few residues. When an animal has to make such a protein precursor of a peptide hormone, it is—as known in the vertebrates (Dockray & Gregory 1980, Hughes et al 1980, Steiner et al 1980)—frequently cut up to produce, also, various kinds of transmitters to be used at various places.

Berridge: The mammalian pancreas has receptors for ACh and for the peptide hormone pancreozymin. However, the information the cell gets is identical whether the cholinergic system or the peptide system is being activated. The same applies to substance P and ACh in the mammalian salivary gland. The cell doesn't get more information by having a more complex peptide. There may be other reasons why these molecules exist.

Mordue: The time course of the signal may be important too. The arthropod peptides of the AKH family are blocked and not charged, for some reason. We were able to isolate AKH because it doesn't behave like a peptide at all. We don't know yet whether its peculiar structure is significantly related to the interactions with its receptors, transport in the blood or resistance to degradation. We now have some labelled material, a tritiated leucine derivative, with a high specific activity and we should be able to look at receptor interactions in a range of different systems. There are modifications during the course of evolution not only in the structure of peptides but also in a variety of receptors and tissues, so even in common systems there must be evolution in the receptor systems too.

Usherwood: If insects produce molecules like neuropeptides that signal cellular events with a high degree of specificity, could these compounds be used as a starting point to develop insect-specific insecticides?

Donnellan: The insecticide industry is certainly aware of the potential of targets related to the known neuropeptides. The biggest drawback may be the slowness of action of either analogues or antagonists. The industry is still primarily geared to looking for fast-acting nerve poisons. We also have the perennial problem of getting intrinsically active substances into the insect. We can make molecules for the target sites but they may never see those sites when applied to the insect's exterior. Peptides don't pose specific problems in that respect.

Mordue: The dual component in some of the systems, particularly the Malpighian tubules which have both aminergic and peptidergic systems, may mean that it is not essential to get the peptide to the target. The peptide may be useful for indicating how the target is responding but it may not be essential to concentrate purely on a peptidergic system. The tse-tse fly diuretic hormone can be blocked very nicely with phentolamine. If locust tubules are challenged with tse-tse fly hormone, fluid secretion increases markedly, as does muscle contraction, which is blocked also by phentolamine. This hormone may be working more on the aminergic side than on the peptidergic side, as in *Rhodnius*.

Narahashi: To be an effective insecticide a compound must be stable. Could a peptide be made very stable?

Dudai: Some snake toxins are quite stable.

Hall: If you tried to feed large polypeptides such as α-bungarotoxin to the insect it would just digest the polypeptide, destroying its activity, so delivery is a major problem.

Dudai: I don't know if it would digest the peptide right away, but the toxin probably doesn't penetrate the insect equivalent of the blood-brain barrier. α-Bungarotoxin should be quite toxic, on the basis of its affinity *in vitro* to nicotinic binding sites, but if you inject it into a fruitfly almost nothing happens (Dudai et al 1980). Even if it enters the circulation there is the blood–brain barrier.

M. E. Eldefrawi: I once injected 10 LD_{50} of α-bungarotoxin for a mouse into a cockroach and the insect actually grew fatter for a few days. But good toxic peptides are available. The important question is whether insects degrade or metabolize these peptides, hormones or toxins if they feed on them. If they do, a good delivery system could be produced that would encapsulate these chemicals so that the insects would be poisoned by them.

Neumann: Whether a peptide is open to enzymic attack depends on its amino acid sequence. I would assume that some peptides are stable enough by themselves. With others, changing or modifying the amino acids should increase their stability. Certainly the pyrethroids provide good examples of strongly increased stability due to chemical modification of a labile natural compound.

Hildebrand: Another point is the possible difference between one group of insects and another. I have the impression that proctolin has proved to be difficult to detect in some groups of insects. Can anyone confirm that proctolin is found only in certain insect orders?

Mordue: There are said to be large amounts of proctolin in the corpus cardiacum in cockroaches (order Dictyoptera) and locusts (order Orthoptera). The activity has to be chemically characterized but it may well be Gersch's neurohormone D, which is certainly not proctolin (Baumann & Gersch 1981). Flies and Lepidoptera may have something else which is acting like proctolin but is not identical to it.

Penzlin: We have some results on the distribution of proctolin in the terminal ganglion of the cockroach, *Periplaneta americana*. In iontophoretic studies we filled the anterior and posterior branches of the unilateral proctodeal nerve with cobalt and succeeded in labelling several cells in the terminal ganglion of *Periplaneta*. At light- and electron-microscopic levels about 30 somata were characterized. Most of these cells were located in the dorsal and ventral midline of the terminal ganglion. We then asked which cells with connections with the proctodeal nerve contain proctolin or a proctolin-like peptide. Dr M. Eckert of our group isolated a specific antibody against proctolin by affinity chromatography. Among the 17–22 cells stained after unilateral axonal cobalt iontophoresis and arranged along the main fibre tract in the dorsal and ventral midline of the terminal ganglion, only the large so-called type 2 cells posterior to the arborization domain stained with anti-proctolin antibodies. Proctolin-like immunoreactivity was localized within neurons of the terminal ganglion projecting into the proctodeal nerve on one hand, and in nerve cells without a direct connection to this system on the other hand. Therefore, we must take into account that proctolin can act as a neuromuscular transmitter at the hind-gut musculature and as a neurotransmitter or neuromodulator on structures in the central nervous system as well. A detailed description of our results has been published (Eckert et al 1981).

Some preliminary experiments by our group on the effect of proctolin on the hyperneural muscle of *Periplaneta americana* show that this muscle is sensitive to proctolin in threshold concentrations lower than 10^{-10} M. The response is fairly independent of the ionic composition of the bath fluid. However, calcium is necessary for the response. We assume that proctolin acts on the hyperneural muscle of *Periplaneta* via the adenylate cyclase system.

Piek: You used a concentration of 10^{-7} M but this is much higher than is needed for contraction of the proctodeal muscle in the cockroach (Brown 1975) and much higher than we used for getting effects on the tonic bundle of the extensor muscle of the locust hindleg (Piek & Mantel 1977). I can't believe that proctolin is present at 10^{-7} M in the insect, because that concentration would have enormous effects on visceral as well as on skeletal muscles. What is the significance of such high concentrations?

Penzlin: The response of the muscle begins to become visible at 10^{-10} M. In our experiments we deliberately used much higher concentrations to induce maximal responses.

Piek: We found 10^{-10} M and May et al (1979) found 10^{-11} to be the threshold for tonic muscle; 10^{-9} M gives physiological effects and 10^{-8} gives enormous contractions.

Evans: O'Shea & Adams (1980) have also identified proctolin-containing neurons in several insect species. They have called them the lateral giant peptide neurons. These represent another very exciting proctolin system that could be worked on in the future.

Usherwood: What do we know about the effects of octopamine and proctolin on insect central neurons?

Evans: It is difficult to obtain good evidence for central synaptic actions of biogenic amines on identified neurons in insects. When they are ionotophoresed into the neuropil we don't know where the sites of action are going to be. If there are any responses we don't know whether they are primary effects or secondary effects induced by stimulation of some parallel pathway.

Penzlin: Walker et al (1980) showed that dorsal median neurons in the cockroach VIth abdominal ganglion responded after addition of proctolin, but I imagine that was an indirect effect.

Sattelle: Are the midline proctolin-containing cells clearly distinct from the dorsal unpaired median (DUM) cells, which contain octopamine?

Penzlin: I can't say that yet.

Usherwood: To sum up, at least two types of chemical signal operate in the nervous systems of insects and other animals. Some signals, such as the traditional neurotransmitters, act mainly on a specific part of the target cell—that is, the postsynaptic membrane. These spatially restricted signals are usually of short duration. Other chemical signals, such as the neuropeptides, act more diffusely and over longer time periods and it is, therefore, more difficult to gain an understanding of their pharmacological roles. The term neuromodulator has been used to describe compounds such as neuropeptides but this does not provide much insight into the functional significance of these compounds.

REFERENCES

Baumann E, Gersch M 1981 Purification and identification of neurohormone D, a heart accelerating peptide from the corpora cardiaca of the cockroach *Periplaneta americana*. Insect Biochem, in press

Brown BE 1975 Proctolin: a peptide transmitter candidate in insects. Life Sci 17:1241-1252

Christensen M, Carlsen J, Josefsson L 1979 Structure and function studies on red pigment concentrating hormone. II. The significance of the C-terminal tryptophan amide. Hoppe-Seyler's Z Physiol Chem 360:1051-1060

Dockray GJ, Gregory RA 1980 Relations between neuropeptides and gut hormones. Proc R Soc Lond B Biol Sci 210:151-164

Dodd J, Kelly JS 1981 The actions of cholecystokinin and related peptides on pyramidal neurones of the mammalian hippocampus. Brain Res 205:337-350

Dudai Y, Nahum-Zvi S, Haim-Granot N 1980 Cholinergic pharmacology of *Drosophila*: comparison of *in vivo* to *in vitro* studies. Comp Biochem Physiol C Comp Pharmacol 65:135-138

Eckert M, Agricola H, Penzlin H 1981 Immunocytochemical identification of proctolin-like immunoreactivity in the terminal ganglion and hindgut of the cockroach *Periplaneta americana* (L.). Cell Tissue Res 217:633-645

Havrankova J, Roth J, Brownstein MJ 1978 Insulin receptors are widely but unevenly distributed throughout the central nervous system of the rat. Nature (Lond) 272:827-829

Hughes J, Beaumont A, Fuentes JA, Malfroy B, Unsworth C 1980 Opioid peptides: aspects of their origin, release and metabolism. J Exp Biol 89:239-255

Maddrell SHP, Pilcher DEM, Gardiner BOC 1971 Pharmacology of the malpighian tubules of *Rhodnius* and *Carausius*: the structure–activity relationship of tryptamine analogues and the role of cyclic AMP. J Exp Biol 54:779-804

May TE, Brown BE, Clements AN 1979 Experimental studies upon a bundle of tonic fibres in the locust extensor tibialis muscle. J Insect Physiol 25:169-181

O'Shea M, Adams M 1980 Identification of giant peptidergic (proctolin containing?) neurons in insects. Soc Neurosci Abstr 6:704

Piek T, Mantel P 1977 Myogenic contractions in locust muscle induced by proctolin and by wasp, *Philanthus triangulum*, venom. J Insect Physiol 23:321-325

Rafaeli A, Mordue W 1981 The responses of the Malpighian tubules of *Locusta* to hormones and other stimulants. Gen Comp Endocrinol, in press

Robinson N, Goldsworthy GJG 1977 A possible site of action for adipokinetic hormone on the flight muscle of locusts. J Insect Physiol 23:153-158

Steiner DF, Patzelt C, Shan SJ et al 1980 Formation of biologically active peptides. Proc R Soc Lond B Biol Sci 221:45-59

Walker RJ, James VA, Roberts CJ, Kerkut GA 1980 Neurotransmitter receptors in invertebrates. In: Sattelle DB et al (eds) Receptors for neurotransmitters, hormones and pheromones in insects. Elsevier/North-Holland Biomedical Press, Amsterdam

Isolation of receptors from the central nervous system

E. A. BARNARD

Department of Biochemistry, Imperial College, London, SW7 2AZ, UK

Abstract The principles of, and some salient findings in, the successful isolation of some vertebrate neurotransmitter receptors are briefly reviewed. These form a logical starting-point for this approach in the study of the molecular neurobiology of the insect nervous system. A case of particular relevance to the latter, namely a nicotinic acetylcholine receptor of the vertebrate brain, is taken as an illustrative example.

I, and other biochemists working in the field of neuroscience, have a certain bias in that we must admit to being vertebrates, and this has obviously influenced our viewpoint. Both this and reasons of abundance of material have led to a high degree of emphasis on the vertebrate nervous system as a source. Yet it has often been pointed out that it makes good sense, in studies of brain mechanisms, to start with the simpler nervous systems. Of course, 'simple' is a term to be used only in a very relative sense in describing the nervous system of insects. Nevertheless, the insect nervous system is undeniably much less complex than the vertebrate brain, and offers cases where identified neurons and synapses can be reproducibly located (as illustrated by others in this symposium) and where several pathways can be readily mapped and related to activity. It is logical, therefore, to enquire into the functioning and control of individual types of neurotransmitter receptors at these synapses, as a necessary stage in the understanding of such neural pathways. For this, the experience gained in the isolation of receptors for those same neurotransmitters from vertebrate sources should obviously be exploited.

In making this statement, I am relying on some important homologies: (a) the great majority of the neurotransmitters which are known or believed to be used in insects have been found also in the vertebrates (see Hildebrand, this symposium p 5–11); (b) the principal physiological and ultrastructural

1982 Neuropharmacology of insects. Pitman, London (Ciba Foundation symposium 88) p 109-117

characteristics of mammalian chemical synapses are found in insect synapses; and (c) protein structure and membrane structure, and the genetic mechanisms for specifying them, are known to be uniform in all their main features in all phyla.

Taking the vertebrate nervous system as a prototype, then, we should first note that the massive body of information accumulated on it at the biochemical and physiological levels leads to the recognition of two general classes of neurotransmitter receptor: (A) receptors that are intimately and permanently associated with fixed ionic channels across the membrane, controlling the opening of the latter, and (B) receptors linked to mobile elements (e.g. adenylate cyclase, Ca^{2+}-linked systems, etc.) which generate the transmembrane signal by some secondary reaction. The latter class, B, comprises several different types (as discussed by Dr Berridge in this symposium p 260–269). Analysis at the molecular level has proceeded much further with some cases of class A than of class B and I shall confine myself here to class A.

The best-characterized example so far in class A is, of course, the vertebrate nicotinic acetylcholine (ACh) receptor, although several others are at present also becoming elucidated. A model of that receptor is shown in Fig. 1. This model summarizes the evidence from a wealth of investigations of

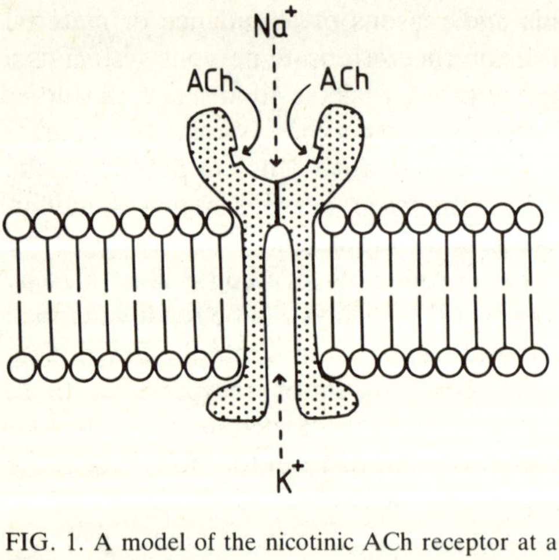

FIG. 1. A model of the nicotinic ACh receptor at a peripheral synapse. Two binding sites for ACh are known to be present per molecule of about 250 000 M_r, and these face the synaptic cleft (top), into which protein segments (carrying polysaccharide side-chains) are known to project. The ion channel is shown as controlled by a gate, which is believed to be opened by a conformational change induced by the binding of two agonist molecules at their sites as shown, and closed by the binding of antagonists (including α-BGTX) at the same sites. In the electric organ receptor the protein structure shown is built up from five subunits, of type $\alpha_2\beta\gamma\delta$. The molecular means of constructing the channel from them has not, however, so far been specified.

the ACh receptor of the electric organ of electric fish and of vertebrate skeletal muscle. It is shown schematically as a fixed, ion-selective channel in the membrane with a protein structure containing a recognition site for ACh. Some kind of gate to the ionic channel can be opened by means of a conformational change in that protein when the ACh is bound, but the gate will remain tightly closed when an antagonist is bound.

While some of the evidence for this structure has come from studies of channel conductance or membrane ligand binding or ultrastructure, the evidence also rests very heavily on the results of the biochemical approach of extraction and purification of this receptor. The former types of study indicate significant features of the mechanism of action, but require the latter to give it concrete structure; molecular knowledge of the system could not have become definite and specific without analysis of the isolated receptor. I shall confine this discussion to the question of isolation and analysis.

Requirements for receptor isolation

The first requirement for successful receptor purification is, obviously, a suitable source, which initially has meant a tissue with an abundance of the receptor. The first receptor to be purified was from fish electric organs, where the nicotinic ACh receptor is exceptionally concentrated (~1000 pmol/g wet weight of tissue); this gave a great advantage to the studies a decade or so ago on this receptor. However, it subsequently proved possible to improve the methods so that the receptor could be obtained in equally pure form (Dolly & Barnard 1975, 1977, Froehner et al 1977) from vertebrate denervated skeletal muscle (down to ~50 pmol/g). Recently it has been purified also from innervated skeletal muscle and from vertebrate brain (see below), in both of which the concentration is much lower still, in some cases down to 1 pmol/g. The experience with the more abundant sources led to the finding that affinity chromatography with high affinity ligands, and the use of protease inhibitors, could be applied in purifying from contents down to ~1 pmol/g tissue. A number of receptor types are found in vertebrate or insect brain at this order of magnitude.

The second requirement is a suitable probe, usable both for assay in solution and for immobilization for affinity chromatography. The availability of the now-celebrated postsynaptic toxins of elapid snake venoms, such as α-bungarotoxin (α-BGTX), of course conferred a great advantage in this respect in the isolation of the peripheral ACh receptor. This was due to their very slow or negligible dissociation from that receptor. However, dissociation is much faster from brain ACh receptors, which—again with the benefit of the methods developed on the more favourable cases—can now be purified, as will be detailed.

A third requirement, which must not be forgotten, is that it is also necessary to demonstrate that such a probe, in any new case of its use, is actually recognizing the functional receptor. Problems in this respect will be exemplified below.

A fourth requirement is a medium in which the receptor can be extracted fully, in stable and active form. For the vertebrate nicotinic receptor, a variety of detergent solutions meet this criterion, showing that a transmembrane receptor protein *can* be removed from its lipid contacts and yet retain its binding activity and also its ability to re-enter a lipid phase to reconstitute its ion channel (Miller & Racker 1980). For other receptors, a compatible detergent may be found by trial-and-error testing: for example, a class of the GABA receptors of the mammalian brain which is linked to the binding sites for benzodiazepines is stable in 0.7% Triton X-100 medium (Lang et al 1979, Sherman-Gold & Dudai 1980), and the erythrocyte β-adrenergic receptor has been purified in digitonin media (Shorr et al 1981a). Failing this, stabilization by added phospholipid in solution (Agnew & Raftery 1979, Schramm 1979, Anholt et al 1981) or other such means may be feasible. Related to the stability requirement is the need to prevent inactivating proteolytic attack during homogenization and extraction of the tissue, by having suitable protease inhibitors present (Dolly & Barnard 1977, Shorr et al 1981b).

Isolation of a brain ACh receptor

Since a suitable probe exists for nicotinic ACh receptors of the vertebrate peripheral nervous system, namely the postsynaptic neurotoxins of the type of α-BGTX, it is instructive to see how this could be applied to the brain, as a new source and one where the nicotinic ACh receptors are in relatively small numbers.

The binding of α-BGTX in radioactive form to sites in mammalian brains has been described by various authors (reviewed by Schmidt et al 1980). However, the third requirement above has not generally been met, in that no physiological blockade by the α-BGTX of synaptic transmission or associated receptor function has been demonstrated in the mammalian brain, spinal cord or autonomic ganglia where nicotinic receptors can be recognized pharmacologically (reviewed by Brown 1979, Schmidt et al 1980). Although a binding component for α-BGTX can be extracted, to ensure that a functional receptor is being isolated it has been necessary to select a site in the brain of lower vertebrates where functional blockade by applied α-BGTX can be demonstrated electrophysiologically (Freeman et al 1980), namely the optic lobe of the visual pathway. In parallel, saturable specific binding sites for α-BGTX have been shown in membranes of the optic lobe of fish (Oswald & Freeman 1979) and chick (Wang et al 1978, Barnard et al 1979).

Another important factor here is the relative concentration of such binding sites: in vertebrate brains α-BGTX binding sites of all types are generally at a low level, but the optic lobe of the newly hatched chick has about a 10-fold higher concentration than any other brain source (Wang et al 1978, Barnard et al 1980), which has been a useful aid in purification.

The binding of α-BGTX to its specific sites in vertebrate brain membranes, including the optic lobe, differs from the peripheral sites in being reversible, with a K_d value of 10^{-9}–10^{-10}M (Wang et al 1978, Barnard et al 1979). This is true, too, for the purified receptor, but the half-time for dissociation of the toxin-receptor complex is ~3 h, sufficiently long for accurate assay.

The procedures for purifying the optic lobe ACh receptor completely are summarized in Table 1. The overall purification from the starting membrane fraction, achieved by R. I. Norman in this laboratory, was 15 600-fold at the maximum, and the overall yield was 15–20% from that fraction. The receptor was shown to be a pure protein by a variety of criteria (Table 2).

Specific properties of the isolated ACh receptor

The following features were used to confirm that the protein isolated is the nicotinic ACh receptor of the chick optic lobe:

(1) It reacts with α-BGTX at the fast rate characteristic of muscle or electric organ ACh receptors, e.g. $k_{on} = 4.6 \times 10^5\ M^{-1}\ s^{-1}$ at 25 °C, pH 8, with mono-^{125}I-α-BGTX (Norman et al 1981), and this reaction is saturable and completely blocked by a large excess of α-BGTX or *d*-tubocurarine.

(2) It shows a high affinity for nicotinic cholinergic ligands, and not other types of ligand tested (Barnard et al 1980, Norman et al 1981).

(3) It reacts specifically with an affinity reagent based on the ACh structure, namely bromoacetylcholine (Norman et al 1981). This reaction occurs also in the electric organ (Damle et al 1978, Moore & Raftery 1979) or muscle (Lyddiatt et al 1979) ACh receptors, in each case (as also here) requiring prior reduction of a disulphide bond in the receptor protein. The rate is exceptionally fast, due to assistance from the binding, for all three sources.

(4) Immunological cross-reactivity has been demonstrated between the pure ACh receptors from cat muscle and from the chick optic lobe (Norman et al 1981). An antibody (anti-muscle) titration curve was obtained and showed 1.0% cross-reactivity, a low but highly significant value, indicating

TABLE 1 Purification procedure (at 4 °C) for optic lobe receptor

Chick optic lobe membranes[a]
↓
5% Lubrol PX + protease inhibitors[b]
↓
α-BGTX-gel (overnight, 84% bound)
↓
1 M-NaCl + buffer washes[c]
↓
1 M-carbamylcholine[d]
↓
G-25 gel filtration (removal of all carbamylcholine)
↓
GTP- agarose[d,e]
↓
Lentil lectin gel[d,e] (100% bound)
↓
0.4 M glycosides[e] + EDTA (84% eluted)
↓
Gel filtration

[a] Lobes (from 150 1-day-old chicks) are homogenized at pH 7, in the presence of protease inhibitors + 0.5 mM-*N*-ethylmaleimide, and the particulate fraction is collected.
[b] The protease inhibitors used are: 1mM-EDTA, 1 mM-EGTA, 100 μg bacitracin/ml, 10 μg soybean trypsin inhibitor/ml and 0.1 mM-phenylmethylsulphonyl fluoride (PMSF).
[c] All procedures from here onwards are in 0.2% Lubrol PX, buffered at pH 7.4, plus 0.1 mM-PMSF.
[d] These stages are combined by cycling the carbamylcholine elution medium in a closed circuit through columns of each of the gels, for 12 h, so that all the eluted receptor accumulates bound to the lectin gel (because of its carbohydrate content).
[e] Details of these materials or other procedures are as in Shorr et al (1981b) and Norman et al (1981).

TABLE 2 Criteria of purity of the ACh receptor isolated from chick optic lobe

Property[a]	*Value or observation*
(1) $s_{20,w}$	9.1S (in 0.5% Lubrol PX medium; one peak)
(2) Electrophoresis (in non-denaturing conditions, in 4% gel)	One band only
(3) Isoelectric point	5.4 (one peak of [^{125}I]-labelled protein in gel isoelectric focusing)
(4) Isoelectric point of complex with [^{125}I]α-BGTX	5.9 (one peak)
(5) Subunit content	M_r = 54 000 (one band only)
(6) Bromo-[^{3}H]acetylcholine-labelled subunit	M_r = 54 000 (one band only)

[a] Properties 1,2,3,5,6 were examined on receptor finally labelled by ^{125}I iodination and recovered by gel filtration and re-binding to the α-BGTX-gel. Results are from Norman et al (1981) and further unpublished observations.

that, despite the large phylogenetic difference, there are structural similarities between the two proteins.

(5) In the intact tissue, localization of the α-BGTX-binding sites has been accomplished at the light and electron microscope levels. For the former, dry-mount autoradiography of $[^3H]\alpha$-BGTX shows highly localized binding in the layers of the chick optic tectum where cholinergic terminals can be expected, in work by J. Cavanagh in this laboratory. In the electron microscope, binding of a 1:1 conjugate of α-BGTX and peroxidase was demonstrated in work with J. Cavanagh, B. Lang and R. I. Norman here. This showed, as illustrated elsewhere (Barnard et al 1979), that labelling occurs and is confined to certain synapses only, and to their postsynaptic membranes.

Hence, a brain receptor, of low abundance in the brain as a whole, has been purified using a reversible probe, both for assay and for affinity chromatography. It should be noted that this CNS receptor already shows some illuminating differences from the related peripheral receptor. Thus, its subunit composition is different. For comparison, the *Torpedo* ACh receptor has subunits with M_r values of about 40 000, 50 000, 58 000 and 65 000 (Karlin 1980) and the chicken muscle ACh receptor has subunits with M_r values of about 41 000, 50 000 and 54 000 (Sumikawa et al 1981), in each case these being only apparent M_r values from gel electrophoresis. Further, as noted, the dissociation of α-BGTX from its complex is much faster, and also its specificity for different postsynaptic toxins, such as those from the mamba *Dendroaspis viridus* (Barnard et al 1979), is different, in comparison to the peripheral receptors. Again, the relative affinities of nicotinic ligands can differ quantitatively between the two classes; decamethonium is bound 150-fold more weakly by the chick optic lobe receptor (membrane-bound or pure) and nicotine five times more strongly than by chick muscle receptor (Norman et al 1981). These differences will be important for an understanding of the detailed structural basis of the activity of the CNS receptor.

Whether the ion channel structure is intrinsic to the ACh recognition protein that has been purified will need to be determined by reconstitution into artificial membranes, as has been achieved with the electric organ ACh receptor. These general approaches should be relevant to the study of at least some of the receptors of the insect nervous system.

Acknowledgements

The work in this laboratory was supported by the Medical Research Council.

REFERENCES

Agnew WS, Raftery MA 1979 Solubilized tetrodotoxin binding component from the electroplax of *Electrophorus electricus*. Stability as a function of mixed lipid-detergent composition. Biochemistry 18:1912-1919

Anholt R, Lindstrom J, Montal M 1981 Stabilization of acetylcholine receptor channels by lipids in cholate solution and during reconstitution in vesicles. J Biol Chem 256:4377-4387

Barnard EA, Dolly JO, Lang B, Lo M, Shorr RG 1979 Application of specifically acting toxins to the detection of functional components common to peripheral and central synapses. Adv Cytopharmacol 3:409-435

Barnard EA, Norman RI, Lang B, Sumikawa K, Dolly, JO 1980 Nicotinic acetylcholine receptors of the CNS, in comparison to peripheral receptors. In: Littauer UZ et al (eds) Neurotransmitters and their receptors. Wiley, New York p 271-292

Brown DA 1979 Neurotoxins and the ganglionic (C6) type of nicotinic receptor. Adv Cytopharmacol 3:225-230

Damle V, McLaughlin M, Karlin A 1978 Bromoacetylcholine as an affinity label of the acetylcholine receptor from *Torpedo californica*. Biochem Biophys Res Commun 84:845-851

Dolly JO, Barnard EA 1975 Complete purification of the acetylcholine receptor from skeletal muscle. FEBS (Fed Eur Biochem Soc) Lett 57:267-271

Dolly JO, Barnard EA 1977 Purification and characterization of an acetylcholine receptor from mammalian skeletal muscle. Biochemistry 16:5053-5060

Freeman JA, Schmidt JR, Oswald RE 1980 Effect of α-bungarotoxin on retino-tectal synaptic transmission in goldfish and toad. Neuroscience 5:929-942

Froehner SC, Reiness CG, Hall ZW 1977 Subunit structure of the acetylcholine receptor from denervated rat skeletal muscle. J Biol Chem 252:8589-8596

Karlin A 1980 Molecular properties of nicotinic acetylcholine receptors. In: Cotman CW et al (eds) Cell surface reviews. Elsevier, Amsterdam, vol 6:192-242

Lang B, Barnard EA, Chang LR, Dolly JO 1979 Putative benzodiazepine receptor: a protein solubilized from brain. FEBS (Fed Eur Biochem Soc) Lett 104:149-153

Lyddiatt A, Sumikawa K, Wolosin JM, Dolly JO, Barnard EA 1979 Affinity labelling by bromoacetylcholine of a characteristic subunit in the acetylcholine receptor from muscle and *Torpedo* electric organ. FEBS (Fed Eur Biochem Soc) Lett 108:20-24

Miller C, Racker E 1980 Reconstitution of membrane transport functions. In: O'Brien RD (ed) The receptors. Plenum Press, New York, vol 1:1-31

Moore HH, Raftery MA 1979 Studies of reversible and irreversible interactions of an alkylating agonist with *Torpedo californica* acetylcholine receptor in membrane-bound and purified states. Biochemistry 18:1862-1867

Norman RI, Mehraban F, Barnard EA, Dolly JO 1981 Nicotinic acetylcholine receptor from chick optic lobe: subunit structure and molecular relationships to the α-neurotoxin binding components of muscle and brain. Proc Natl Acad Sci USA, in press

Oswald RE, Freeman JA 1979 Characterization of the nicotinic acetylcholine receptor isolated from goldfish brain. J Biol Chem 254:3419-3426

Schmidt J, Hunt S, Polz-Tejera G 1980 Nicotinic acetylcholine receptors from brain. In: Essman WB (ed) Neurotransmitters, receptors and drug action. Spectrum Publications, New York, p 1-65

Schramm M 1979 Transfer of glucagon receptor from liver membranes to a foreign adenylate cyclase by a membrane fusion procedure. Proc Natl Acad Sci USA 76:1174-1178

Sherman-Gold R, Dudai Y 1980 Solubilization and properties of a benzodiazepine receptor from calf cortex. Brain Res 198:485-490

Shorr RG, Lefkowitz RJ, Caron MG 1981a Purification of the β-adrenergic receptor. J Biol Chem 256:5820-5826

Shorr RG, Lyddiatt A, Lo MMS, Dolly JO, Barnard EA 1981b Acetylcholine receptor from mammalian muscle. Oligomeric forms and their subunit structures. Eur J Biochem 116:143-153

Sumikawa K, Barnard EA, Dolly JO 1981 Similarity of acetylcholine receptors of denervated, innervated and embryonic chick muscles. Eur J Biochem, in press

Wang CF, Molinaro S, Schmidt J 1978 Ligand responses of α-bungarotoxin binding sites from skeletal muscle and optic lobe of the chick. J Biol Chem 253:8507-8512

Pharmacological characteristics of a putative nicotinic acetylcholine receptor from *Musca domestica*

CHRISTINE S. MARCH, KENNETH J. CATTELL and JOHN F. DONNELLAN*

Department of Science, Bristol Polytechnic, Coldharbour Lane, Bristol, BS16 1QY and **Sittingbourne Research Centre, Shell Research Ltd, Sittingbourne, Kent, ME9 8AG, UK*

Abstract Insect central nervous system extracts appear to contain at least three types of putative cholinergic receptors whose molecular and pharmacological properties can be distinguished and studied by biochemical techniques. In this study housefly head extracts are shown to contain high and low affinity sites for ^{125}I-labelled α-bungarotoxin (a nicotinic cholinergic antagonist) and competition studies indicate that the high affinity sites have a distinctly nicotinic binding profile. The high affinity sites have been extensively purified using differential centrifugation and chromatography on sequential α-toxin and concanavalin A affinity columns. Another radiolabelled nicotinic ligand, [^{3}H]4-(*N*-maleimido)-α-benzyltrimethylammonium iodide, has been used to covalently label the reduced form of the purified receptor as an initial step in characterizing the molecular properties of the putative cholinergic receptor protein.

There is a growing body of both biochemical and electrophysiological evidence to suggest that acetylcholine (ACh) is a major excitatory neurotransmitter in the insect central nervous system (CNS) (Sattelle 1980). An integral component of the cholinergic synapse is the postsynaptic ACh receptor, whose interaction with the neurotransmitter liberated into the synaptic cleft triggers the transduction of the signal between the neurons involved in the cholinergic synapse. The study of cholinergic receptor molecules is, therefore, of fundamental importance to an understanding of the mechanism of cholinergic transmission in the insect CNS. Moreover, insects could be ideal species in which to study nervous system development (Hall 1980). Besides the obvious possibilities of comparative studies of insect

1982 Neuropharmacology of insects. Pitman, London (Ciba Foundation symposium 88) p 118-136

and vertebrate ACh receptors, information on the insect ACh receptor could help in the designing of new classes of insecticides capable of interacting with CNS cholinergic receptors in a similar manner to nicotine, an established insecticide and a defined cholinergic agonist.

The introduction of specific radiolabelled ligands with high affinities for cholinergic receptors present in mammalian tissues, and in particular in the electroplax organ of electric fishes, has led to a marked advance in our knowledge of ACh receptors (see Barnard, this volume). Comparable studies with insect ACh receptors have been more limited but considerable progress has been made in the past four years.

Biochemical studies designed to characterize insect ACh receptors have the advantage of avoiding the problem of drug accessibility, which is the major problem in electrophysiological studies. However, binding studies done with cell-free extracts of insect nervous tissue have to satisfy a number of criteria before one can conclude that a physiological receptor is being studied (Birdsall & Hulme 1976). In essence the receptor molecule should show saturable binding of physiologically active molecules; the receptor should bind such ligands specifically and with high affinity; and there should be a correlation with the apparent affinity derived from complementary electrophysiological assays. The receptor should also be localized at the appropriate subcellular site in tissues known to respond to the neurotransmitter. The satisfaction of these criteria with insect ACh receptors poses a severe challenge because the complementary electrophysiological information is only now starting to become available, with the introduction of methods for studying single neurons (Sattelle et al 1980, Goodman & Spitzer 1980).

Binding studies with insect CNS extracts have identified three populations of putative ACh receptors, as reviewed by Sattelle (1980). The first experiments designed to look for ACh receptors in insect nervous tissue used aqueous extracts of housefly heads. Competitive binding studies using reversible cholinergic ligands, such as decamethonium, revealed a putative ACh receptor with a pronounced nicotinic–muscarinic pharmacology (Cattell et al 1980, Eldefrawi & Eldefrawi 1980, Tripathi et al 1979). The putative receptor has been extensively purified. It appears to be a glycoprotein with an isoelectric point between pH 4.8 and 5.0 and a relative molecular mass (M_r) of 320 000–400 000, with subunit populations of 83 000 and 91 000 (Cattell et al 1980), although Tripathi et al (1979) have described a molecule with subunits of 94 000 and 64 000.

A second type of putative ACh receptor appears to be muscarinic, with ligand-binding characteristics closely related to those of the muscarinic receptors of mammalian CNS extracts. Binding studies with [^{3}H]quinuclidinyl benzilate (QNB), a muscarinic antagonist, have revealed a high affinity QNB-binding site in membrane fractions isolated from *Drosophila melano-*

gaster heads (Dudai & Ben-Barak 1977), housefly heads (Jones & Sumikawa 1981) and cricket ganglia (Meyer & Edwards 1980).

Finally, a putative nicotinic ACh receptor has been identified by using α-bungarotoxin (α-BGTX), the high affinity nicotinic antagonist, to characterize putative receptor sites in extracts of *Drosophila* heads (Hall 1980, Dudai 1980, Rudloff et al 1980), housefly heads (Harris et al 1979, March et al 1980, Eldefrawi & Eldefrawi 1980, Jones et al 1981), *Manduca sexta* brains (Sanes et al 1977) and cockroach ganglia (Gepner et al 1978). Evidence linking the α-BGTX-binding site with a physiological ACh receptor has been obtained in electrophysiological studies. These have demonstrated that α-BGTX blocks cholinergic processes in identified neurons involved in synaptic transmission between the cercal afferents and the giant interneurons in the cockroach sixth abdominal ganglion (Harrow et al 1980, 1982, Sattelle 1980).

This paper reviews experiments designed to characterize and purify the α-BGTX-binding component, thought to be the nicotinic ACh receptor, from housefly heads (*Musca domestica*).

Subcellular localization of α-BGTX-binding sites

Specific α-BGTX-binding components could readily be demonstrated in a $500 \times g$ supernatant derived from flyhead homogenates (Harris et al 1979, March et al 1980). Fig. 1 is a Scatchard plot of typical α-BGTX-binding data and clearly shows the presence of high affinity sites with a dissociation constant in the low nanomolar (<10 nM) range and with a site concentration (B_{max}) also similar to the value reported for other insect α-BGTX-binding sites (Sattelle 1980). At higher concentrations of free α-BGTX (>50 nM), however, there appears to be another specific binding site of low affinity but of higher capacity. It is difficult to say whether this low affinity site is due to a distinct site on a separate protein molecule or on a separate subunit of the same molecule. Another possibility is that the two sites represent different configurations of essentially the same site.

Hill plots of the binding data of the high affinity site (Fig. 1) gave Hill coefficients of 0.96–1.0, indicating lack of cooperativity in the binding of the toxin to this site. The pharmacological profile of the high affinity site was investigated by measuring the abilities of different concentrations of ligands to inhibit the binding of α-BGTX. The concentration of α-BGTX (11.25 nM) used was just sufficient to saturate the high affinity site. Table 1 shows that nicotinic ligands were the most effective in blocking the binding of ^{125}I-labelled α-BGTX (^{125}I-α-BGTX).

Differential centrifugation of homogenates of flyheads, prepared in sucrose, showed a low total recovery of the high affinity α-BGTX binding sites

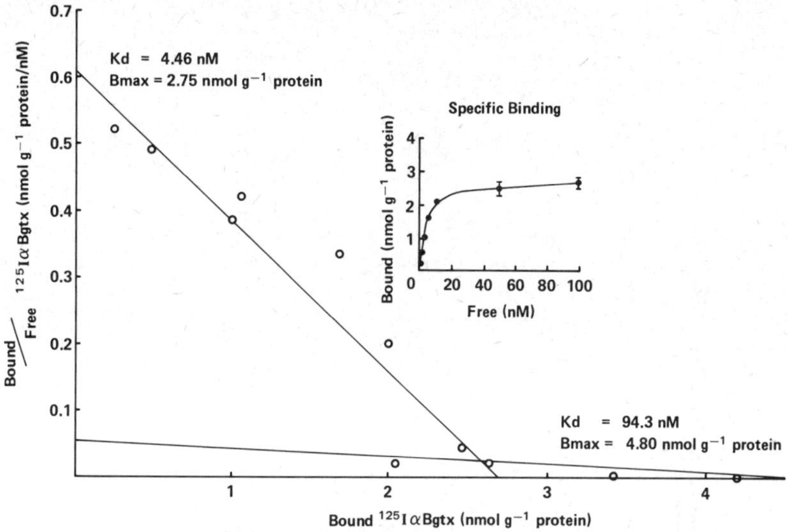

FIG. 1. Scatchard plot of α-BGTX binding to a 500 × *g* supernatant fraction from flyheads. Aliquots of the supernatant fraction were incubated with a concentration range of ^{125}I-α-BGTX and assayed for α-BGTX-binding ability as described by March et al (1980). Non-specific binding was measured in the presence of unlabelled α-BGTX (10^{-5} M).

from the various fractions. We therefore studied the subcellular localization and recovery of the sites as a function of the medium in which the flyheads were homogenized. Table 2 shows that the total recovery was enhanced, in an additive manner, by the presence of phosphate, phenyl methyl sulphonyl fluoride (PMSF, a protease inhibitor) and EDTA (a protease inhibitor and a chelating agent).

We also noted that the composite medium led to a substantial proportion of the α-BGTX-binding sites being concentrated in the soluble fraction (S_3) material not sedimented by centrifugation for 90 min at 150 000 × *g*. Other workers using frozen *Drosophila* heads have reported that most α-BGTX-binding sites are membrane-bound, sedimenting at 20 000–40 000 × *g* (Schmidt-Nielsen et al 1977). We homogenized frozen and freshly isolated housefly heads in the medium described by these workers (5 mM-phosphate buffer, pH 7.4, containing 1% sucrose) and found that the major portion of the α-BGTX was recovered in a 20 000 × *g* pellet. The solubilization of α-BGTX-binding activity in our composite homogenization medium appears therefore to be due to detachment of the α-BGTX sites from a presumed membrane locale. As we were primarily interested in purifying the α-BGTX-binding component, the adventitious solubilization was an advantage because

TABLE 1 Pharmacological specificity of the high affinity α-bungarotoxin-binding component in flyhead subcellular functions

Ligand	I_{50}[a] *(M)* 500 × **g** *supernatant* (S_1)	150 000 × **g** *supernatant* (S_3)	*Purified extract*	*Drosophila (purified receptor)*[b]	*Cockroach electrophysiology*[c]
Nicotine	1.0×10^{-8}	2.0×10^{-8}	2.3×10^{-7}	3.5×10^{-7}	0.8×10^{-8}
d-Tubocurarine	6.2×10^{-8}	7.9×10^{-9}	5.6×10^{-8}	6.6×10^{-7}	0.3×10^{-6}
α-Toxin (*Naja naja siamensis*)	—	1.8×10^{-10}			
Acetylcholine[d]	—	3.6×10^{-6}	5.6×10^{-6}	3.0×10^{-6}	1.3×10^{-6}
Decamethonium	—	4.0×10^{-5}	—	6.3×10^{-5}	$>5.0 \times 10^{-4}$
Hyoscyamine	—	1.6×10^{-5}	—	—	—
(+)Dexetimide	9.5×10^{-3}	1.8×10^{-5}	1.8×10^{-5}	6.0×10^{-5}	—
QNB	$>10^{-3}$	$>10^{-3}$	1.3×10^{-5}	—	—
Muscarine	$>10^{-3}$	$>10^{-3}$	3.2×10^{-4}	1.4×10^{-3}	—

[a] I_{50}, the concentration required for inhibition by 50% of the binding of α-BGTX, was determined by competition studies with 11.25 nM-^{125}I-α-BGTX. Ligands were preincubated with extract for 30 min at room temperature before incubation with ^{125}I-α-BGTX in a final volume of 0.25 ml. 100 µl samples were assayed for α-BGTX binding by the DEAE-cellulose disk method of Schmidt & Raftery (1972).

[b] Hall (1980).

[c] Sattelle (1980). I_{50} is defined as the concentration estimated to produce half-maximal physiological action at cercal afferent, giant-interneuron synapses.

TABLE 2 Effect of homogenization medium on the subcellular distribution of [^{125}I]-α-BGTX binding

	% recovered α-BGTX binding[a]						
Homogenization medium (pH 7.4)	P_1	S_1	P_2	S_2	P_3	S_3	*Total recovery (%)*[b]
0.32 M-sucrose	16.5	82.9	5.0	73.4	11.1	6.0	37.6
0.32 M-sucrose/10 mM-phosphate	14.5	35.4	9.2	23.7	1.5	24.7	49.9
10 mM-phosphate	15.5	54.2	8.0	50.2	18.3	26.2	68.0
0.32 M-sucrose/10 mM-phosphate/1 mM-PMSF	13.0	61.1	17.6	56.0	26.0	30.1	86.7
0.32 M-sucrose/10 mM-phosphate/2 mM-EDTA	11.1	113.1	12.2	50.2	5.9	50.0	79.2
0.32 M-sucrose/10 mM-phosphate/1 mM-PMSF/ 2 mM-EDTA	3.5	101.1	11.8	82.9	17.7	63.8	96.8

Homogenates of flyheads (40 ml^{-1}) in the various buffers were filtered through nylon net and centrifuged for 10 min at 500 × ***g*** to give a pellet (P_1) and a supernatant (S_1). S_1 was centrifuged for 20 min at 20 000 × ***g*** to give a pellet (P_2) and a supernatant (S_2) which was further centrifuged for 90 min at 150 000 × g to give a pellet (P_3) and a supernatant (S_3) (n = 3).

[a] DEAE assay using 10 nM-[^{125}I]-α-BGTX.

[b] Homogenate = 100%.

it removed the necessity of using detergents to solubilize the α-BGTX sites for further chromatographic purification.

Properties of the S_3 fraction

Binding studies with the soluble fraction revealed a biphasic binding for α-BGTX (Fig. 2) similar to that shown for S_1 in Fig. 1. Studies with the P_3 and

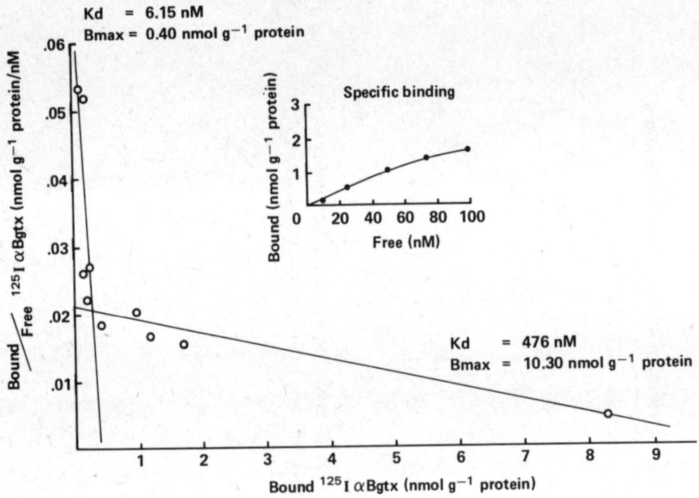

FIG. 2. Scatchard plot of α-BGTX binding to a 150 000 × *g* supernatant fraction (S_3) from flyheads. The binding assay used was that described in Fig. 1.

P_2 membrane fractions (Table 2) also show a similar biphasic binding. Competition studies with the S_3 fraction revealed the same nicotinic profile as exhibited by S_1 (Table 1).

An estimate of the molecular weight of the soluble α-BGTX-binding component in S_3 was obtained by gel filtration chromatography on Sephadex G-200 after labelling by incubation with 1 nM-^{125}I-α-BGTX (Fig. 3). Comparison with standard proteins indicated that the bound ^{125}I-α-BGTX is associated with a protein fraction of M_r 200 000, assuming that it is a globular protein. Free ^{125}I-α-BGTX was well separated from the bound species and in experiments in which ^{125}I-α-BGTX alone was eluted through the column only a single peak of low relative molecular mass was observed. Comparable M_r values of 460 000 to 500 000 have been reported for the detergent-solubilized α-BGTX receptors purified from *Drosophila* heads (Rudloff et al 1980, Hall

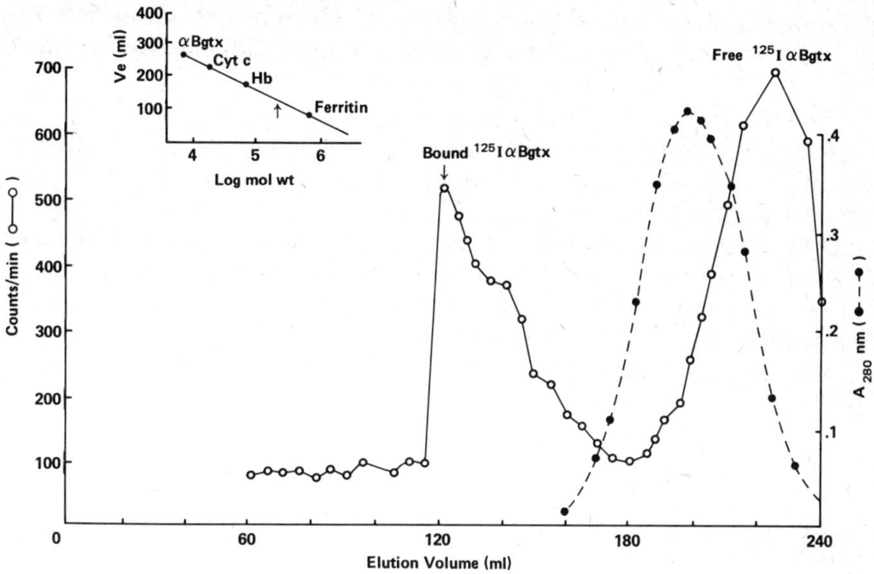

FIG. 3. Gel filtration chromatography of the ^{125}I-α-BGTX-binding component present in the 150 000 × ***g*** supernatant fraction (S_3). An aliquot (1 ml) of the S_3 fraction was incubated with 1 nM-^{125}I-α-BGTX for 1 h at 20 °C and then chromatographed on a Sephadex G-200 column (2.6 × 40 cm) equilibrated and eluted with 0.1 M-Tris-Cl^- (pH 7.5) containing 0.1 M-NaCl and 0.5 mM-NaN_3. The column was calibrated with ferritin, haemoglobin, cytochrome *c* and ^{125}I-α-BGTX.

1980). Values for the solubilized nicotinic ACh receptors purified from various electroplax and mammalian muscles range from 250 000 to 300 000 (Heidmann & Changeux 1978).

The solubilized decamethonium-binding protein from flyheads (the putative mixed muscarinic–nicotinic receptor) has a reported M_r of 320 000–340 000 (Cattell et al 1980, Tripathi et al 1979).

Purification by affinity chromatography

Affinity chromatography has been a powerful tool for the purification of nicotinic ACh receptors, using as affinity ligands either snake toxins (Hall 1980) or lectins which interact with glycoproteins (Salvaterra et al 1977). We investigated three alternative procedures of affinity chromatography: (i) a single step on a *Naja naja siamensis* α-toxin–Sepharose 4B column with elution by carbamylcholine chloride, (ii) a single step on a concanavalin

A(Con A)–Sepharose 4B column with elution by α-methyl D-glycopyranoside, and (iii) sequential elution through the α-toxin and Con A columns.

After the single chromatographic separations on α-toxin or Con A columns the bulk of the S_3 protein, including all the measurable acetylcholinesterase activity, was eluted in the pre-washing stage. Specific α-BGTX-binding activity was eluted only with 0 to 0.5 M gradients of either carbamylcholine or α-methyl D-glycopyranoside with purifications of binding activity of 350- and 280-fold, respectively. The purification achieved by sequential chromatography on the two columns was 700-fold and this procedure has been adopted for routine use (Fig. 4). The purified extract contained less than 1% of the recovered protein, no acetylcholinesterase activity and 75% of the high affinity α-BGTX-binding sites present in S_1. Control experiments showed that

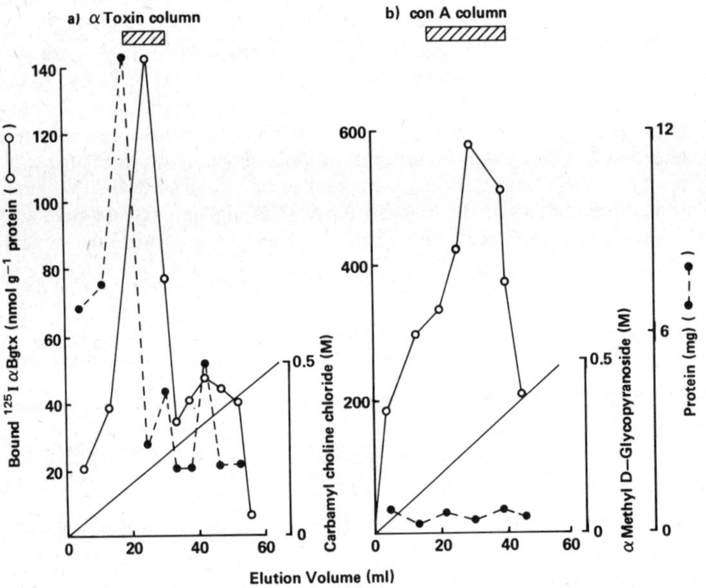

FIG. 4. Purification of α-BGTX-binding component by affinity chromatography. S_3 (~200 mg protein) was loaded onto a 0.5 × 10 cm α-toxin-Sepharose 4B column and equilibrated for 30 min before elution with 50 mM-Tris-Cl⁻ buffer (pH 7.5) containing 0.1 M-NaCl and 0.5 mM-NaN_3 with a flow rate of 10 ml h^{-1} until the absorbance at 280 nm returned to zero. The α-BGTX-binding protein peak (o–o) was eluted with a 0 to 0.5 M gradient of carbamylcholine chloride (CCC). Pooled peak fractions (▨) were dialysed against the Tris buffer and then chromatographed on a 0.5 × 10 cm Con A column with elution by a 0 to 0.5 M gradient of α-methyl D-glycopyranoside with a flow rate of 10 ml h^{-1}. AChE activity was detected only in the pregradient eluant of the α-toxin step.

no purification was achieved without the affinity arms of the cyanogen bromide-activated Sepharose 4B columns.

The affinity column fraction showed the same nicotinic profile, as judged by competition studies, as that shown by the earlier S_1, P_3 and S_3 fractions and by other insect preparations (Table 1, purified extract). Saturation studies also showed the same specific biphasic binding profile for α-BGTX as was exhibited by the other fractions (Table 3). The high affinity site had a Hill

TABLE 3 α-Bungarotoxin binding constants

	K_d *(nM)*		B_{max} *(nmol g⁻¹ protein)*	
	High affinity	*Low affinity*	*High affinity*	*Low affinity*
S_1 (500 × *g* supernatant)	4.46[a]	94.3	2.73	4.8
S_3 (150 000 × *g* supernatant)	6.15	476.2	0.40	10.3
P_3 (150 000 × *g* pellet)	17.86	961.5	1.90	24.0
Purified fraction obtained by sequential affinity chromatography (two batches)	6.25[b]	1572.3	25.0	2170.0
	9.9[c]	ND	173[c]	ND

[a] Hill coefficient 0.95.
[b] Hill coefficient 1.06.
[c] Considerable variation in B_{max} of the high affinity site was observed from batch to batch. These values were obtained with the fraction used in the MBTA work (Fig. 5).
ND, not determined.

coefficient approaching unity and therefore showed no signs of cooperativity in α-BGTX binding. Also of interest was the gradual decrease in the dissociation constant of the low affinity site with increasing purity of the receptor, although the concentration of the low affinity sites (B_{max}) showed a far bigger increase than the equivalent B_{max} values for the high affinity sites. An explanation of this trend is not readily apparent. One possibility could be an interconversion of the two sites.

Preliminary sodium dodecyl sulphate (SDS)–polyacrylamide gel electrophoresis of the purified fraction revealed major polypeptides with M_r values of 42 000 and 25 000. As mentioned earlier, the major polypeptide subunits of the flyhead decamethonium-binding protein, which has a mixed pharmacology, appear to have M_r values of 91 000 and 64 000. The much-studied electroplax nicotinic ACh receptor has an $\alpha_2\beta\gamma\delta$ subunit composition, and it is thought that the α polypeptide (M_r = 40 000) is the acetylcholine-binding receptor/ionophore (Heidmann & Changeux 1978). In the mammalian muscle receptor the α subunit is a 43 000 M_r polypeptide (Barnard et al 1980).

Specific labelling with [^{3}H]MBTA

An alternative procedure for irreversibly labelling the nicotinic ACh receptor *in vitro* was developed by Karlin (1969) from electrophysiological experiments with intact electroplax from *Electrophorus electricus*. In summary, a disulphide bond at the periphery of the ACh-binding site was shown to be susceptible to reduction by low concentrations of dithiothreitol, and one of the SH groups so formed could be reacted with a tritiated affinity alkylating agent, [^{3}H]4-(*N*-maleimido)-α-benzyltrimethylammonium iodide ([^{3}H]MBTA; Karlin et al 1971). Using carbamylcholine chloride, α-BGTX or α-neurotoxin as protectants to correct for non-specific SH labelling we have demonstrated specific [^{3}H]MBTA labelling of the purified flyhead protein using the assay described by Karlin et al (1976).

From preliminary experiments, the labelling is linear with protein concentrations up to 5 μg per sample and the optimum reduction of disulphide bonds, protectable by cholinergic ligands, was achieved with 0.2 mM-dithiothreitol. Binding studies indicate that [^{3}H]MBTA binding is saturated at approximately 1 μM with a site concentration in the region of 150 nmol g^{-1} protein when carbamylcholine (Fig. 5) or α-BGTX are used to monitor

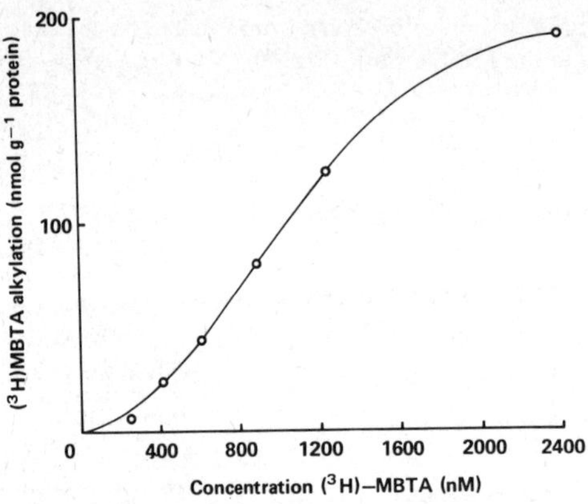

FIG. 5. Site-specific alkylation of the purified fraction by [^{3}H]MBTA using carbamylcholine as protectant. Samples (up to 50 μg of protein) of the α-BGTX-binding component purified by affinity chromatography were assayed for their ability to bind [^{3}H]MBTA as described by Karlin et al (1976). Non-specific binding was measured in the presence of 1 mM-carbamylcholine chloride or, in some experiments, 3 μg ml^{-1} α-toxin or α-BGTX.

non-specific labelling. It is interesting that the B_{max} for the direct binding of α-BGTX to the same extract was 173 nmol g^{-1} protein, suggesting a 1:1 ratio of α-BGTX- and MBTA-binding sites. This differs from the 2:1 ratio commonly observed for the electroplax nicotinic receptor (Weill et al 1974) but is similar to the ratio obtained after the alkylation by bromoacetylcholine of reduced ACh receptors from both electroplax and cat muscle (Barnard et al 1980).

We have also investigated the potential of [^{3}H]MBTA as a specific label of polypeptides of the purified flyhead protein by using the SDS–polyacrylamide gel electrophoretic procedure described by Rübsamen et al (1978). Specific labelling was demonstrated for the 42 000 M_r species, which suggests that MBTA can be used to further characterize the α-BGTX-binding protein.

Conclusions

(1) Flyhead homogenates contain a specific high affinity binding site for α-BGTX which has the pharmacological characteristics expected of a nicotinic ACh receptor.

(2) The α-BGTX site has been solubilized by homogenization in detergent-free medium before differential centrifugation. A 1.35×10^7 g min^{-1} supernatant has then been subjected to sequential affinity chromatography on α-toxin and Con A columns to give a 700-fold enrichment in α-BGTX-binding capacity.

(3) Crude and purified fractions, membrane-bound or soluble, show biphasic specific binding of α-BGTX. The high affinity site has a similar K_d (<10 nM) to other putative nicotinic receptors isolated from *Drosophila* heads, electroplax and mammalian muscle.

(4) Preliminary experiments indicate that the α-BGTX-binding protein has a relative molecular mass of approximately 200 000 and consists of polypeptide subunits of M_r values 42 000 and 26 000.

(5) The purified α-BGTX-binding protein also reacts specifically with the covalent affinity ligand [^{3}H]MBTA, which appears to label the 42 000 M_r subunit.

REFERENCES

Barnard EA 1982 Isolation of receptors from the central nervous system. This volume, p 109-117

Barnard EA, Dolly JO, Lyddiatt A, Sumikawa K, Wolosin JM 1980 Current views on the structure of nicotinic acetylcholine receptors from different sources. In: Brzin M et al (eds) Synaptic constituents in health and disease. Pergamon Press, Oxford, p 143-145

Birdsall NJ, Hulme EC 1976 Biochemical studies on muscarinic acetylcholine receptors. J Neurochem 27:7-16

Cattell KJ, Harris R, Donnellan JF 1980 The identification and characterisation of acetylcholine receptors from housefly brain—is it possible? In: Sattelle DB et al (eds) Receptors for neurotransmitters, hormones and pheromones in insects. Elsevier/North-Holland Biomedical Press, Amsterdam, p 71-83

Dudai Y 1980 Cholinergic receptors of *Drosophila*. In: Sattelle DB et al (eds) Receptors for neurotransmitters, hormones and pheromones in insects. Elsevier/North-Holland Biomedical Press, Amsterdam, p 93-110

Dudai Y, Ben-Barak J 1977 Muscarinic receptor in *Drosophila melanogaster* demonstrated by binding of (^{3}H)-quinuclidinyl benzilate. FEBS (Fed Eur Biochem Soc) Lett 81:134-136

Eldefrawi AT, Eldefrawi ME 1980 Putative acetylcholine receptors in housefly brain. In: Sattelle DB et al (eds) Receptors for neurotransmitters, hormones and pheromones in insects. Elsevier/North-Holland Biomedical Press, Amsterdam, p 59-70

Gepner JI, Hall LM, Sattelle DB 1978 Insect acetylcholine receptors as a site of insecticide action. Nature (Lond) 276:188-190

Goodman CS, Spitzer NC 1980 Embryonic development of neurotransmitters. In: Sattelle DB et al (eds) Receptors for neurotransmitters, hormones and pheromones in insects. Elsevier/North-Holland Biomedical Press, Amsterdam, p 195-207

Hall LM 1980 Biochemical and genetic analysis of an α-bungarotoxin-binding receptor from *Drosophila melanogaster*. In: Sattelle DB et al (eds) Receptors for neurotransmitters, hormones and pheromones in insects. Elsevier/North-Holland Biomedical Press, Amsterdam, p 111-124

Harris R, Cattell KJ, Donnellan JF 1979 Identification of a putative nicotinic acetylcholine receptor in fractions from housefly brain. Biochem Soc Trans 7:136-138

Harrow ID, Hue B, Gepner JI, Hall LM, Sattelle DB 1980 An α-bungarotoxin sensitive acetylcholine receptor in the CNS of the cockroach *Periplaneta americana*. In: Insect neurobiology and pesticide action. Society of Chemical Industry, London, p 37-144

Harrow ID, David JA, Sattelle DB 1982 Acetylcholine receptors of identified insect neurons. This volume, p 12-27

Heidmann I, Changeux JP 1978 Structural and functional properties of the acetylcholine receptor protein in its purified and membrane-bound states. Annu Rev Biochem 47:317-357

Jones SW, Sumikawa K 1981 Quinuclidinyl benzilate binding in housefly heads and rat brain. J Neurochem 36:454-459

Jones SW, Sudershan P, O'Brien RD 1981 α-Bungarotoxin binding in housefly heads and *Torpedo* electroplax. J Neurochem 36:447-453

Karlin A 1969 Chemical modification of the active site of the acetylcholine receptor. J Gen Physiol 54:245S-264S

Karlin A, Prives J, Deal W, Winnik M 1971 Affinity labelling of the acetylcholine receptor in the electroplax. J Mol Biol 61:175-188

Karlin A, McNamee MG, Cowburn DA 1976 Assay of the acetylcholine receptor by affinity labelling. Anal Biochem 76:442-451

March CS, Cattell KJ, Donnellan JF 1980 The characterisation of a solubilised α-bungarotoxin-binding component from *Musca domestica*. Biochem Soc Trans 8:742-743

Meyer MR, Edwards JS 1980 Muscarinic cholinergic binding sites in an Orthopteran central nervous system. J Neurobiol 11:215-219

Rübsamen H, Eldefrawi AT, Eldefrawi ME, Hess GP 1978 Characterization of the calcium-binding sites of the purified acetylcholine receptor and identification of the calcium-binding subunit. Biochemistry 17:3818-3825

Rudloff E, Jiminez F, Bartels J 1980 Purification and properties of the nicotinic receptor of *Drosophila melanogaster*. In: Sattelle DB et al (eds) Receptors for neurotransmitters, hormones and pheromones in insects. Elsevier/North-Holland Biomedical Press, Amsterdam, p 85-92

Salvaterra PM, Gurd, JM, Mahler HR 1977 Interactions of nicotinic acetylcholine receptor from rat brain with lectins. J Neurochem 29:345-348

Sanes JR, Prescott DJ, Hildebrand JG 1977 Cholinergic neurochemical development of normal and deafferented antennal lobes during metamorphosis of the moth *Manduca sexta*. Brain Res 119:389-402

Sattelle DB 1980 Acetylcholine receptors of insects. Adv Insect Physiol 15:215-315

Sattelle DB, Harrow ID, Hue B 1980 α-Bungarotoxin blocks acetylcholine responses in identified neurons of the cockroach *Periplaneta americana* (L). In: Sattelle DB et al (eds) Receptors for neurotransmitters, hormones and pheromones in insects. Elsevier/North-Holland Biomedical Press, Amsterdam, p 125-139

Schmidt J, Raftery MA 1972 A simple assay for the study of solubilised acetylcholine receptors. Anal Biochem 52:349-354

Schmidt-Nielsen BK, Gepner JI, Teng NNH, Hall LM 1977 Characterization of an α-bungarotoxin binding component from *Drosophila melanogaster*. J Neurochem 29:1013-1029

Tripathi RK, Tripathi HL, O'Brien RD 1979 Properties and affinity purification of the mixed-type putative acetylcholine receptor from wild and a mutant strain of houseflies. Biochim Biophys Acta 586:624-631

Weill CL, McNamee MG, Karlin A 1974 Affinity-labelling of purified acetylcholine receptor from *Torpedo californica*. Biochem Biophys Res Commun 61:997-1003

DISCUSSION

Hall: Did you use any detergents for the purification?

Donnellan: No. We were surprised to find the binding sites in the soluble fraction, but this meant we did not have to use detergents. There might be a contribution from 'endogenous detergents'. Something peculiar happens to membrane proteins in insect head homogenates.

M. E. Eldefrawi: We saw something similar in 1976 but we didn't pursue it (unpublished results). When we took the supernatant at 105 000 ***g*** and passed it once on Sephadex G-200, the very early fractions came out cloudy, even though the fraction that went on the column was crystal clear. It looked as if something was separating on the column and the proteins were aggregating. The particulate material could be precipitated by centrifugation at 100 000 ***g***. The protein concentration in the particulate fraction was very low, and it bound much more [^{3}H]decamethonium than the soluble protein that remained in the supernatant. We wondered whether an endogenous material with a detergent action separated from the proteins on the column. You also said that you tried protease inhibitors such as EDTA and PMSF that we use in isolating the ACh receptor of electric organs. But these two inhibitors are not sufficient to inhibit all protease activity. Did you test the effects of other protease inhibitors in preventing release of α-BGTX-binding proteins into the supernatant?

Donnellan: Bacitracin should be very effective, but its use did not seem to alter the picture.

Breer: Did you ever try to isolate this receptor from insect tissue other than housefly heads?

Donnellan: We haven't, but Dr Lunt has done some work with the CNS of the locust.

Lunt: We have looked at the α-toxin-binding component in cerebral ganglia dissected out from locusts, and fractioned in either the presence or absence of the full range of proteases that Professor Barnard has described. The most we have ever seen in a supernatant fraction is perhaps 5% of the total binding activity.

Breer: We found a similar distribution of α-BGTX- and quinuclidinyl benzylate-binding sites in ganglionic tissue from *Locusta migratoria* (Breer 1981).

Lunt: The rest of the binding activity is always associated with membranous fractions which we can then solubilize with Triton.

Breer: Since a soluble fraction of binding sites has been found only when total insect heads instead of freshly dissected ganglionic tissue were used as starting material, the postulated 'endogenous detergent' may be a substance in the cuticle or on the surface of the insect heads.

Sattelle: How readily does the housefly α-BGTX receptor dissociate?

Donnellan: I think our rate constants and K_d values are similar to those published for *Drosophila* (see Sattelle 1980).

Olsen: The changes in binding characteristics as you purify the protein are a potential problem with which all workers in membrane protein biochemistry can sympathize. They may indicate conformational changes or denaturation. Even though SDS gel chromatography indicates that you are getting near the end of purification, and the use of affinity methods should provide good purification, the specific activity you measure is too low for a pure receptor. Do you think you have to purify it much further?

Donnellan: I don't know. I would be happier if we could load more proteins onto our gels. The final specific activity of the high affinity α-BGTX-binding site is routinely about 200 nmol g^{-1} protein, which is far below what we would expect.

Barnard: Either there must be a lot of inactivated receptor or there is much other protein still present. Did you have a high specific activity when you calculated the MBTA-labelled fraction?

Donnellan: No, it was about 200 nmol MBTA/g protein too. We get the increase in B_{max} of the low affinity site, which is the puzzle.

Dudai: In the gel filtration experiment the peak for labelled α-BGTX looked as if it was in the void volume of the column.

Donnellan: Yes, it was almost there. G-200 wasn't the best column material to use. The bound α-BGTX was however eluted between haemoglobin and ferritin, which indicated that it was not eluted in the void volume.

M. E. Eldefrawi: Professor Barnard, in the final organization of the nicotinic ACh receptor molecule in the membrane, do you favour the monomer arrangement? I had the impression that you were dealing with an aggregate with an M_r of about 250 000. Did you ever detect in a sucrose gradient anything that would indicate the presence of a dimer as well?

Barnard: No, but neither has a dimer ever been seen of the comparable ACh receptor from vertebrate skeletal muscle (Lo et al 1981). The dimer, of sedimentation coefficient 13S, may be something special to the electric organ of *Torpedo* and other rays. It has not been shown for that receptor from the electric organ of the eel (Meunier et al 1974, Lindstrom et al 1980), only for *Torpedo*. I don't know that it has a fundamental significance. The direct answer to your question is that from the chick optic lobe we always find that the ACh receptor is of the 'monomer' size with sedimentation coefficient 9S, unless we get the artifact I described in which extraneous protein complexes with it in certain conditions, in which case it goes to the 12S form, which is *not* a dimer of the 9S form.

Hall: Like Dr Donnellan, we have been working on purification of the acetylcholine receptor, but using *Drosophila*. T. Schmidt-Glenewinkel, T. R. Venkatesh and I used a similar purification procedure except that our binding activity does not go into a high-speed supernatant. Our battery of protease inhibitors includes aprotinin, iodoacetamide, pepstatin, phenyl methyl sulphonyl fluoride (PMSF), EDTA and EGTA. We have purified the receptor 3728-fold and when we iodinate the purified receptor and put it on a non-denaturing gel we see a single component. If we dissociate this receptor and put it on SDS gels we see reproducible peaks at 100 000, 79 000, 65 000, 57 000, 42 000 and 24 000 M_r. The problems to consider are whether all of these are *bona fide* receptor subunits and what the relationship is between the different subunits.

One way to look at this is to make a heterobifunctional derivative of α-bungarotoxin. We derivatize ^{125}I-labelled α-BGTX with methyl 4-azidobenzoimidate, so one part of this cross-linker is covalently attached to the α-BGTX molecule. We then incubate the α-BGTX with a membrane preparation that contains the receptor. This is a photoactivatable derivative, so once we wash off unbound toxin we shine light on the preparation and cross-link the toxin to the receptor. We can then solubilize the membranes directly, using SDS, and run that material on SDS–polyacrylamide gels. Then we get a profile with peaks at 102 000, 78 000, 61 000, 57 000, 42 000 and 18 000, after correction for the M_r of α-BGTX (8000). That is in good agreement with the subunits seen in the purified material.

I think John Donnellan is getting proteolytic degradation in his preparation. The material probably appears soluble because endogenous proteases are present that partially digest the receptor, releasing it from the membrane. Our purified receptor has a specific activity of 2.71 μmol/g protein. If we let this material stand, we see conversion of the high molecular weight peaks into material of about 24 000 M_r. We believe that is due to proteolysis, and I think that the 42 000 and 24 000 subunits Dr Donnellan sees are analogous to what we see. We are trying more protease inhibitors now.

We are interested in the relationship between the subunits. We plan to take the subunits from purified receptor, slice them off gels, subject them to limited proteolytic digestion and compare their peptide maps. There are various proteases that can be used on SDS gel slices. So far we have compared degradation fragments from the 100 000 and the 65 000 M_r subunits, using limited papain degradation. We see exactly the same banding pattern for both the 100 000 and the 65 000 subunits, so we feel that the 100 000 subunit is a dimer of the 65 000 subunit. Using this same kind of technique and a battery of different proteases, one should be able, with an insect preparation, to look at the relationship between the various subunits. This same approach can be very useful in analysing mutants that affect the receptors.

Usherwood: Even when proteolysis is inhibited the purified receptor still breaks down to the 24 000 subunit on standing. What is the explanation for this?

Hall: Different proteases are inhibited by different inhibitors, and we haven't yet tested every known inhibitor. We shall use additional inhibitors for our next purification.

Usherwood: So the 24 000 M_r material might be just a degradation product?

Hall: I feel it is a degradation product. The relationship between the proteolytic fragments should tell us whether some of the polypeptides are degradation products or dimers of other polypeptides. From all of our work to date, we feel that the α-BGTX receptor from *Drosophila* is composed of four different polypeptide subunits with M_r values of 42 000, 57 000, 65 000 and 79 000.

Donnellan: Your purified protein seems to fall apart and you imply that this is due to protease. Are proteases co-purified with the α-BGTX receptor? Can you show protease activity in that 'pure' protein?

Hall: There is something in the preparation that is degrading it.

M. E. Eldefrawi: Perhaps one or more subunits of the receptor have protease activity.

Dudai: Which subunit does the labelled toxin bind to?

Hall: I don't know yet. We could test that using radioactively labelled 4-(*N*-maleimido)benzyltrimethylammonium (MBTA) and bromoacetylcholine.

Donnellan: We have just done an experiment using MBTA to label 20 000 *g* membranes. Binding, as judged by SDS–polyacrylamide gel electrophoresis, seems to be to a 42 000 M_r subunit.

Dudai: Kalderon & Silman (1977) purified the nicotinic receptor from the electric eel without detergents by using autolysis and proteolysis. They found a major component which stayed in solution without detergent. It had a sedimentation coefficient of about 20S and one of the major subunits had an M_r of about 30 000. That is another, vertebrate, source, but again proteolysis is involved.

Lunt: Did you measure the size of your purified component, either the S value or the relative molecular mass?

Hall: On a sucrose gradient we got a 9S and a 13S peak. Most of the time we got 9S material with Triton X-100 present. We are now trying to do an H_2O–D_2O experiment to see if we can get a better estimate of the M_r value. My guess would be that it is similar to the acetylcholine receptor from *Torpedo* and other vertebrate preparations, that is, about 250 000.

Barnard: There may be some similarity between the BGTX-sensitive nicotinic ACh receptor of the vertebrate brain and the receptor in the fly. In size, both are 9S. If the major labelling at a 61 000 M_r subunit after cross-linking is because the toxin is primarily binding to that subunit, that would give an M_r of 53 000 for the free subunit, not really different from the 54 000 size that we found as the subunit with the binding site in the chick brain, or the 52 000 M_r subunit found recently by Seto et al (1981) to bind α-BGTX in mouse brain. The 61 000 subunit is a candidate for the ligand-bound subunit which should be looked at more closely.

The best way to counter proteolysis of the binding-site subunit may be to put the covalent label into the membrane. When it is feasible to use the source material essentially intact and obtain reaction in it of a radioactive highly specific ligand which is modified so as to bind irreversibly at the ligand site, whether through photo-activation as in your case or through direct affinity alkylation, as with bromoacetylcholine in our case, it doesn't matter what other proteins were present. One can plunge it into heated SDS solution immediately, to kill the proteases, and hence see the labelled subunit in a gel. Of course, one must subtract any non-specific labelling, shown by a control containing also a large excess of the reversibly bound ligand or toxin. We have done that and get the same size subunit specifically labelled with bromoacetylcholine from the ACh receptor in the membrane as from the purified receptor (Wolosin et al 1980). This same procedure was earlier used successfully in *Torpedo* membranes by Karlin et al (1975) with MBTA (see comments above) as the reagent and by Witzemann & Raftery (1978) using the method of photoactivating a derivative of bound α-BGTX.

Hall: That was what we did with the photoactivatable derivative of bungarotoxin. The membrane preparation was immediately solubilized in SDS. That is why I have some confidence that these receptor subunits are *bona fide* receptor subunits.

M. E. Eldefrawi: The fact that you have so many subunits or peptides labelled with the toxin reminds me of the finding of Raftery and co-workers (Witzemann et al 1979) that the photolabile derivative of ^{125}I-labelled α-BGTX labelled the 40 K subunit as well as the 65 K subunit of the *Torpedo* ACh receptor. Labelling of the latter subunit was suggested to be due to the large size of the toxin and the proximity of the two subunits. Knowing this now, would you still pursue labelling with the toxin or would you go back and favour a smaller ligand such as MBTA or bromoacetylcholine to identify the site of binding?

Hall: We know those results. We chose the toxin first because we wanted to label all the subunits in the membrane-bound receptor to define what was present *in situ*. We wanted to know whether the subunits that we had purified were real receptor subunits or were fortuitously associated with the receptor. This has suggested that we are dealing with a complex macromolecule with different subunits in it. To identify the active site we plan to use these smaller ligands that you mention.

REFERENCES

Breer H 1981 Properties of putative nicotinic and muscarinic cholinergic receptors in the central nervous system of *Locusta migratoria*. Neurochem Int 3:43-52

Kalderon N, Silman I 1977 Water-soluble acetylcholine receptor from *Torpedo californica*. Solubilization, purification and characterization. Biochim Biophys Acta 465:331-340

Karlin A, Weill CL, McNamee MG, Valderrama R 1975 Facets of the structures of acetylcholine receptors from *Electrophorus* and *Torpedo*. Cold Spring Harbour Symp Quant Biol 40:203-210

Lindstrom J, Cooper J, Tzartos S 1980 Acetylcholine receptors from *Torpedo* and *Electrophorus* have similar subunit structures. Biochemistry 19:1454-1458

Lo MMS, Dolly JO, Barnard EA 1981 Molecular forms of the acetylcholine receptor from vertebrate muscles and *Torpedo* electric organ. Interactions with specific ligands. Eur J Biochem 116:155-163

Meunier JC, Sealock R, Olsen R, Changeux J-P 1974 Purification and properties of the cholinergic receptor from *Electrophorus electricus* electric tissue. Eur J Biochem 45:371-394

Sattelle DB 1980 Acetylcholine receptors of insects. Adv Insect Physiol 15:215-315

Seto A, Arimatsu Y, Amano T 1981 Subunit structure of α-bungarotoxin binding component in mouse brain. J Neurochem 37:210-216

Witzemann V, Raftery MA 1978 Specific molecular aggregates of *Torpedo californica* acetylcholine receptor. Biochem Biophys Res Commun 81:1025-1031

Witzemann V, Muchmore D, Raftery MA 1979 Affinity-directed cross-linking of membrane-bound acetylcholine receptor polypeptides with photolabile α-bungarotoxin derivatives. Biochemistry 18:5511-5518

Wolosin JM, Lyddiatt A, Dolly JO, Barnard EA 1980 Stoichiometry of ligand binding sites in the acetylcholine receptor oligomer from muscle and from electric organ. Measurements by affinity alkylation with bromoacetylcholine. Eur J Biochem 109:495-505

Binding of acetylcholine receptor/channel probes to housefly head membranes

AMIRA T. ELDEFRAWI, NADER SHAKER and MOHYEE E. ELDEFRAWI

Department of Pharmacology and Experimental Therapeutics, University of Maryland School of Medicine, Baltimore, Maryland 21201, USA

Abstract Perhydrohistrionicotoxin (H_{12}-HTX) and phenycyclidine (PCP) bound to ionic channel sites of the acetylcholine (ACh) receptor/channel molecule in motor end-plates and *Torpedo* electric organs. Binding of [^{3}H]H_{12}-HTX and [^{3}H]PCP was inhibited by ionic channel drugs (such as H_{12}-HTX, PCP and amantadine) that alter the time course of receptor-regulated end-plate currents. The apparent rate of binding was slow but was increased up to 100-fold by receptor agonists in a dose-dependent manner. This increase was inhibited by receptor antagonists or by persistence of agonists in the medium, which led to receptor desensitization. Antagonists alone also increased the apparent rate of binding of the radiolabelled probes, but to a much lower degree than agonists. The allosteric interactions and coupling between receptor and channel sites proved the identity of the binding macromolecules.

Both [^{3}H]H_{12}-HTX and [^{3}H]PCP, at nanomolar concentrations, bound reversibly to membrane preparations from housefly heads; their binding was inhibited by local anaesthetics but was not stimulated by ACh receptor agonists. The concentrations of the binding sites for the two probes in housefly head membranes were much higher than the concentration of nicotinic and muscarinic cholinergic receptor sites. ACh receptor agonists and antagonists failed to affect the extent or rate of binding of these probes to housefly membranes. It is concluded that most of these binding sites in housefly heads are not associated with cholinergic receptors. Several phenothiazine derivatives were poor inhibitors of [^{3}H]PCP binding, but chlorpromazine was potent. Since there was no relationship between inhibition by phenothiazines and their insecticidal action it is suggested that PCP-binding sites in housefly head membranes may not be targets for the insecticidal action of phenothiazines.

The regulation of ion transport across cellular membranes is a vital process for cell functioning and occurs mostly through ionic channel macromolecules, which are proteins that traverse the membrane lipid bilayer. They may be cation-specific and voltage-dependent like the Na^+, K^+ and Ca^{2+} channels of nerve cell membranes, or anion-selective like the Cl^- channel of the red

1982 Neuropharmacology of insects. Pitman, London (Ciba Foundation symposium 88) p 137-152

blood cell membrane and the ionic channel associated with γ-aminobutyric acid (GABA) receptors. Certain receptor-regulated ionic channels, such as the acetylcholine (ACh) receptor/channel complex of motor end-plates, are cation-selective but have very little cation specificity (Dwyer et al 1980).

Ion channels are primary targets for the action of specific toxins and drugs, and they might serve well as targets for insecticides. Thus, the Na^+ channel of vertebrate nerve and muscle is the primary target for tetrodotoxin, batrachotoxin, scorpion toxin, anemone toxin and local anaesthetics (Catterall & Ray 1978b). The Ca^{2+} channel of nerve and smooth muscle is the primary target for Ca^{2+} antagonists, such as verapamil (Rosenberger & Triggle 1978), which are now being used to treat cardiac arrhythmias. There is enough evidence to suggest that several drugs affect GABA- and glutamate-mediated conductance by interacting with the ionic channel rather than with the receptor sites (Nistri & Constanti 1978, Ticku & Olsen 1978). The receptor/channel system that we have most information on in vertebrates is the ACh receptor/channel molecule of motor end-plates and fish electric organs. It has also been studied in insects by both physiological and biochemical methods (Sattelle et al 1980, Eldefrawi & Eldefrawi 1980).

On the basis of their specific binding of α-bungarotoxin (α-BGTX), putative nicotinic ACh receptors have been identified in housefly brain (*Musca domestica*) (Eldefrawi & Eldefrawi 1980, Jones et al 1981), in the antennal lobes of the moth *Manduca sexta* (Hildebrand et al 1979), in the nerve cord of the American cockroach (*Periplaneta americana*) (Sattelle et al 1980) and in *Drosophila melanogaster* (Schmidt-Nielsen et al 1977, Haim et al 1979). These putative insect receptors have much lower affinities than the ACh receptor of the electric organ of the electric ray *Torpedo* for decamethonium and other bis-quaternary drugs as well as for ACh and other choline esters (Eldefrawi & Eldefrawi 1980, Jones et al 1981, Schmidt-Nielsen et al 1977).

In addition, putative muscarinic ACh receptors have been identified in *Drosophila* (Dudai & Ben-Barak 1977) and housefly brain (Jones & Sumikawa 1981, Shaker & Eldefrawi 1981). Again there are some differences between these insect receptors and mammalian brain muscarinic receptors. They have higher affinity for *d*-tubocurarine, and the Hill coefficients for their agonist and antagonist binding are <1, while in mammals the Hill coefficient is 1 for antagonists and <1 for agonists (Birdsall & Hulme 1976). Most interesting is the high ratio of nicotinic to muscarinic receptors in insect brain and of muscarinic to nicotinic receptors in mammalian brain (Dudai & Ben-Barak 1977, Shaker & Eldefrawi 1981).

Another protein that bound cholinergic drugs was identified in housefly heads, but it had nicotinic as well as muscarinic drug specificity, and it could be solubilized without the use of detergents (Mansour et al 1977). Although it is found in much higher concentrations than the membrane-bound putative receptors (50-fold more than the nicotinic and 1000-fold more than the muscarinic), its function is still not known. The properties of ionic channels associated with the cholinergic insect receptors described here are also unknown.

The best-known inhibitor of the ACh receptor/channel of motor end-plates of vertebrates is histrionicotoxin (Albuquerque et al 1980a), which is a spiropiperidine compound found in the skin of the Colombian frog *Dendrobates histrionicus*. Its tritiated derivative, $[^3H]$perhydrohistrionicotoxin ($[^3H]H_{12}$-HTX), was used as a probe for the identification of channel sites on the nicotinic ACh receptor/channel molecule of the electric organ of the electric ray *Torpedo* sp. (Eldefrawi et al 1980a). A large number of non-cholinergic drugs with various pharmacological actions interacted with this ionic channel without inhibiting the binding of receptor agonists or antagonists (Albuquerque et al 1980a, Eldefrawi & Eldefrawi 1980). These 'channel' or 'allosteric' drugs caused voltage-dependent postsynaptic inhibition of neuromuscular transmission, whereas a receptor antagonist such as α-BGTX caused voltage-independent inhibition. An example of these channel drugs is the general anaesthetic and hallucinogen, phencyclidine (PCP). It has two advantages over H_{12}-HTX, one its commercial availability in tritiated form (Eldefrawi et al 1980b) and the other the availability of several derivatives that have different potencies in modifying animal behaviour. We found that these channel probes bound reversibly to membrane preparations from housefly heads and with high affinities. Comparison of the binding of these probes to membranes from *Torpedo* electric organ and housefly head is the subject of this paper.

Methods used

The electric organ membrane preparation was a receptor-enriched preparation (1 nmol/mg protein), obtained by differential centrifugation of homogenates of the electric organ of *Torpedo ocellata*, as described by Eldefrawi et al (1978). Housefly membranes were obtained by centrifugation of head homogenates (20% w/v) at 3000 × ***g*** for 10 min, then recentrifugation of the supernatant fraction at 70 000 × ***g*** for 60 min (Shaker & Eldefrawi 1981). The resulting pellets were suspended in 50 mM-phosphate buffer, pH 7.4, so that 1 ml represented 0.2 g of original head weight. Supernatant fractions after 100 000 × ***g*** centrifugation of housefly head homogenates in glass-distilled water were obtained as previously described (Mansour et al 1977).

Binding of the ^3H-labelled channel probes to *Torpedo* membranes and housefly head membranes was measured by a filter assay, as described by

Aronstam et al (1981). In this assay, the tissue (≈50 μg of protein) was added to an incubation mixture (950 μl) containing 2 nM-[^{3}H]H$_{12}$-HTX (21 Ci/mmol), 50 mM-Tris buffer, pH 7.4, or [^{3}H]PCP (48 Ci/mmol; New England Nuclear), and any other drugs as indicated for the experiment. Incubation times also varied, from 30 s to 4 h. After this the mixture was suction-filtered over a Whatman GF/B glass fibre filter and washed with 7–10 ml of 50 mM-Tris buffer. The radioactive content of the filters was counted 10 h after they had been placed in 5 ml of a toluene-base liquid scintillation solution. Specific binding was the difference between binding with and without amantadine (5 mM)—a concentration that inhibited competitively all specific binding. Binding of [^{3}H]decamethonium (specific activity 559 mCi/mmol; Radiochemical Centre, Amersham) to the supernatant fraction of housefly heads was measured by equilibrium dialysis as described by Mansour et al (1977).

Binding of [^{3}H]H$_{12}$-HTX and [^{3}H]PCP to Torpedo electric organ membranes

Both [^{3}H]H$_{12}$-HTX and [^{3}H]PCP bound reversibly to equal numbers of sites in *Torpedo* membranes (≈0.7 nmol/g tissue) with K_d values of 0.11 μM for [^{3}H]H$_{12}$-HTX (Aronstam et al 1981) and 0.10 and 3.1 μM for the two affinities of [^{3}H]PCP (Eldefrawi et al 1980b). ACh receptor-regulated $^{22}Na^+$ flux was inhibited totally by concentrations of H$_{12}$-HTX that did not inhibit the binding of [^{3}H]ACh or of [^{125}I]BGTX to the receptor sites (Fig. 1B). The binding of both radiolabelled probes was inhibited by ionic-channel drugs, such as H$_{12}$-HTX, amantadine and PCP, though with somewhat different potencies (Eldefrawi et al 1980b).

A major advantage in using channel probes to study an ionic channel associated with a receptor is that their binding has proved to be conformation-dependent. Thus, the channel probe not only identifies the binding sites but also acts as a reporter molecule indicating the conformational state of the receptor/channel. The binding of [^{3}H]H$_{12}$-HTX and [^{3}H]PCP to *Torpedo* membranes in the absence of any receptor ligand was fairly slow ($T_{1/2} < 25$ min). But, in the presence of carbamylcholine, the binding was potentiated in a dose-dependent manner such that the initial rate of binding (measured after 30 s) increased dramatically (Fig. 2A). Various agonists potentiated binding of [^{3}H]H$_{12}$-HTX to different degrees that correlated well with their potencies in depolarizing cholinergic postsynaptic membranes in electric organs (Aronstam et al 1981). Decamethonium was the only agonist known to have channel-blocking properties (Adams & Sakmann 1978), and its stimulatory effect on [^{3}H]H$_{12}$-HTX binding was inhibited at higher concentrations. The effect of agonists on [^{3}H]H$_{12}$-HTX binding is due to binding to the receptor

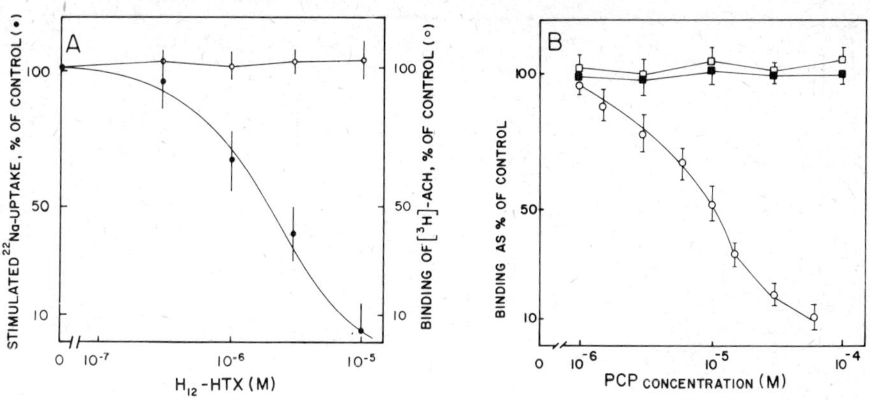

FIG. 1. Effect of ionic channel drugs (H_{12}-HTX and PCP) on binding of ligands to membranes from *Torpedo* electric organ and on ACh receptor-regulated ionic flux. Each symbol plus vertical line represents the mean ±SD of three experiments.

A. Effect of H_{12}-HTX on [^{3}H]ACh (1 μM) binding (○) and the carbamylcholine-stimulated $^{22}Na^+$ influx (●), which was measured by incubating 200 μl of *Torpedo* microsacs for 2 min with 1 μM-carbamylcholine and H_{12}-HTX; then 0.1 μCi of $^{22}Na^+$ in 200 μl of buffer and 100 μM-carbamylcholine were added and after 30 s the microsacs were passed through a Dowex 50W-X8 column, the free $^{22}Na^+$ was adsorbed, while the influxed $^{22}Na^+$ was collected in the filtered microsacs and counted. A low concentration of carbamylcholine (1 μM) was included during preincubation with H_{12}-HTX to stimulate the slow interaction. (From Eldefrawi et al 1980a.)

B. Effect of PCP on binding of 0.1 μM-[^{3}H]ACh (□) and 10 nM-^{125}I-α-BGTX (■) to ACh receptor sites and on binding of 2 nM-[^{3}H]H_{12}-HTX (○) to ionic channel sites in the membranes. (From Albuquerque et al 1980b.)

site, since α-BGTX inhibits it (Fig. 2A). However, this agonist-induced increase in the apparent binding rate of channel probe, which may reach up to 100-fold, cannot be explained by the twofold increase in affinity for channel drug and is believed to be controlled by factors other than the dissociation rate of [^{3}H]H_{12}-HTX. Accessibility may be a factor, as well as the rate constant of decay of the activated channel, since H_{12}-HTX seems to bind better to the activated states (Albuquerque et al 1980a, Eldefrawi et al 1980b).

Preincubation of the tissue with carbamylcholine reduced the stimulated [^{3}H]H_{12}-HTX binding (Fig. 2B) and $^{22}Na^+$ flux (Eldefrawi et al 1980a) in a concentration and time-dependent manner. This was suggested to be due to receptor desensitization. The presence of quaternary antagonists alone stimulated binding of [^{3}H]H_{12}-HTX to *Torpedo* membranes, but to a much lower extent, and this increase had a much slower time course than that caused by agonists, suggesting that antagonists induced conformational changes to a non-conducting state that was different from the resting and desensitized states (Aronstam et al 1981).

Therefore, the allosteric effects of the receptor sites on [^{3}H]H_{12}-HTX or

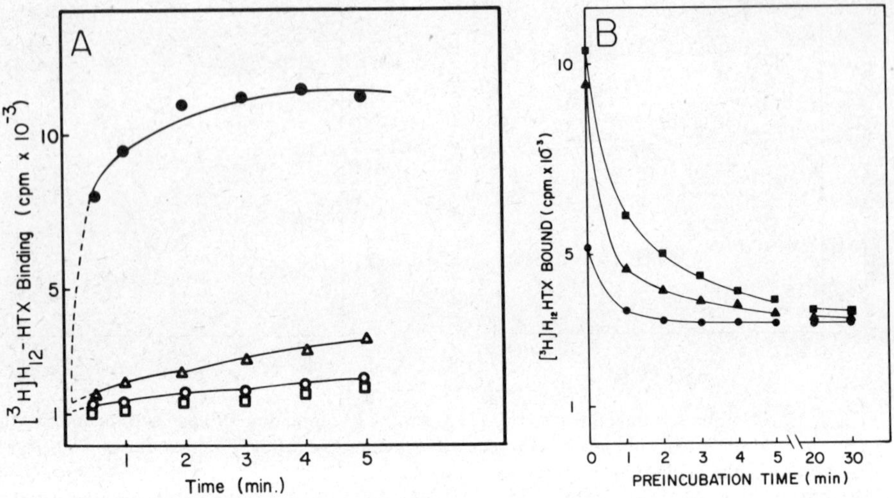

FIG. 2. Effect of receptor ligands on the kinetics of binding of $[^3H]H_{12}$-HTX to *Torpedo* membranes. Each symbol is the mean of three experiments, and the standard deviation is <10%.

A. Binding of $[^3H]H_{12}$-HTX (2 nM) alone (○) and with the simultaneous addition of carbamylcholine (20 μM) (●), carbamylcholine (20 μM) and *d*-tubocurarine (100 μM) (△) to untreated membranes as well as to membranes pretreated with 10 μM-α-BGTX for 60 min before the addition of $[^3H]H_{12}$-HTX (2 nM) and carbamylcholine (20 μM) (□).

B. Binding of $[^3H]H_{12}$-HTX (2 nM) to membranes that were preincubated with carbamylcholine (1 μM, ●; 10 μM, ▲; or 100 μM, ■) for the times indicated on the *x* axis. Binding was measured after 30 s. (From Aronstam et al 1981.)

[³H]PCP binding prove that this binding is indeed to the channel component of the ACh receptor of the *Torpedo* membrane and that the receptor and channel sites are coupled. When the membranes are first exposed for one hour to α-BGTX the allosteric effect is cancelled, and the interaction of drugs with the channel site can be studied without interference from the receptor sites.

Binding of $[^3H]H_{12}$-HTX and [³H]PCP to housefly head membranes

Like *Torpedo* electric organ membranes, those from housefly heads also bound $[^3H]H_{12}$-HTX and [³H]PCP reversibly, with high affinities (Fig. 3A,B). The binding of [³H]PCP to housefly head membranes showed two affinity constants (Fig. 3C): 250 pmol/g protein with a K_d of 0.023 μM and 500 pmol/g protein with a K_d of 0.19 μM; equivalent to 46.5 pmol/g heads and 93.1 pmol/g heads, respectively. The binding of both probes was inhibited equally

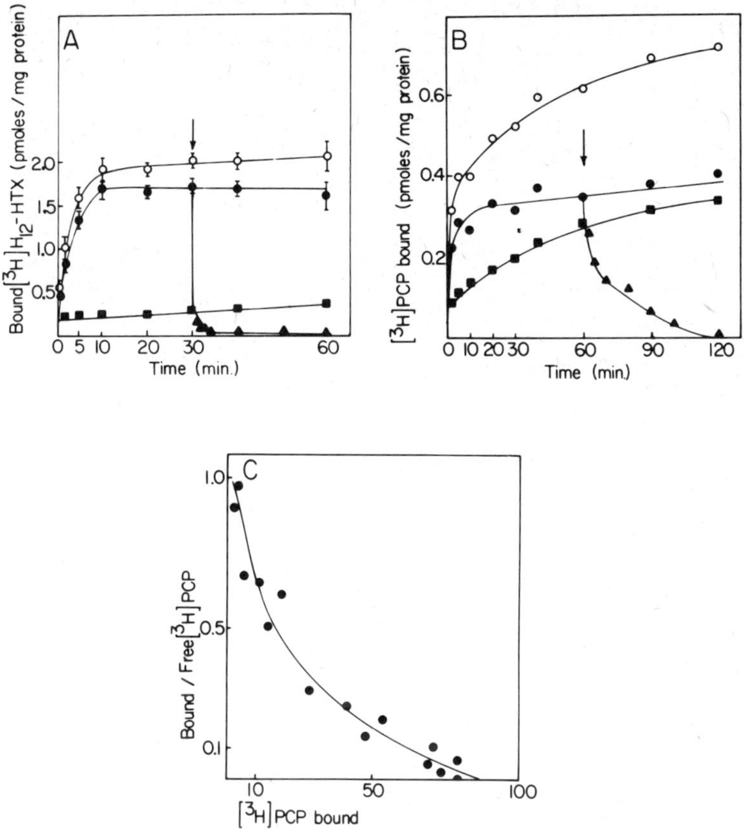

FIG. 3. Binding of two radiolabelled channel probes (at 2 nM) to housefly head membranes as a function of time. A. $[^3H]H_{12}$-HTX binding. B. $[^3H]$PCP binding. C. Scatchard plot of the binding of $[^3H]$PCP at different concentrations. Total binding (○), binding in presence of 5 mM-amantadine (■) and specific binding (●) calculated as the difference between the two. Amount bound is in pmol/0.1 g protein. The arrows point to the time at which 5 mM-amantadine was added to part of the incubation mixture. The bound $[^3H]$ligand that remained bound was then determined (▲). Symbols and bars represent means ±SD of triplicate assays.

effectively by ionic channel drugs (Fig. 4). At the concentrations used, H_{12}-HTX and PCP did not inhibit $[^3H]\alpha$-BGTX binding to housefly head membranes. However, unlike binding to *Torpedo* membranes, binding of $[^3H]$PCP and $[^3H]H_{12}$-HTX to housefly head membranes was not activated by cholinergic agonists or antagonists, or by other neurotransmitters. This suggests that the binding in insect brain may not be to the ionic channel sites of the ACh receptor, or that the insect ACh receptor/channel complex is different from that of vertebrate peripheral ones. However, the concentration

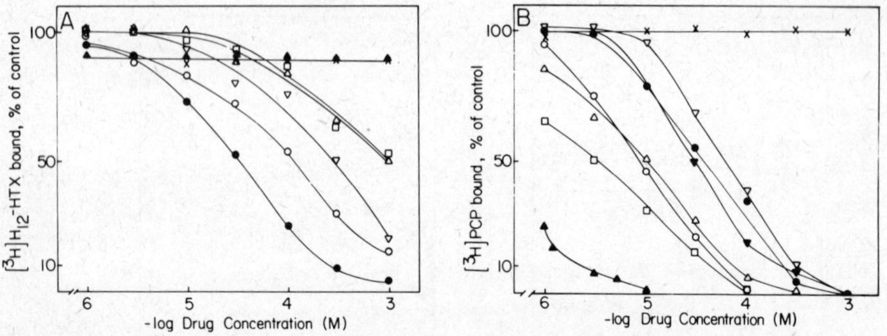

FIG. 4. The effect of drugs on the binding of (A) $[^3H]H_{12}$-HTX and (B) $[^3H]$PCP to housefly head membranes. Tissue (50 μl) was added to an incubation tube containing 950 μl of 50 mM-phosphate buffer, pH 7.4, containing the probe and drug; after 60 min the amount of probe bound was determined by a filter assay. Symbols represent means of three experiments with SD <10%. △, amantadine; ○, dibucaine; ●, quinacrine; ▽, tetracaine; ▼, piperocaine; ▲, nicotine in A and PCP in B; X, α-BGTX in A and carbamylcholine in B; □, *d*-tubocurarine in A and chlorpromazine in B.

of the $[^3H]$PCP binding sites in housefly head membranes (80 pmol/g heads) (Fig. 3C) was much higher than the concentration of putative nicotinic and muscarinic sites (Shaker & Eldefrawi 1981), which ranged from 20 (Eldefrawi & Eldefrawi 1980) to 23 (Jones et al 1981) pmol/g heads for the former and from 0.8 (Shaker & Eldefrawi 1981) to 1 (Jones & Sumikawa 1981) pmol/g heads for the latter. Thus, it appears likely that the observed $[^3H]$PCP binding to housefly head membranes may be to several proteins with close affinities, particularly since PCP has been shown to interact with muscarinic receptors (Aronstam et al 1980), K^+ channels (Albuquerque et al 1980b, Tsai et al 1980) and biogenic amine uptake carriers (Smith et al 1977) with similarly high affinities. Thus, if binding to the cholinergic channel represents 2% of total binding observed and agonists increase this binding by up to fivefold, the stimulation may not be detectable by the assay used.

Toxicity of channel probes to insects

Since binding of the two channel probes to housefly head preparations was inhibited best by local anaesthetics and drugs with a tricyclic ring structure (Fig. 4), we determined the toxicities of several such drugs to insects and compared them to their potencies in displacing the binding of $[^3H]$PCP to housefly head membranes. As shown in Table 1, there was some correlation between the toxicities of local anaesthetics to houseflies and their potencies in inhibiting $[^3H]$PCP binding. It is interesting to note that dibucaine, the only local anaesthetic tested which is an amide, was more toxic and had higher affinity for [^{3}H]PCP binding sites than the other three, which are benzoic acid esters. Among the tricyclic compounds, chlorpromazine and quinacrine were fairly potent in inhibiting [^{3}H]PCP binding to housefly membranes but were not very toxic to mosquito larvae (Table 2). On the other hand, another

TABLE 1 Toxicity of local anaesthetics to houseflies and their inhibition of binding of [^{3}H]PCP and [^{3}H]decamethonium to housefly head preparations

Local anaesthetic	*K_i for inhibition of binding (M)*		*Toxicity to houseflies*[c] *(LD_{50} μg/female)*
	[^{3}H]PCP[a]	*[^{3}H]Decamethonium*[b]	
Procaine	>1 × 10^{-3}	2.9 × 10^{-6}	140
Piperocaine	3.1 × 10^{-5}	7.4 × 10^{-7}	192
Tetracaine	4.9 × 10^{-5}	2.3 × 10^{-6}	34
Dibucaine	8.9 × 10^{-6}	1.6 × 10^{-7}	26

[a] Binding of [^{3}H]PCP at 2 nM to housefly head membranes was measured by a filter assay. Two affinities were detected with K_d values of 2.3 × 10^{-8} and 1.9 × 10^{-7} M.

[b] Binding of [^{3}H]decamethonium at 0.1 μM was measured by equilibrium dialysis to the 'soluble' fraction of housefly head membranes (from Mansour et al 1977).

[c] Toxicity to houseflies was determined by injecting 5 μl of the aqueous solution into the abdomen. For each of seven concentrations tested, 25 female houseflies were tested. Mortality counts were made after 24 h and LD_{50} values were obtained from probit analysis.

TABLE 2 Toxicity of phenothiazine and other tricyclic compounds to mosquito larvae and their ability to inhibit binding of [^{3}H]PCP to membranes from housefly head and *Torpedo* electroplax

Compound	*K_i for inhibition of [^{3}H]PCP binding (M)*[a]		*LD_{50} to Aedes aegypti*[b] *(p.p.m.)*
	Housefly	*Torpedo*	
Phenothiazine	>1 × 10^{-4}	>1 × 10^{-4}	0.6
2-Chlorophenothiazine	>1 × 10^{-4}	>1 × 10^{-4}	7.5
2-Trifluoromethylphenothiazine	>1 × 10^{-4}	>1 × 10^{-4}	7.8
2-Acetylphenothiazine	1.0 × 10^{-4}	>1 × 10^{-4}	>250
10-Acetylphenothiazine	>1 × 10^{-4}	2.1 × 10^{-5}	7.5
Phenoxazine	>1 × 10^{-4}	>1 × 10^{-4}	6.0
Chlorpromazine	2.9 × 10^{-6}	1.7 × 10^{-7}	30
Amantadine	1.0 × 10^{-5}	5.2 × 10^{-5}	>300

[a] Binding of [^{3}H]PCP at 2 nM was measured by a filter assay. Two affinities were detected: K_d values for [^{3}H]PCP binding to housefly head membranes were 2.3 × 10^{-8} and 1.9 × 10^{-7}M and to *Torpedo* membranes 10^{-7} and 3.1 × 10^{-6}M.

[b] Toxicity was determined on 4th instar larvae, 25 in 100 ml for each of seven concentrations tested. Mortality counts were made after 24 h and LD_{50} values were obtained from probit analysis.

tricyclic compound, phenothiazine, is an insecticide and was toxic to mosquito larvae but did not inhibit [^{3}H]PCP binding. Substitutions on the phenothiazine nucleus that improved the ability of the compound to inhibit [^{3}H]PCP binding (Fig. 5) also reduced its toxicity. Thus, we may conclude

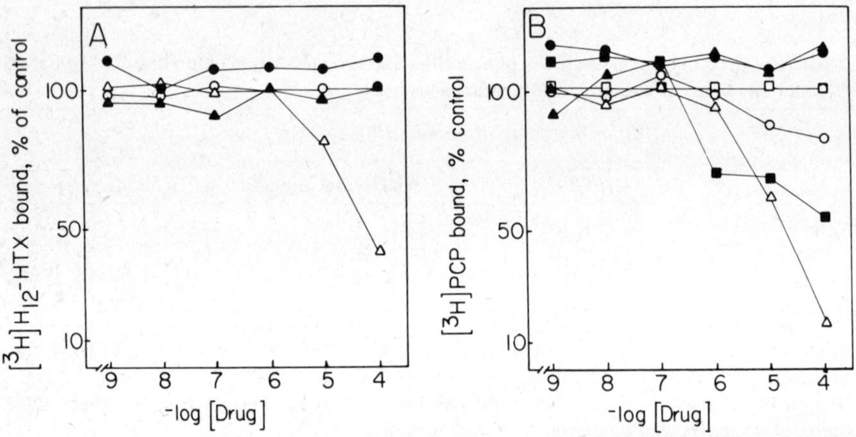

FIG. 5. The effect of phenothiazine derivatives on binding of (A) [^{3}H]H_{12}-HTX and (B) [^{3}H]PCP to housefly head membranes (solid symbols) and *Torpedo* membranes (open symbols). Phenothiazine (○, ●), 10-acetylphenothiazine (△, ▲) and 2-acetylphenothiazine (□, ■). Other phenothiazine derivatives tested but with no effect on binding of either ligand to either tissue included 2-chlorophenothiazine, 2-trifluoromethylphenothiazine and phenoxazine. Symbols are means of three experiments with SD $<10\%$.

that the target of the insecticidal action of phenothiazines is not the same target in housefly heads that binds [^{3}H]PCP. PCP-HCl was not very toxic to mosquito larvae (LD_{50} above 100 p.p.m.), and H_{12}-HTX was non-toxic to houseflies up to concentrations of 25 μg/female. We have no information on the effect of H_{12}-HTX or PCP on the function of insect nerve and muscle. Consequently, we have been unable to characterize their binding site(s) and identify it (them) with known regulatory moieties. It is possible that a compound like PCP, which significantly modifies the behaviour of humans and other mammals, such as rat and monkey, may adversely affect insect behaviour without being toxic. In humans, PCP is a powerful psychotropic agent. Thus, channel inhibitors may be good candidates for testing on the behaviour of highly organized societies such as those of honey bees and ants.

Acknowledgement

This work was supported in part by a grant from the National Institutes of Health, ES 02594.

REFERENCES

Adams PR, Sakmann B 1978 Decamethonium both opens and blocks endplate channels. Proc Natl Acad Sci USA 75:2994-2998

Albuquerque EX, Adler M, Spivak CE, Aguayo L 1980a Mechanism of nicotinic channel activation and blockade. Ann NY Acad Sci 358:204-252

Albuquerque EX, Tsai M-C, Aronstam RS, Witkop B, Eldefrawi AT, Eldefrawi ME 1980b Phencyclidine interactions with ionic channels of the acetylcholine receptor and electrogenic membranes. Proc Natl Acad Sci USA 77:1224-1228

Aronstam RS, Albuquerque EX, Eldefrawi AT, Eldefrawi ME, Jim KF, Triggle DJ 1980 Sites of action of phencyclidine. III. Interactions with muscarinic receptors. Mol Pharmacol 18:179-184

Aronstam RS, Eldefrawi AT, Pessah IN, Daly JW, Albuquerque EX, Eldefrawi ME 1981 Regulation of [^{3}H]perhydrohistrionicotoxin binding to *Torpedo* electroplax by effectors of the acetylcholine receptors. J Biol Chem 256:2843-2850

Birdsall NJM, Hulme EC 1976 Biochemical studies on muscarinic acetylcholine receptors. J Neurochem 27:7-16

Catterall WA, Ray R 1976 Interactions of neurotoxins with the action potential Na^+ ionophore. J Supramolec Struct 5:397-407

Dudai Y, Ben-Barak J 1977 Muscarinic receptor in *Drosophila melanogaster* demonstrated by binding of [^{3}H]quinuclidinyl benzilate. FEBS (Fed Eur Biochem Soc) Lett 81:134-136

Dwyer TM, Adams DJ, Hille B 1980 The permeability of the endplate channel to organic cations in frog muscle. J Gen Physiol 75:469-492

Eldefrawi AT, Eldefrawi ME 1980 Putative acetylcholine receptors in housefly brain. In: Sattelle DB et al (eds) Receptors for neurotransmitters, hormones and pheromones in insects. Elsevier/North-Holland Biomedical Press, Amsterdam, p 59-70

Eldefrawi ME, Eldefrawi AT, Mansour NA, Daly JW, Witkop B, Albuquerque EX 1978 Acetylcholine receptor and ionic channel of *Torpedo* electroplax: binding of perhydrohistrionicotoxin to membrane and solubilized preparations. Biochemistry 17:5474-5484

Eldefrawi ME, Aronstam RS, Bakry NM, Eldefrawi AT, Albuquerque EX 1980a Activation, inactivation and desensitization of acetylcholine receptor channel complex detected by binding of perhydrohistrionicotoxin. Proc Natl Acad Sci USA 77:2309-2313

Eldefrawi ME, Eldefrawi AT, Aronstam RS, Maleque MA, Warnick JE, Albuquerque EX 1980b [^{3}H]Phencyclidine: A probe for the ionic channel of the nicotinic receptor. Proc Natl Acad Sci USA 77:7458-7462

Haim N, Nahum S, Dudai Y 1979 Properties of a putative muscarinic cholinergic receptor from *Drosophila melanogaster*. J Neurochem 32:543-552

Hildebrand JG, Hall LM, Osmond BC 1979 Distribution of binding sites for ^{125}I-labeled α-bungarotoxin in normal and deafferented antennal lobes of *Manduca sexta*. Proc Natl Acad Sci USA 76:499-503

Jones SW, Sumikawa K 1981 Quinuclidinyl benzilate binding in house fly head and rat brain. J Neurochem 36:454-459

Jones SW, Sudersham P, O'Brien RD 1918 α-Bungarotoxin binding in house fly heads and *Torpedo* electroplax. J Neurochem 36:447-453
Mansour NA, Eldefrawi ME, Eldefrawi AT 1977 Isolation of putative acetylcholine receptor proteins from housefly brain. Biochemistry 16:4126-4132
Nistri A, Constanti A 1978 Effects of flurazepam on amino acid-evoked responses recorded from the lobster muscle and the frog spinal cord. Neuropharmacology 17:127-135
Rosenberger L, Triggle DJ 1978 Calcium, calcium translocation, and specific calcium antagonists. In: Weiss GB (ed) Calcium in drug action. Plenum, New York, p 3-31
Sattelle DB, David JA, Harrow ID, Hue B 1980 Actions of α-bungarotoxin on identified insect central neurones. In: Sattelle DB et al (eds) Receptors for neurotransmitters, hormones and pheromones in insects. Elsevier/North-Holland Biomedical Press, Amsterdam, p 125-139
Schmidt-Nielsen BK, Gepner JI, Teng NNH, Hall LM 1977 Characterization of an α-bungarotoxin binding component from *Drosophila melanogaster*. J Neurochem 29:1013-1029
Shaker N, Eldefrawi A 1981 Muscarinic receptor in house fly brain and its interaction with chlorobenzilate. Pestic Biochem Physiol 15:14-20
Smith RC, Meltzer HY, Arora RC, Davis JM 1977 Effects of phencyclidine on [^{3}H]catecholamine and [^{3}H]serotonin uptake in synaptosomal preparations from rat brain. Biochem Pharmacol 26:1435-1437
Ticku MK, Olsen RW 1978 Interaction of barbiturates with dihydropicrotoxinin binding sites related to the GABA receptor-ionophore system. Life Sci 221:1643-1652
Tsai M-C, Albuquerque EX, Triggle DJ, Aronstam RS, Eldefrawi AT, Eldefrawi ME 1980 Sites of action of phencyclidine. I. Effects on the electrogenic and chemosensitive properties of the neuromuscular junction of skeletal muscle. Mol Pharmacol 18:159-166

DISCUSSION

Usherwood: Since histrionicotoxin (HTX) binds to *Torpedo* membrane in the absence of receptor ligand, does this binding depress the physiological response when you subsequently add the cholinergic ligand?

M. E. Eldefrawi: From experiments on the mechanism of action of HTX, Albuquerque and coworkers (Lapa et al 1975) concluded that its inhibition was stimulation-dependent. When they added it to the frog nerve-muscle preparation and then stimulated, the higher the stimulation frequency, the more potent was the inhibition obtained. The first twitch response of the muscle was almost normal and then it decayed very quickly, as if the muscle was desensitizing. If the system was washed, HTX was added and then the nerve was stimulated, the first response was again normal and then decayed very quickly. This stimulation dependency indicates that the toxin, if added in the resting state, reacts very slowly but after stimulation the action is exaggerated and becomes very fast.

Usherwood: Does HTX bind to the same sites in *Torpedo* membranes in the absence of receptor ligand as when the receptors are activated by cholinergic ligand?

M. E. Eldefrawi: Yes.

Barnard: In support of your conclusions, the results of Kato & Changeux (1976) should be recalled. They showed that the binding at the ACh receptor (in *Torpedo* membranes) of the cholinergic agonist itself is reciprocally affected by HTX. This gives a further insight into the mechanism of action of this drug.

M. E. Eldefrawi: There is reciprocal allosteric coupling between the receptor site and the site on the ionic channel where HTX binds. As you say, Kato & Changeux (1976) showed that the affinity for ACh was increased when HTX was present. They even predicted that the toxin binding may cause allosteric transformation to a desensitized state. The same effect is produced with local anaesthetics, which compete very well with the toxin binding. They identified that site as a local anaesthetic binding site. There is a reciprocal interaction between the two sites and basically we believe we are dealing with a single molecule.

Barnard: What about the results of single-channel conductance measurements in the presence of HTX?

M. E. Eldefrawi: Our analysis is similar to that presented by Dr Cull-Candy.

Barnard: So it is a pure channel-blocking effect?

M. E. Eldefrawi: Yes.

Narahashi: There are other end-plate channel-blocking agents. We have studied *N*-alkylguanidine derivatives (Farley et al 1981), which have a similar channel-blocking effect to that of HTX. However, the situation is more complicated with the guanidine derivatives because there are two types of blocking action. One is the blocking of open channels, just as you mentioned, and the other is a blocking of closed channels. The closed-channel block is highly voltage-dependent, being intensified by hyperpolarization, yet the open-channel block is totally voltage-independent. Is there any voltage-dependent binding of H_{12}-HTX or phencylidine, or any of the chemicals you have studied?

M. E. Eldefrawi: We are doing biochemical studies now. We are using [^{3}H]methyl triphenyl phosphonium to determine the membrane potential in our preparation. We hope to be able to manipulate that potential and see the effect. Then we shall be able to answer this question, but I think that Albuquerque has done enough, in looking at the voltage dependency of the action of HTX and PCP, to show that the action of these drugs is definitely voltage-dependent and time-dependent (Albuquerque et al 1973, 1980a). Agents with very different chemical structures, ranging from amantadine to tetraethylammonium and tricyclic compounds, affect the ionic channel activity. They have different mechanisms. H_{12}-HTX and PCP, for instance, produce non-linearity in the current-voltage relationship as well as a change in the time course of the end-plate current. Imipramine, on the other hand, has no effect on the time course of the channel though it causes the non-linearity (Eldefrawi et al 1981). The interpretation is that imipramine binds to the closed as well as the activated but non-conducting receptor/channel conformation, while HTX and PCP interact with the closed, the open, and the activated but non-conducting conformations.

Narahashi: Depending on the rates of association and dissociation, one could have conditions in which the time course of the end-plate current is not changed during channel block. The time course of the end-plate current does not necessarily tell you whether the effect is a channel block or a receptor block.

A. T. Eldefrawi: The channel can be blocked from either inside or outside the membrane. Thus when Albuquerque et al (1980b) and Katz & Miledi (1980) applied intracellularly charged molecules large enough *not* to pass through the end-plate channel and not to get out of the membrane, they blocked the channel. So we know there are different binding sites on the ionic channel of the ACh receptor in the end-plate.

Narahashi: Recently we have done an experiment with a cultured myoball preparation from chick embryo using internal perfusion and voltage clamp techniques. The ACh-associated channels are blocked only when a guanidine derivative is applied from outside. This is in keeping with what we believed in the past. Local anaesthetics can block voltage-dependent sodium channels of the myoball membrane only from inside. This also agrees with what has been thought previously.

Barnard: I liked very much, Professor Eldefrawi, the idea that instead of having to kill insect pests you could just drive them crazy. I wonder if the correlation with lethality that you were looking for—and did not find—in discussing the comparative data in insects might in fact be unobtainable. You were looking at the lethality of these channel-blocking toxins, but if HTX is injected into mammals there is a poor correlation between its lethality and its channel-blocking effectiveness: its lethality, in fact, is very low. It seems that using a postsynaptic channel blocker is a poor way to kill an animal. Presumably it is too lipophilic, gives too much non-specific binding and so doesn't get efficiently distributed to the channels, and so forth. The lack of correlation with the LD_{50} values doesn't mean that the substances do not have a channel-blocking effect. However, can you make use of the rank order of potency of the various channel blockers you have looked at, to deduce whether the channel in the insect nicotinic cholinergic receptor is different from the channel in the vertebrate?

M. E. Eldefrawi: It is quite possible that the cholinergic receptor of the insect CNS has a different mechanism for stimulating the cellular response. The channel may have a different structure from that of *Torpedo*. We have no evidence either way yet.

Cull-Candy: Is there a site for each channel binder or are many of them binding to the same site?

M. E. Eldefrawi: H_{12}-HTX and PCP share a binding site and their interaction is competitive. The interaction of imipramine with both PCP and H_{12}-HTX is not competitive, so it binds to a different site. Imipramine and chlorpromazine are competitive. It seems to me that the channel represents the bulk of the molecule, which apparently has many possible sites, whose binding of drugs changes the properties of the channel.

A. T. Eldefrawi: I should point out that the acetylcholine receptor channel probably binds only a small proportion of the total detectable binding of $[^3H]$PCP and $[^3H]H_{12}$-HTX to housefly heads. Thus, if activation of $[^3H]$PCP or $[^3H]H_{12}$-HTX binding by the cholinergic agonist is small, this change would not be detectable in the total binding. But, if it were large, we should have detected it.

Barnard: Has anybody tried adding HTX or PCP in electrophysiological experiments on the nicotinic cholinergic junction in insects?

M. E. Eldefrawi: Our hope is that somebody will do that.

Narahashi: Has anyone investigated the effect of HTX on glutamate-activated ionic channels?

Barnard: Glavinovic et al (1974), on the basis of intracellular recording from a known pathway in the spinal cord of the frog, reported that HTX blocked an amino acid receptor channel. I have not seen a further investigation using HTX of such a system in vertebrates.

Usherwood: Is recovery from the toxin enhanced by the presence of the agonist?

M. E. Eldefrawi: The agonist increases the rates of association and of dissociation.

Usherwood: Does that tell us anything about the structure of the ionic channel?

M. E. Eldefrawi: I don't know yet. It tells us that the site becomes accessible. It is more exposed in the activated state, maybe because it happens to be on the part of the channel that opens up.

Sattelle: The discussion has ranged over vertebrate and invertebrate studies. It is important not to generalize too widely.

M. E. Eldefrawi: That ties in with Eric Barnard's question about whether the channel of the insect nicotinic receptor is pharmacologically different from the vertebrate channel. We can't answer that question yet.

Hall: What is the relationship between the number of saturable binding sites for PCP, for HTX and for α-BGTX receptors in an insect preparation?

M. E. Eldefrawi: The electric organ has an equal number of binding sites for PCP, α-BGTX and ACh. The insect has 10 times more binding sites for $[^3H]H_{12}$-HTX than for $[^3H]$PCP, and the number of binding sites for PCP is probably 3–4 times that for α-BGTX. Even though we have saturable binding sites their number is much higher than the number of α-BGTX-binding sites.

REFERENCES

Albuquerque EX, Barnard EA, Chiu TH et al 1973 Acetylcholine receptor and ion conductance modulator sites at the murine neuromuscular junction: evidence from specific toxin reactions. Proc Natl Acad Sci USA 70:949-953

Albuquerque EX, Tsai M-C, Aronstam RS, Witkop B, Eldefrawi AT, Eldefrawi ME 1980a Phencyclidine interactions with the ionic channel of the acetylcholine receptor and electrogenic membrane. Proc Natl Acad Sci USA 77:1224-1228

Albuquerque EX, Adler M, Spivak CE, Aguayo L 1980b Mechanism of nicotinic channel activation and blockade. Ann NY Acad Sci 358:204-238

Eldefrawi ME, Warnick JE, Schofield GG, Albuquerque EX, Eldefrawi AT 1981 Interaction of imipramine with the ionic channel of the acetylcholine receptor of motor endplate and electric organ. Biochem Pharmacol 30:1391-1394

Farley JM, Yeh JZ, Watanabe S, Narahashi T 1981 Endplate channel block by guanidine derivatives. J Gen Physiol 77:273-293

Glavinovic M, Henry J, Kato G, Krnjevic K, Pail E 1974 Histrionicotoxin: effects on some central and peripheral excitable cells. Can J Physiol Pharmacol 52:1220-1226

Kato G, Changeux J-P 1976 Studies on the effect of histrionicotoxin on the monocellular electroplaque from *Electrophorus electricus* and on the binding of [^{3}H]acetylcholine to membrane fragments from *Torpedo marmorata*. Mol Pharmacol 12:92-100

Katz B, Miledi R 1980 Blockade of endplate responses by intracellular application of procaine. In: Taxi J (ed) Ontogenesis and functional mechanisms of peripheral synapses. Elsevier/North-Holland Biomedical Press (INSERM Symp 13) p 171-178

Lapa AJ, Albuquerque EX, Sarvey JM, Daly J, Witkop B 1975 Effects of histrionicotoxin on the chemosensitive and electrical properties of skeletal muscle. Exp Neurol 47:558-580

Subcellular fractionation of invertebrate nervous tissue

RICHARD W. OLSEN

Division of Biomedical Sciences and Department of Biochemistry, University of California, Riverside, California 92521, USA

Abstract The usefulness of subcellular fractions from vertebrate nervous tissue in neuropharmacological research has been profound. The preparation and properties of subcellular fractions enriched in pinched-off nerve terminals (synaptosomes) and pre-synaptic and postsynaptic membrane fragments are reviewed briefly. Attempts to produce similar fractions from invertebrate animals are summarized. A method for preparing a sarcolemma-enriched fraction from crayfish muscle has been described. This method is suitable for the biochemical analysis of the action of neurotransmitters, drugs, toxins and insecticides in invertebrates.

Research on neuropharmacology in insects would undoubtedly be greatly aided by the availability of subcellular fractions from synaptic regions of the nervous system, like those used successfully in studies of higher organisms. The preparation of pinched-off nerve endings, or synaptosomes, and synaptic membrane fractions from the central and peripheral nervous systems of vertebrates has contributed significantly to the study of neurotransmission. These fractions have served as enriched sources of neurotransmitter substances and of the macromolecules involved in the biosynthesis, secretion, degradation, re-uptake, and receptor–effector response mechanisms of synaptic transmission. Such fractions can be used for cell-free model studies of the mechanisms of neurotransmitter release and uptake, and of the regulation of the biosynthesis of synaptic proteins and neurotransmitter-synthesizing enzymes. They can also be used for the isolation of, and structure–function studies on, the macromolecules involved.

Despite numerous research advantages and the practical interest in insect and other invertebrate neuropharmacology, studies on the subcellular fractionation of nervous tissue in invertebrates have been too few and only

1982 Neuropharmacology of insects. Pitman, London (Ciba Foundation symposium 88) p 153-157

modestly successful. Techniques used with higher organisms do not appear to be generally successful in work with invertebrates, although some approaches are promising. We have prepared a sarcolemma-enriched fraction of crayfish muscle suitable for the biochemical study of excitatory and inhibitory peripheral neurotransmission (Meiners et al 1979). Preparations of synaptic membrane fractions from the insect central nervous system (Breer, this volume) and muscle (Briley et al, this volume) likewise appear very useful.

Applications of subcellular fractions from vertebrate nervous tissue to research in neuropharmacology

The specialized secretory portions of nerve endings and other synaptic regions (boutons and/or varicosities), can be isolated from gently homogenized vertebrate nervous tissue (De Robertis 1971, Dowdall et al 1975). The plasma membrane of these specialized regions apparently pinches off and then reseals (in a medium of appropriate osmolarity) to form a cell-free (nucleus-free) but cell-like nerve-ending vesicle. This nerve-ending cytoplasm bound by the cellular synaptic membrane—the synaptosome—usually contains some portions of the synaptic contact with the postsynaptic cell, including part of the postsynaptic membrane (presumably complete with neurotransmitter-responsive receptor–effector complexes) and the synaptic 'glue' which holds the two cells together at the synapse. Other macromolecules in the postsynaptic 'density' structure near the membrane may or may not remain attached. Further enrichment of the synaptic connection itself, at the expense of the presynaptic cytoplasmic vesicle, can be obtained by osmotic disruption and/or mild detergent treatment, plus additional density gradient centrifugation (De Robertis 1971). This approach has proved useful in identifying major proteins associated with the synaptic connections (Cotman 1974).

With the acceptance that the synaptosome-enriched fraction represents a portion of the tissue involved in synapse formation came the conclusion that any molecule specifically involved in synaptic transmission ought to be enriched in such fractions. This could be a very oversimplified conclusion in many cases. Nevertheless, application of this fractionation technique to the vertebrate neuromuscular junction or to the electric organ of electric fish (a tissue analogous to vertebrate muscle and receiving nicotinic cholinergic innervation) resulted in a synaptosome fraction highly enriched in the putative neurotransmitter, acetylcholine (ACh) (Dowdall et al 1975). These cholinergic synaptosomes contain high concentrations of cytoplasmic ACh and also considerable amounts bound to or enclosed in vesicles, which are abundant in the presynaptic cytoplasm adjacent to the synaptic specialization.

The presence of neurotransmitter-containing vesicles, some of which have been observed in exocytotic contact with the presynaptic membrane (De Robertis 1971, Heuser & Reese 1973), appears to confirm the quantal hypothesis of neurotransmitter release (Katz 1971). These advances in ultrastructural and neurochemical knowledge, while not crucially dependent on the synaptosomal preparation, were greatly aided by the development of this tissue fraction.

Although clearly identifiable synaptosome fractions were readily prepared from the vertebrate central nervous sytem, enrichment of putative neurotransmitter substances was not as clear-cut as for the ACh nerve ending in the peripheral nervous system (Dowdall et al 1975). Nevertheless, the central tissue synaptosomes have been extremely valuable in cell-free studies of the uptake of putative neurotransmitters, of Ca^{2+}-dependent release of neurotransmitters, and of the enzymic synthesis of neurotransmitters (Bradford 1972). The enrichment of neurotransmitter synthetic enzymes in nerve endings has been used for the histochemical mapping of specific synaptic contacts in the vertebrate brain. Synaptosomes have also been used for studies of ion transport and channels involved in action potentials and secretion, as well as for a few studies on protein synthesis, metabolic regulation, gene expression and cytoskeleton function (Bradford 1972, Cobb & Pentreath 1978).

The electric organ of electric fish has also provided an excellent cell-free vesicle membrane preparation suitable not only for the isolation of the ACh receptor protein but also for biochemical studies on the functional mechanism by which the neurotransmitter regulates ion channel permeability *in vitro* (Kasai & Changeux 1971).

Advantages and special properties of invertebrates for research in neuropharmacology

Invertebrates, including insects, have played a big role in neurobiological research and should continue to do so. For example, the giant axon of the squid has provided an excellent model for studies of electrically excitable membranes and ion channels (Hodgkin & Huxley 1952). Invertebrates, including insects, have a relatively easily defined neuroanatomy, with identified neurons that can be repeatably located in each individual organism (Strang 1979). Insects, including *Drosophila melanogaster* and also some other species, are particularly amenable to genetic research, which is now being applied to neurobiology (Strang 1979). Insects have a highly active central cholinergic system that should be extremely useful for studying ACh (Strang 1979), in addition to the practical aspect of the action of insecticides at this level. The striated muscles of arthropods, including insects and crustaceans, are excellent tissues in which to study the putative amino acid neurotransmitters such as glutamate and γ-aminobutyric acid (GABA)—compounds under justifiably intensive study today in the vertebrate nervous system. It was less than 20 years ago that GABA was shown to be the peripheral inhibitory neurotransmitter substance in crustaceans (Roberts et al 1976). This result was obtained in the lobster by microdissection and chemical analysis of the inhibitory neurons and their secretion product. Since this identification, insect and crustacean muscle has provided considerable neuropharmacological information about glutamate and GABA, studied by both electrophysiological and neurochemical approaches (Roberts et al 1976). We measured the response to GABA in crayfish abdominal muscle fibres using a radioisotope of chloride, $^{36}Cl^-$ (Ticku & Olsen 1977). Subcellular fractionation of this tissue gave us a plasma membrane-enriched fraction, as determined by marker enzyme localization (Meiners et al 1979). Microscopic examination revealed vesicular membrane fragments, 300 nm in diameter. Whether any of these membrances were derived from muscle or nerve cells was not determined. Since the membrane fraction was enriched in the receptor sites for GABA (Meiners et al 1979) and the GABA Cl^- channel-blocking drug, picrotoxinin (Olsen et al 1978), it appears to be derived principally from sarcolemma. Thus, it could also be used for studies of the uptake of and functional responses to GABA and the excitatory amino acids.

Subcellular fractionation of insect nervous tissue

The homogenization and density fractionation scheme used to produce synaptosomes from vertebrate tissue does not seem to work the same way in invertebrates. This is probably due to differences in synaptic ultrastructure, both in molecular composition and in three-dimensional organization (Cobb & Pentreath 1978). The preparative methods used in vertebrates produced synaptosome-like structures from the squid (Dowdall et al 1975) but very few such structures from the cockroach (Telford & Matsumura 1970) and the honey-bee (Fox & Larsen 1972). Modifications in methods have led to some success in the production of cholinergic nerve terminals from fleshfly heads (Donnellan et al 1976), and numerous synaptosomal profiles were observed in a flotation pellicle fraction from the locust nervous system (Breer & Jeserich 1980).

The fractionation of insect muscle has also been attempted (Donnellan et al 1974), with little success in the production of synaptosomes, but some enrichment in membrane fractions showing receptor-like binding activities (Briley et al, this volume).

Acknowledgements

Supported by NIH Research Career Development Award NS-00224.

REFERENCES

Bradford HF 1972 Cerebral cortex slices and synaptosomes: *in vitro* approaches to brain metabolism. In: Fried R (ed) Methods of neurochemistry. Dekker, New York, vol 2:155-202

Breer H 1982 Properties of synaptosomes from the central nervous system of insects. This volume p 158-169

Breer H, Jeserich G 1980 A microscale floatation technique for the isolation of synaptosomes from nervous tissue of *Locusta migratoria*. Insect Biochem 10:457-463

Briley PA, Filbin MT, Lunt GG, Donnellan JF 1982 Binding and uptake of glutamate and γ-aminobutyric acid in membrane fractions from locust muscle. This volume, p 176-187

Cobb JLS, Pentreath VW 1978 Comparison of the morphology of synapses in invertebrate and vertebrate nervous systems: analysis of the significance of the anatomical differences and interpretation of the morphological specializations. Prog Neurobiol (Oxf) 10:231-252

Cotman CW 1974 Isolation of synaptosomal and synaptic plasma membrane fractions. Methods Enzymol 31A:445-454

De Robertis E 1971 Molecular biology of synaptic receptors. Science (Wash DC) 171:963-971

Donnellan JF, Jenner DW, Ramsey A 1974 Subcellular fractionation of fleshfly flight muscle in attempts to isolate synaptosomes and to establish the location of glutamate enzymes. Insect Biochem 4:243-265

Donnellan JF, Alexander K, Chendlik R 1976 The isolation of cholinergic nerve terminals from fleshfly heads. Insect Biochem 6:419-423

Dowdall MJ, Fox G, Wächtler K, Whittaker VP, Zimmerman H 1975 Recent studies on the comparative biochemistry of the cholinergic neuron. Cold Spring Harbor Symp Quant Biol 40:65-81

Fox PM, Larsen JR 1972 Glutamic acid decarboxylase and the GABA shunt in the supra-oesophageal ganglion of the honey-bee, *Apis mellifera*. J Insect Physiol 18:439-457

Heuser JE, Reese TS 1973 Evidence for recycling of synaptic vesicle membrane during transmitter release at the frog neuromuscular junction. J Cell Biol 57:315-344

Hodgkin AL, Huxley AF 1952 A qualitative description of membrane current and its application to conduction and excitation in nerve. J Physiol (Lond) 117:500-512

Kasai M, Changeux JP 1971 *In vitro* excitation of purified membrane fragments by cholinergic agonists. J Membr Biol 6:1-80

Katz B 1971 Quantal mechanism of neural transmitter release. Science (Wash DC) 173:123-126

Meiners BM, Kehoe P, Shaner DM, Olsen RW 1979 γ-Aminobutyric acid receptor binding and uptake in membrane fractions of crayfish muscle. J Neurochem 32:979-990

Olsen RW, Ticku MK, Miller T 1978 Dihydropicrotoxinin binding to crayfish muscle sites possibly related to γ-aminobutyric acid receptor-ionophores. Mol Pharmacol 14:381-390

Roberts E, Chase TN, Tower DB (eds) 1976 GABA in nervous system function. Raven Press, New York

Strang RHC 1979 In praise of insects. Trends Neurosci 1:172-174

Telford JN, Matsumura F 1970 Dieldrin binding in subcellular nerve components of cockroaches. An electron microscopic and autoradiographic study. J Econ Entomol 63:795-800

Ticku MK, Olsen RW 1977 γ-Aminobutyric acid-stimulated chloride permeability in crayfish muscle. Biochim Biophys Acta 464:519-529

Properties of synaptosomes from the central nervous system of insects

HEINZ BREER

Zoophysiologie/Zellphysiologie, University of Osnabrück, D-4500 Osnabrück, FRG

Abstract Nerve terminals can be isolated from the cerebral ganglia of insects (*Locusta migratoria*) by means of a microscale flotation technique. Electron microscopic analysis revealed morphologically well-preserved synaptosomal profiles containing numerous synaptic vesicles, mainly spherical and electron-lucent. The synaptosomes were also characterized by their high energy reserves (glycogen). Gangliosides, typical constituents of synaptosomal membranes of vertebrates, were totally lacking in locust synaptosomes, which however contained large amounts of galactolipids. Nevertheless, locust synaptosomes had calcium-binding sites, as shown fluorometrically and electron microscopically with terbium ions. Cholinergic activities were concentrated in the synaptosomal fraction. The cholinergic nature as well as the functional integrity of isolated nerve endings were also confirmed by the detection of a high affinity choline uptake system. The accumulated choline was mostly acetylated and could be released by depolarizing conditions.

In contrast to the enormous progress that has been made in exploring the neurophysiological correlates of insect behaviour, our knowledge of insect neurochemistry and especially of the molecular details of chemical transmission in insects is still very limited. Neurochemical research on synaptic events in insects, however, may give us some valuable new insights into general neurobiological principles as well as into neuropharmacological features of arthropods that may help us to understand the mode of action of certain insecticides.

The investigation of synaptic transmission in vertebrates, in particular of presynaptic processes at the molecular level, has been greatly aided by the discovery that when vertebrate brain tissue is homogenized, presynaptic nerve terminals are pinched off from the axon and form sealed structures (synaptosomes) (Gray & Whittaker 1962, Whittaker 1969). Vertebrate synaptosomes, which retain most of the morphology and chemical composi-

1982 Neuropharmacology of insects. Pitman, London (Ciba Foundation symposium 88) p 158-175

tion of the original nerve endings, are an excellent *in vitro* preparation for studying many aspects of synaptic function as well as an enriched source of various synaptosomal components (e.g. synaptic vesicles, plasma membranes, intraterminal mitochondria). Unfortunately, all attempts to isolate synaptosomes from insect nervous tissue with standard techniques have failed. However, the use of density gradients of Ficoll, which forms density solutions of low osmotic pressure, has produced subcellular fractions enriched in synaptosome-like structures in step gradients (Donnellan et al 1976, Breer & Jeserich 1980) and in flotation gradients (Breer & Jeserich 1980).

Preparation of synaptosomes

Ganglia from *Locusta migratoria* were prepared and homogenized as previously described (Breer 1981b). Briefly, head and thoracic ganglia were excised and chopped with a tissue chopper to disintegrate the peripheral sheath of the connective tissue; the tissue was then homogenized in a Tris-buffered iso-osmolar sucrose solution. A crude mitochondrial fraction (P_2) was isolated by differential centrifugation and resuspended in an iso-osmolar buffered Ficoll solution. The mixture was centrifuged for 40 min at 10 000 ***g*** in a swing-out rotor. A shallow continuous density gradient was formed, causing the particulate material to separate into a pellet and a pellicle fraction.

Comparison of pellet and pellicle fractions

Analysis of the protein distribution in the various subcellular fractions showed that the pellet and pellicle each contained about 10% of the original protein (Breer 1981b). Fig. 1 shows the relative specific activity of marker enzymes in the fractions of the flotation gradient and demonstrates that acetylcholinesterase and Na^+, K^+-ATPase accumulated mainly in the pellicle, whereas succinate dehydrogenase was concentrated in the pellet. After detergent treatment, lactate dehydrogenase was shown to be enriched in the pellicle, but was largely occluded, apparently representing cytoplasm sequestered within the synaptosomes during preparation.

A high proportion of cholesterol is said to be typical of neuronal membranes. The cholesterol:phospholipid ratio for the pellicle material was 1:1.7, that for the pellet being 1:4.3; furthermore, cardiolipin, a mitochondrial phospholipid, constituted about 10% of the pellet phospholipids but was only a minor component in the pellicle (Breer 1981b).

Electron micrographs showed that the pellet fraction consisted mainly of

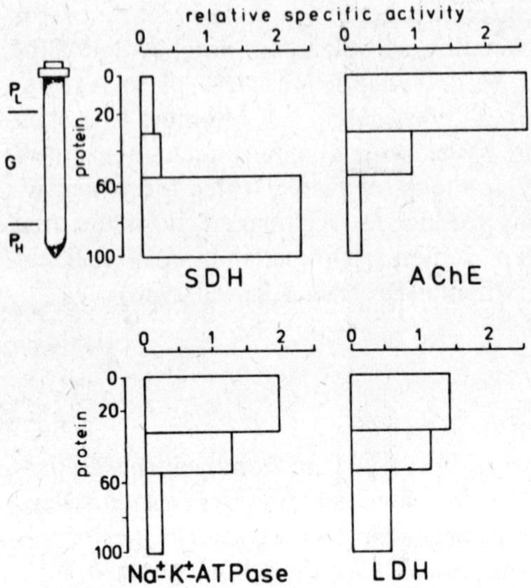

FIG. 1. De Duve plots of the relative specific activities of marker enzymes in the fractions (pellet, P_H; pellicle, P_L; gradient, G) isolated from flotation gradients. The areas of the columns are proportional to the amounts of activity and the heights proportional to the relative enrichment of the activity in the various fractions. SDH, succinate dehydrogenase; AChE, acetylcholinesterase; LDH, lactate dehydrogenase.

essentially intact mitochondria of different sizes and shapes, whereas the pellicle fraction consisted of numerous synaptosomal profiles containing synaptic vesicles and occasional intraterminal mitochondria (Fig. 2). The results of biochemical and ultrastructural analysis thus demonstrate the marked separation of free mitochondria and nerve endings and characterize the pellicle as a typical synaptosomal fraction.

Characterization of synaptosomes

Several of their properties showed that insect synaptosomes are well-sealed structures. Soluble cytoplasmic enzymes such as lactate dehydrogenase and choline acetyltransferase became fully active only when the external membrane was ruptured; and small molecules, such as choline, were taken up and retained by synaptosomes (see below). In cytochemical studies continuous staining of the external membrane surface could be demonstrated (Breer & Jeserich 1981). In experiments with the fluorescent dye 3,3′-dipentyl-2,2′-oxocarbocyanine (generously provided by Dr A. S. Waggoner), which is an accurate indicator of membrane polarization (Waggoner 1976), we used the

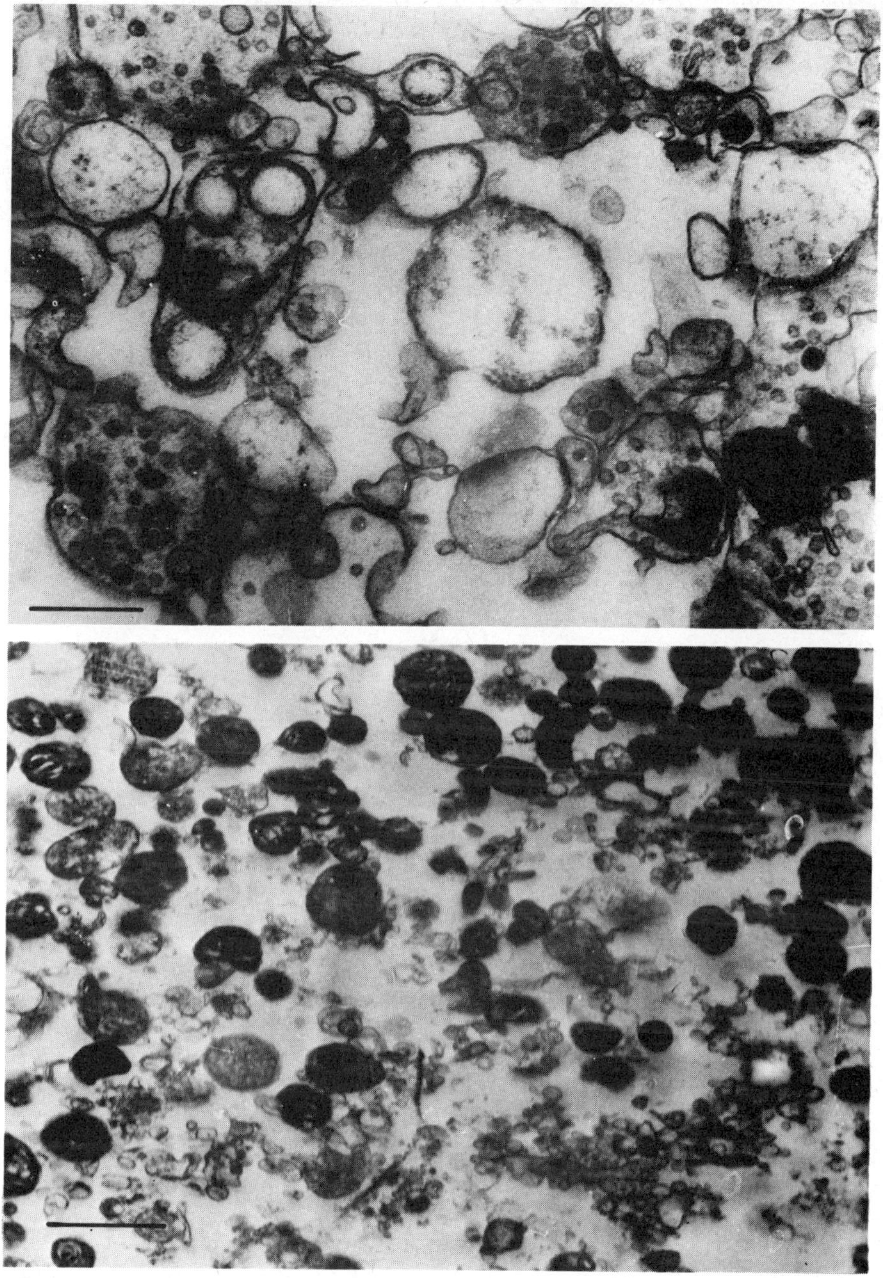

FIG. 2. Representative electron micrographs from pellet and pellicle fractions of the flotation gradient. *Upper:* pellicle (synaptosomal profiles). *Lower:* pellet (structurally intact mitochondria). Bars represent 1 μm.

technique of Blaustein & Goldring (1975) to see whether insect synaptosomes retained a membrane potential. A maximal signal was obtained when synaptosomes were treated with gramicidin D to depolarize the membranes fully. Successive increases in the external potassium concentration, however, induced an increase in fluorescence intensity which was linearly proportional to the log of the K^+ concentration (Fig. 3). The results show that isolated

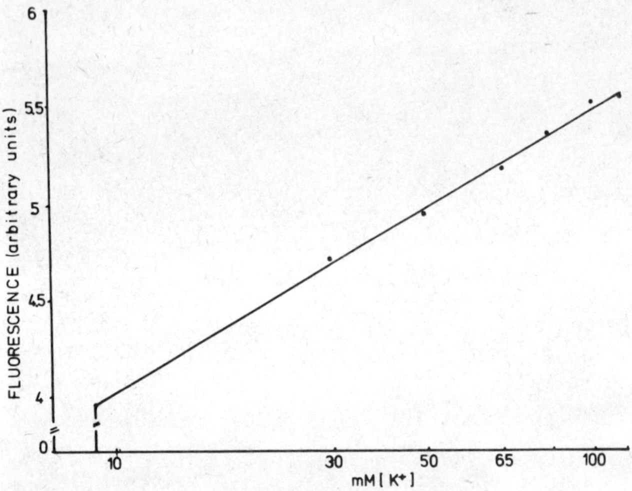

FIG. 3. Changes in cyanine dye fluorescence in suspensions of insect synaptosomes in relation to various potassium concentrations. Dye concentrations: 1 µg/ml. Fluorescence intensity at a wavelength of 510 nm was measured with excitation at 450 nm.

nerve terminals from insects maintain an intact membrane potential that varies with the external K^+ concentration. This is an indication of the viability of insect synaptosomes.

Detailed ultrastructural analysis showed that insect synaptosomes were about 1 µm in diameter and contained spherical vesicles about 50 nm in diameter, mostly of the electron-lucent type (Fig. 4). A few synaptosomes were filled with large electron-dense vesicles, probably neurosecretory granules (Breer & Jeserich 1980).

In contrast to mammalian synaptosomes, nerve terminals from the insect central nervous system (CNS) often contained conspicuous glycogen granules. We found a glycogen content of 1.38 µmol/mg protein, an unusually high energy reserve for synaptosomes.

Locust synaptosomes, suspended in locust Ringer solution, pH 7.4, and incubated at 20 °C in a microvessel of an oxygen electrode, used oxygen in a linear fashion at about 81.6 nmol O_2/(mg protein h) without addition of

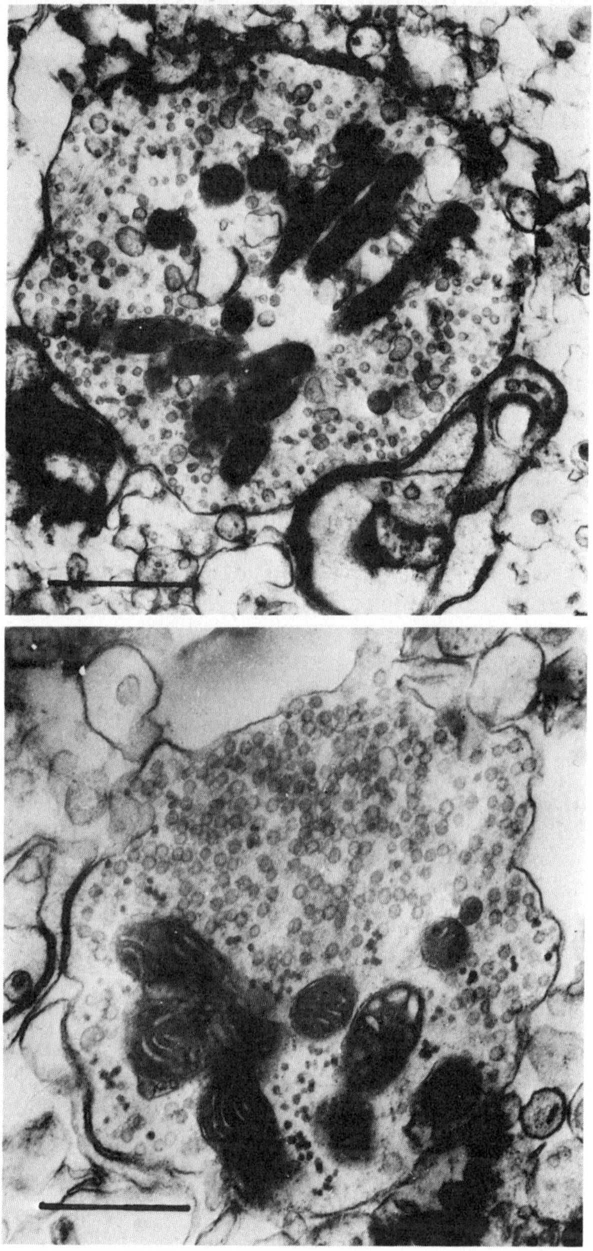

FIG. 4. Various synaptosome profiles from the pellicle fraction. The synaptosomes are well preserved and contain numerous electron-lucent synaptic vesicles and mitochondria. Bars represent 0.5 μm.

substrate to the medium. With freshly prepared synaptosomes no stimulation of oxygen consumption was observed with glucose or pyruvate (10 mM), indicating the availability of carbohydrate reserves in the synaptosomes. Studies of the survival of insect synaptosomes showed that respiration continued at a linear rate for several hours. These experimental results all emphasize the structural and functional integrity of nerve endings isolated from the CNS of an insect.

Synaptosomal membranes

Since the membranes of nerve terminals play a specific part in synaptic transmission, the biochemical properties of insect synaptosomal membranes are of great interest. We found that synaptosomal membranes from insects and vertebrates contained similar proportions of cholesterol and phospholipids but there were basic differences in the glycolipids. Whereas vertebrate synaptosomal membranes contain large concentrations of sialoglycolipids (gangliosides), none of these membrane constituents was found in synaptosomal preparations from locusts. The insect neuronal membranes, however, contained a high proportion of galactolipids, while negligible amounts of these were found in vertebrate synaptosomes (Breer 1981b).

Gangliosides are likely candidates for the calcium binding sites of synaptic membranes (Rahmann et al 1976). We therefore studied the binding properties of locust synaptosomes using the terbium fluorescence technique. Terbium ions bind to insect neuronal membranes with a K_m of 4.2 μM and this high affinity binding was competitively inhibited by calcium ions. As judged electron microscopically, terbium ions produced electron-dense patches in regular arrays on the outer surface of synaptosomal membranes and, occasionally, particularly strong cytochemical staining could be demonstrated at the synaptic contact region. Since terbium monitors calcium-binding sites (Martin & Richardson 1979), the ganglioside-free insect membranes evidently contain specific calcium-binding sites. Inside the synaptic terminals terbium ions induced a marked aggregation of vesicles, often into a single cluster, which possibly reflects a physiological intrasynaptosomal action of calcium ions (Breer & Jeserich 1981).

Cholinergic activities of insect synaptosomes

Neuropharmacological observations and measurements have indicated that acetylcholine (ACh) is probably the major excitatory transmitter at synapses in the insect CNS (Gerschenfeld 1973, Pichon 1974, Breer 1981a). Subcellular fractionation using the flotation technique greatly enriched cholinergic activities (ACh, choline acetyltransferase, acetylcholinesterase, ACh receptors) in the synaptosomal fraction (Breer 1981b, c). The high specific activities of these enzymes indicate that insect synaptosomal preparations contain a high proportion of cholinergic nerve endings.

These results indicate that insect synaptosomes are provided with the enzymic machinery for the synthesis of ACh and with organelles for its storage. Furthermore, they are surrounded by a plasma membrane which is thought to maintain many of the specific properties of presynaptic membranes *in situ*. Isolated nerve endings seem therefore to provide a suitable system in which to study the uptake, storage and release of neurotransmitters as well as to analyse presynaptic feedback mechanisms controlling transmitter synthesis and release.

Choline uptake

Vertebrate nervous tissue contains a high affinity uptake system for choline, unique to cholinergic nerve terminals (Kuhar & Murrin 1978). Synaptosomal preparations from locusts were therefore studied for their ability to transport exogenous choline. The results showed that isolated locust nerve endings accumulate choline in a linear, temperature-sensitive process. Analysis of the transport kinetics revealed two linear components, a high affinity system ($K_T \approx 1.0$ μM) and a low affinity system ($K_T \approx 25$ μM). The maximal rate of uptake (V_{max}) of about 100 pmol/(min mg protein) for the high affinity system, which is several times higher than in synaptosome preparations from mammalian brain, indicates the high proportion of cholinergic nerve endings in the synaptosomal fraction from the locust. The high affinity transport of choline, which is thought to be a rate-limiting regulatory step in the synthesis of ACh (Jope 1979), is competitively inhibited by choline analogues (e.g. hemicholinium-3). The high affinity uptake process is essentially sodium-dependent, the optimal Na^+ concentration being about 180 mM. Potassium ions at higher concentrations inhibited high affinity choline transport and the uptake rate was shown to be reciprocally proportional to log [K^+], suggesting that the reduced uptake is caused by a decrease in the synaptosomal membrane potential (see above) (Breer 1982). This view was supported by experiments using veratridine, which depolarizes membranes (Minchin 1980); these experiments gave similar results.

Chromatographic separation of extracts from synaptosome preparations incubated for 5 min with low concentrations of tritiated choline revealed that 60–70% of the accumulated choline was converted to ACh during that period.

Release of acetylcholine

The release of transmitter substances from presynaptic nerve endings is a fundamental step in chemical transmission which is physiologically triggered by the nerve impulse and may be influenced by changes in the ionic environment as well as by endogenous modulators. Since the complex neuronal organization of insect ganglia makes it impractical to use *in vivo* preparations to investigate release mechanisms at the subcellular and molecular level, isolated nerve terminals should be more appropriate for studying these presynaptic events.

When insect synaptosomes were incubated for 10 min with tritiated choline the accumulated label was released by adding a high concentration of potassium ions, which depolarize nerve endings (70 mM-K^+ in locust Ringer solution). The released labelled material was mostly ACh, as demonstrated by thin layer chromatography followed by radioactivity scanning. Maximal release was achieved at about 100 mM-K^+. The specificity of this potassium-induced release was supported by studies in which veratridine was used as a depolarizing agent (Fig. 5). The reduced release induced by veratridine in a sodium-free medium (not illustrated) may indicate that veratridine affected choline release through a sodium-dependent depolarization.

Similar experiments indicated that the evoked ACh release was calcium-dependent. To further specify this calcium-dependent release we incubated synaptosomes with tritiated choline and rapidly perfused this preparation with a calcium-free medium containing depolarizing levels of potassium. Samples of the perfusate were collected at various time intervals. The efflux rate fell, levelling off at a very low rate; the addition of 1 mM-Ca^{2+} had only minor effects but adding 5 mM-Ca^{2+} to the perfusion medium increased the efflux rate markedly. The highest rate of release was attained immediately after calcium was added, after which there was a lower rate of sustained release. When calcium was replaced by Mg^{2+}, no ACh release was induced.

To study the effect of calcium on the release process independently of membrane depolarization we used the calcium ionophore A23187, which specifically facilitates the flux of Ca^{2+} across cell membranes (Pressman 1976). Synaptosomes preloaded with tritiated choline were incubated with A23187 and perfused with a calcium-free solution. The addition of Ca^{2+} to the perfusion medium dramatically increased the rate of efflux of ACh. Thus the combined effects of calcium and depolarization or of calcium and its ionophore are required to induce ACh release from insect synaptosomes. The agents used to stimulate release (potassium, veratridine, A23187) act by completely different mechanisms but lead to a similar action: an increased permeability of synaptosomal membranes to calcium ions.

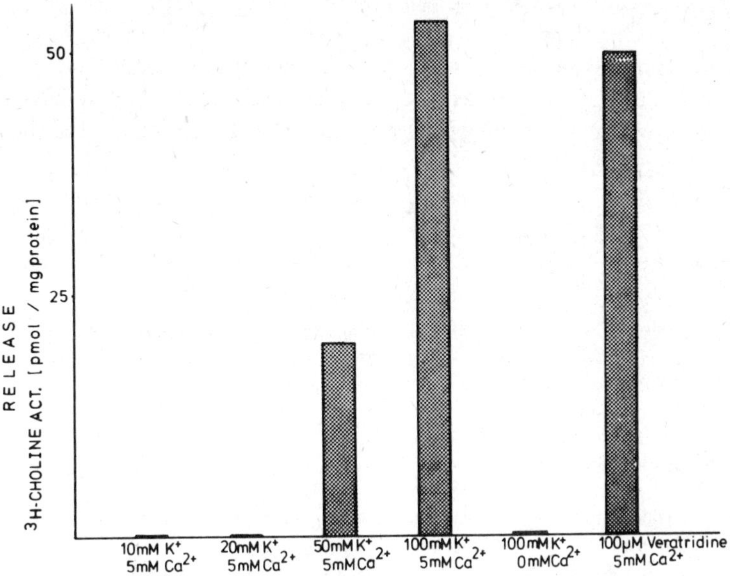

FIG. 5. Potassium- and veratridine-induced release of acetylcholine from insect synaptosomes incubated with [^{3}H]choline for 10 min.

There is increasing evidence that cholinergic synapses in vertebrates have presynaptic muscarinic receptors which regulate the extent of ACh release via a negative feedback mechanism (Szerb 1977). Our preliminary experiments with muscarinic agonists (oxotremorine) and antagonists (atropine) indicate that muscarinic agents can modulate ACh release from insect synaptosomes: whereas atropine enhanced the potassium-evoked release of ACh, the rate of release was reduced by oxotremorine. These results suggest that in insects, muscarinic receptors may be located on central cholinergic terminals that have a physiological role in the autoregulation of ACh release.

Conclusion

We have shown that viable synaptosomes can be isolated from the central nervous tissue of insects. These preparations of pinched-off presynaptic nerve terminals, when partially purified by differential and density gradient centrifugation, provide a useful *in vitro* system for biochemical and pharmacological investigations of presynaptic processes in insects, in particular of central cholinergic synapses.

Acknowledgements

I should like to thank Professor Dr Lueken for support and Dr Jeserich for the electron micrographs. The expert technical assistance of Mrs Düvel is gratefully acknowledged. This work was supported financially by the Deutsche Forschungsgemeinschaft.

REFERENCES

Blaustein MP, Goldring JM 1975 Membrane potentials in pinched-off presynaptic nerve terminals monitored with a fluorescent probe: evidence that synaptosomes have potassium diffusion potentials. J Physiol (Lond) 247:589-615

Breer H 1981a Comparative studies on cholinergic activities in the central nervous system of *Locusta migratoria*. J Comp Physiol 141:271-275

Breer H 1981b Characterization of synaptosomes from the central nervous system of insects. Neurochem Int 3:155-163

Breer H 1981c Properties of putative nicotinic and muscarinic cholinergic receptors in the central nervous system of *Locusta migratoria*. Neurochem Int 3:43-52

Breer H 1982 Uptake of [N-Me-^{3}H]-choline by synaptosomes from the central nervous system of *Locusta migratoria*. J Neurobiol 13:in press

Breer H, Jeserich G 1980 A microscale floatation technique for the isolation of synaptosomes from nervous tissue of *Locusta migratoria*. Insect Biochem 10:457-463

Breer H, Jeserich G 1981 Calcium binding sites of synaptosomes from insect nervous system as probed by trivalent terbium. J Neurochem 37:276-286

Donnellan JF, Alexander K, Chendlik R 1976 The isolation of cholinergic terminals from fleshfly heads. Insect Biochem 6:419-423

Gerschenfeld HM 1973 Chemical transmission in invertebrate central nervous systems and neuromuscular junctions. Physiol Rev 53:1-119

Gray EG, Whittaker VP 1962 The isolation of nerve endings from brain: an electronmicroscopic study of cell fragments derived by homogenisation and centrifugation. J Anat 96:79-88

Jope RS 1979 High affinity choline transport and acetyl CoA production in brain and their roles in the regulation of acetylcholine synthesis. Brain Res Rev 1:313-344

Kuhar MJ, Murrin LC 1978 Sodium-dependent, high affinity choline uptake. J Neurochem 30:15-21

Martin RB, Richardson FS 1979 Lanthanides as probes for calcium in biological systems. Q Rev Biophys 12:181-209

Minchin MCW 1980 *Veratrum* alkaloids as transmitter-releasing agents. J Neurosci Methods 2:111-121

Pichon Y 1974 The pharmacology of the insect nervous system. In: Rockstein M (ed) The physiology of Insecta. Academic Press, New York, vol 4:162-179

Pressman BC 1976 Biological applications of ionophores. Annu Rev Biochem 45:501-530

Rahmann H, Rösner H, Breer H 1976 A functional model of sialo-glyco-macromolecules in synaptic transmission and memory formation. J Theor Biol 57:231-237

Strang RH 1979 In praise of insects. Trends Neurosci 1:172-174

Szerb JC 1977 Characterization of presynaptic muscarinic receptors in central cholinergic neurons. In: Jenden DJ (ed) Cholinergic mechanisms and psychopharmacology. Plenum Press, New York, p 49-60

Waggoner AS 1976 Optical probes of membrane potential. J Membr Biol 27:317-334
Whittaker VP 1969 The subcellular fractionation of nervous tissue. In: Bourne GH (ed) The structure and function of nervous tissue. Academic Press, New York, p 1-24

DISCUSSION

Olsen: You mentioned that the preparation of insect synaptosomes was improved by using 'gentle' homogenization. What is this exactly?

Breer: The homogenization was done in a glass microhomogenizer with a Teflon pestle (12 strokes by hand, 4 °C) (Breer 1981). When standard techniques for vertebrate tissue, which use a motor-driven pestle, were applied, most of the nerve endings from insect tissue were destroyed.

Mordue: We found the same when isolating neurosecretory granules from *Locusta* and *Schistocerca*. We too now use a very gentle homogenization by hand; a motor-driven pestle destroys the granules completely.

Breer: The mechanical and osmotic aspects of the procedure are very important. Insect synaptosomes seem to be very susceptible to osmotic and mechanical perturbations.

Narahashi: Have you tested calcium-channel inhibitors such as cobalt, nickel or other divalent cations in your synaptosomal system? You used the calcium ionophore, which increased the release of transmitter. Was the release inhibited by calcium-channel inhibitors?

Breer: We have not done such experiments yet. We have just looked for the ability of Ca^{2+} and ionophore to induce transmitter release. We now have to do some more pharmacological work to characterize this effect.

Lunt: When you preload locust synaptosomes with labelled choline, what proportion of the choline that has been taken up is released with high potassium concentrations?

Breer: Between 10 and 20%, but the proportion depends on various parameters, such as temperature and time of incubation.

Lunt: Does it 'saturate out' at that value, or can you get further choline release?

Breer: The release depends on the concentration of potassium to some extent, but in normal conditions (70 mM-K^+) about 20% was released, which was almost the maximum that could be released.

Lunt: Do oxotremorine or any of the other muscarinic agents that you tried affect the *spontaneous* release of ACh?

Breer: We have not studied that. So far we have just done these preliminary experiments to see whether there are effects of muscarinic agonists or antagonists on the potassium-evoked release.

Hall: You showed that veratridine inhibited choline uptake. That suggests that you have the voltage-sensitive sodium channel in the synaptosomal material that you purify. Have you measured sodium fluxes, and seen whether sodium flux can be blocked with tetrodotoxin in this preparation?

Breer: We haven't measured ion fluxes in insect synaptosomes, but since veratridine in micromolar concentrations inhibited choline uptake and induced transmitter release, the existence of sodium channels can be assumed.

M. E. Eldefrawi: I was interested in the high proportion of galactolipids in your synaptosomal preparation from the locust, contrary to what is seen in the mammal, as you pointed out. In our work with the muscarinic cholinergic receptor of mammalian brain we found that the best detergent with which to solubilize these receptors from the membrane is the glycoside, digitonin (Aronstam et al 1978). The question was raised earlier whether the housefly head extracts contain an 'endogenous' detergent. The galactolipids, to my mind, would be an interesting detergent-like group of compounds, since they are so abundant in the nerve membranes of insects. Perhaps it is not so far-fetched to think that there is a detergent-like material that is responsible for solubilizing bungarotoxin-binding sites and other material as well.

Berridge: The absence of gangliosides is particularly interesting. I have tried to activate adenylate cyclase in the salivary gland of *Calliphora* using cholera toxin but without success, even though this is a standard procedure with vertebrate tissue. It could be that insects lack gangliosides in their membranes and thus fail to bind the toxin.

Hildebrand: I have been told that attempts to demonstrate gangliosides in insects have been unsuccessful. Perhaps insects simply lack this important class of complex lipids.

Berridge: I understand that it is easy to insert gangliosides into cell membranes. It might be possible to insert GM_1 into insect membranes, thus making them susceptible to cholera toxin.

Hildebrand: If we are right to assume that gangliosides have important physiological functions in the cell membranes in which they are concentrated in vertebrates, and if these lipids really are completely missing in insects, what is taking their place? What is the chemical structure of the galactosphingolipids you detect, Dr Breer? It seems unlikely that the galactosphingolipids in your synaptosomal preparations would replace gangliosides if their molecular structure is much different. The galactolipids apparently lack the charged groups of sialic acid residues, for example.

Breer: We don't know the galactolipid composition of insect synaptosomal membranes, yet. At the moment we can say they are glycosphingolipids, but that does not mean they are identical to cerebrosides and sulphatides—important galactolipids in vertebrate nervous tissue. In fact, there are some indications that these glycolipids are absent from or are only minor components of insect nervous tissue.

Evans: Have you examined the effects of any other putative transmitters, such as GABA or octopamine, on your locust synaptosomes?

Breer: Not yet. We have concentrated on cholinergic receptors and activities so far.

Donnellan: We (Kilpatrick et al 1980) obtained a crude P_2 fraction from locust ganglia and found an octopamine-stimulated adenylate cyclase activity, but we never characterized the morphology of the P_2 fraction.

Breer: Dr E. Zlotkin recently told me that he found a GABA uptake system in insect synaptosomes, using our isolation procedure.

Usherwood: Did he demonstrate GABA release from his synaptosomal preparations?

Breer: I have no information about GABA release.

Usherwood: The fact that ACh is released from the synaptosomal preparations doesn't necessarily mean that they contain cholinergic endings. Have you evidence for the release of any other putative transmitters from your synaptosome preparations?

Breer: Our statement that insect synaptosome preparations contain many cholinergic nerve endings is mainly based on the very high activity of cholinergic parameters in this fraction and the comparatively high V_{max} for the high affinity choline uptake.

Evans: You presumably obtain a mixed population of synaptosomes with your method, some of which are glutamatergic, some cholinergic, and some may even be octopaminergic!

Breer: Yes, that is what I assume.

Narahashi: How homogeneous is a synaptosome preparation? Are the synaptosomes 'inside in' or 'outside in'?

Breer: It seems probable that synaptosome preparations from the CNS of insects, as from other brain tissue, contain various types of nerve endings, depending on the types of terminals existing in this particular tissue. So far there is no appropriate method for separating nerve endings containing different neurotransmitters.

Concerning the membrane polarity, it can be assumed that at least morphologically intact nerve endings are 'right side out': the membrane is not inverted.

Barnard: I have a question related to this topic, on the P_3 fraction—the material remaining in the subcellular fractionation after the usual P_2 fraction is taken, and sedimentable at very high centrifugal force. This fraction is sometimes neglected. Can one consider whether there is perhaps a different population of receptors in the P_3 fraction from the main synaptosome fraction (the P_2 fraction)? In the fractionation of vertebrate brain homogenates one can find electron microscopic evidence for parts of the 'postsynaptic density' structure being sheared off and some of this going elsewhere. In addition, there are a large number of dendritic spines in many regions and these may break off during the homogenization to yield very small membrane fragments which are collected in the P_3 fraction. Thus, when we re-examined the distribution of enkephalin receptors in mammalian brain, a considerable amount was actually found in the P_3 fraction (Glasel et al 1980). That might be attributable to enkephalin receptors which were located on the dendritic spines. Does this division of receptors between the P_2 and P_3 fractions occur in insects? Could one miss a population of receptors which are very small membrane fragments?

Olsen: The method of fractionation used in vertebrates has not been used on insects and there is no P_3 fraction as such in Dr Breer's preparation. It is possible that the binding sites for neurotransmitters are in different types of membranes in brain, but it hasn't been proved. Soon, with irreversible labels and good morphological analysis of the fractions, we shall perhaps answer that question. The experts say that maybe there are small synaptosomes in the P_3 fraction, because when they try to subfractionate the two fractions (P_2 and P_3), the membranes with most receptor binding activity co-migrate (DeBlas & Mahler 1978), but we don't have conclusive evidence.

Donnellan: Thinking back to David Sattelle's cell body receptors, I wonder what happens to those during fractionation, and also in the general receptor studies when we smash up everything to try to get at the α-bungarotoxin-binding sites.

Breer: In our subcellular fractionation procedure a lot of perikarya were found in the P_1 fraction which show considerable binding for α-bungarotoxin as well as for quinuclidinyl benzilate.

Hall: The way that Dr Sattelle prepares his membranes, the membranes from the cell bodies would be in the same membrane fraction as he used for his binding studies.

Barnard: Are there different types of ACh receptor—i.e. separate muscarinic and nicotinic receptors—in the same synaptosome preparation? And is there evidence there for a receptor of mixed (muscarinic/nicotinic) properties?

Breer: We found both nicotinic and muscarinic receptors in the synaptosome fraction, but we have not looked for a receptor type with mixed muscarinic and nicotinic properties.

Barnard: Could I ask generally, then, what is the status in the insect CNS of the receptor with mixed muscarinic and nicotinic properties which has been proposed in certain cases there?

M. E. Eldefrawi: The dorsal median unpaired (DUM) neurons of the grasshopper were found to be sensitive to ACh and this sensitivity was blocked with *d*-tubocurarine and decamethonium but not α-BGTX (Goodman & Spitzer 1980). Whether that receptor is similar to what has been referred to here as the mixed nicotinic/muscarinic 'soluble' receptor protein or not remains a problem. As we once wrote (Mansour et al 1977), the only point we see against it being a receptor molecule is that it is soluble! One expects a receptor protein to be membrane-bound.

Dudai: The mixed receptor was reported to exist in extremely high concentrations in *Musca domestica*.

M. E. Eldefrawi: Yes, there are very high concentrations of these soluble putative ACh receptors in housefly heads but they are only about twofold higher in concentration than the *Torpedo* ACh receptor.

Dudai: We are talking of higher concentrations than that.

Donnellan: We were keen to go back to the 'mixed' receptor but, unfortunately, labelled decamethonium (the probe we used for the 'mixed' receptor) is no longer available.

Dudai: Decamethonium has specific pharmacological effects on *Drosophila*, incidentally. It seems to be unique in its actions, quite different from oxotremorine, mecamylamine or nicotine (Dudai et al 1980). It is the only cholinergic drug tested by us that caused catatonia. The flies stand on their legs for a day without moving. This might be related to your idea, Professor Eldefrawi, that one could drive insects crazy by using the right drugs!

M. E. Eldefrawi: Decamethonium stands out among all the cholinergic depolarizing drugs as a powerful channel blocker. Using electrophysiological methods Adams & Sakmann (1978) showed it to be a channel blocker in muscle end-plates and using biochemical methods we found it to be so in *Torpedo* electric organ (Aronstam et al 1981).

Narahashi: May I ask Professor Barnard about the differential sensitivity to α-BGTX of various receptor preparations from different species? Is there any example in which α-BGTX binds an ACh receptor but does not block its response to agonist?

Barnard: There are many such instances. A good example is that studied by Patrick & Stallcup (1977a, b) with rat sympathetic neurons in culture. A nicotinic-like binding component on these cells was demonstrated pharmacologically, which binds α-BGTX to saturation, with high affinity. Patrick & Stallcup monitored the ACh-induced sodium flux into these cells, which also behaves pharmacologically like a nicotinic receptor, although of somewhat different specificity to the former. When all of the toxin-binding sites are occupied by the toxin, this sodium flux response is totally unaffected. Yet an antibody to the *Torpedo* ACh receptor cross-reacts with this functional nicotinic receptor and blocks the agonist-induced sodium flux; however, it does *not* precipitate the α-BGTX-binding component in an extract of these cells. In other words, there is a functional receptor in such a case which has some structural relationship to known true nicotinic receptors, and yet the toxin-binding site which is present on those cells is not concerned in the function. Another well-documented case of the presence of a specific binding site for α-BGTX and the lack of its effect on agonist action on the nicotinic receptor also present was shown by Carbonetto et al (1978) on chick sympathetic neurons in culture (for other relevant references to such discrepancies, see Barnard, this volume).

There are several possible explanations for this phenomenon. One is that most of the observed α-BGTX-binding sites might be from presynaptic and extrasynaptic receptors, so that their blockade is not readily monitored electrophysiologically *in situ*. Another possibility, that Wang et al (1978) have put forward to explain the dichotomy, derives from the rapid dissociation of α-BGTX from its CNS and ganglionic binding sites. The toxin dissociates two to three orders of magnitude faster there than from the muscle or electric organ receptors, as I have detailed in my paper. Schmidt and co-workers (Wang et al 1978) suggest that ACh is able to bind to part of the α-BGTX-binding site, forming a ternary complex, hence activation is seen in the electrophysiological measurements despite the presence of α-BGTX because the toxin is already in the process of coming off. If these or related proposals are not the correct interpretation, we are left with a mystery: there are, indeed, a large number of such sites in the vertebrate brain and autonomic ganglia with the capacity for binding α-BGTX specifically, yet apparently not on a recognizable, functional ACh receptor.

Lunt: This emphasizes the dangers of considering an α-BGTX-binding site as a nicotinic receptor site. We have been looking at synaptosome preparations from rat striatum (Futerman et al 1981). In this system there is a cholinergic presynaptic receptor on the nerve terminals that modulates the release of dopamine. There is an ACh-induced dopamine release, in other words, which can be demonstrated either in striatal slices or in striatal synaptosomes. We found it impossible to block this ACh-induced release with α-BGTX, but the toxin binds specifically to the striatal synaptosomes, with high affinity. We can block the α-BGTX binding with tubocurarine but, as I say, we cannot block the ACh-induced release of dopamine. What is even more confusing, α-BGTX has itself a very powerful agonist-like effect on these dopaminergic nerve terminals, giving a marked release of dopamine. It appears that α-BGTX is *not* binding to the same site as ACh; it is not competing with ACh, but has an agonist-like effect (Futerman et al 1981).

REFERENCES

Adams PR, Sakmann B 1978 Decamethonium both opens and blocks endplate channels. Proc Natl Acad Sci USA 75:2994-2998

Aronstam RS, Schuessler DC Jr, Eldefrawi ME 1978 Solubilization of muscarinic acetylcholine receptors of bovine brain. Life Sci 23:1377-1382

Aronstam RS, Eldefrawi AT, Pessah IN, Daly JW, Albuquerque EX, Eldefrawi ME 1981 Regulation of [^{3}H]perhydrohistrionicotoxin binding to *Torpedo ocellata* electroplax by effectors of the acetylcholine receptor. J Biol Chem 256:2843-2850

Barnard EA 1982 Isolation of receptors from the central nervous system. This volume, p 109-117

Breer H 1981 Characterization of synaptosomes from the central nervous sytem of insects. Neurochem Int 3:155-163

Carbonetto ST, Fambrough DM, Muller KJ 1978 Non-equivalence of α-bungarotoxin receptors and acetylcholine receptors in chick sympathetic neurons. Prox Natl Acad Sci USA 75:1016-1020

DeBlas AL, Mahler HR 1978 Studies on nicotinic acetylcholine receptors in mammalian brain. Characterization of a microsomal subfraction enriched in receptor function for different transmitters. J Neurochem 30:563-577

Dudai Y, Nahum-Zvi S, Haim-Granot N 1980 Cholinergic pharmacology of *Drosophila*: comparison of *in vivo* to *in vitro* studies. Comp Biochem Physiol C Comp Pharmacol 65:135-138

Futerman AH, Harrison R, Lunt GC, Wonnacott S 1981 Inhibition by anti-(nicotinic acetylcholine receptor) antibodies of acetylcholine-induced dopamine release from striatal nerve terminals. Biochem Soc Trans 9:410-411

Glasel J, Venn RF, Barnard EA 1980 Distribution of stereospecific opiate binding activity between subcellular fractions from ovine corpus striatum. Biochem Biophys Res Commun 95:263-268

Goodman CS, Spitzer NC 1980 Embryonic development of neurotransmitter receptors in grasshoppers. In: Sattelle DB et al (eds) Receptors for neurotransmitters, hormones and pheromones in insects. Elsevier/North-Holland Biomedical Press, p 195-207

Kilpatrick AT, Vaughan PFT, Donnellan JF 1980 Monoamine sensitive adenylate cyclase in *Schistocerca gregaria* nervous tissue. In: Insect neurobiology and pesticide action. Soc Chem Ind (Lond) Monogr, p 341-345

Mansour NA, Eldefrawi ME, Eldefrawi AT 1977 Isolation of putative acetylcholine receptor proteins from housefly brain. Biochemistry 16:4126-4132

Patrick J, Stallcup WB 1977a Immunological distinction between acetylcholine receptors and the α-bungarotoxin-binding component on sympathetic neurons. Proc Natl Acad Sci USA 74:4689-4692

Patrick J, Stallcup WB 1977b α-Bungarotoxin binding and cholinergic receptor function on a rat sympathetic nerve line. J Biol Chem 252:8629-8633

Wang GK, Molinaro S, Schmidt J 1978 Ligand responses of α-bungarotoxin binding sites from skeletal muscle and optic lobe of the chick. J Biol Chem 253:8507-8572

Binding and uptake of glutamate and γ-aminobutyric acid in membrane fractions from locust muscle

PATRICIA A. BRILEY, MARIE T. FILBIN, GEORGE G. LUNT and JOHN F. DONNELLAN*

Department of Biochemistry, University of Bath, Bath, BA2 7AY and **Shell Research Ltd, Sittingbourne Research Centre, Sittingbourne, Kent, UK*

Abstract Membranous subfractions have been prepared from locust muscle by differential centrifugation. One fraction has both glutamate and γ-aminobutyric acid (GABA) associated with it. The fraction is enriched in a number of enzymes that are considered to be associated with plasma membranes. The fraction shows specific, high affinity uptake of L-glutamate and D- and L-aspartate. Glutamate and aspartate appear to be transported by a common mechanism. The fraction also shows specific high affinity binding of glutamate and of GABA. The number of GABA-binding sites is considerably lower than the number of glutamate-binding sites. The findings lend biochemical support to the hitherto predominantly electrophysiological evidence that glutamate and GABA are neurotransmitters at the insect neuromuscular junction.

The nature of the neurotransmitters at the insect neuromuscular junction has been the subject of a considerable amount of research over the past 15 years. A wealth of evidence has been accumulated that supports the identification of L-glutamate as the excitatory transmitter and γ-aminobutyric acid (GABA) as its inhibitory counterpart (Gerschenfeld 1973, Usherwood 1975). The bulk of the evidence, however, comes from electrophysiological studies and there is very little information available on the biochemical characteristics of these transmitter systems. Thus we do not know whether the amino acids are localized in the synaptic vesicles within the nerve terminal or whether the mechanisms of transmitter storage and release are similar to those in the better-characterized vertebrate systems. There are a few reports of receptor binding studies, and there is some evidence that enzymic

1982 Neuropharmacology of insects. Pitman, London (Ciba Foundation symposium 88) p 176-198

destruction of the transmitters has no role in the termination of transmission and that transmitter is removed by uptake systems (Faeder & Salpeter 1970).

Our meagre knowledge of the neurochemistry of the insect neuromuscular systems is in part the result of lack of well-characterized tissue preparations in which to study the cellular and subcellular distribution of the various components of the neurotransmitter systems. This contrasts markedly with the vertebrate nervous system where the characterization of neurotransmitter systems has been greatly aided by the ready availability of preparations of intact nerve terminals. These terminals, or synaptosomes, contain synaptic vesicles and have the receptor-containing postsynaptic membrane still attached. They provide a convenient model system in which to study the many biochemical events in synaptic transmission (Jones 1975). Unfortunately, no such preparations from the arthropod neuromuscular junction have been available until very recently, but Breer & Jeserich (1980) and Breer (this volume) have now described synaptosomal preparations from the insect central nervous system. Donnellan et al (1974) devised an extensive subcellular fractionation scheme for the flight muscles of the fleshfly *Sarcophaga barbata* but were unable to obtain any fraction that was substantially enriched in nerve terminals. We have re-examined the possibility of isolating nerve terminals from insect muscle. Although we have not achieved this goal, the muscle subfractions are proving useful in characterizing the neurotransmitter systems at the neuromuscular junction.

Muscle fractionation

All our studies have been made using the total intrathoracic musculature from two-week-old adult locusts (*Schistocerca gregaria*). In preliminary experiments the muscle was homogenized in 0.25 M-sucrose in Tris-HCl (5.0 mM) containing EGTA (1.0 mM) (pH 7). Examination of such homogenates in the electron microscope revealed that extensive disruption of the tissue had taken place and that only very few intact nerve terminals were present. Centrifugation of these homogenates at 100 000 × ***g*** for 1 h gave a membranous pellet but we could detect no glutamate in such pellets. The homogenization medium was modified and was based on locust saline, a medium that is generally used for electrophysiological preparations of locust muscle. This system gave tissue homogenates that contained a high proportion of membrane-limited structures, and the homogenates were accordingly subfractionated. The procedure that we finally adopted is shown in Fig. 1.

Examination in the electron microscope of the fractions derived from this scheme revealed that fraction P_1 was composed mainly of large fragments of muscle fibres and connectives and was not obviously enriched in any

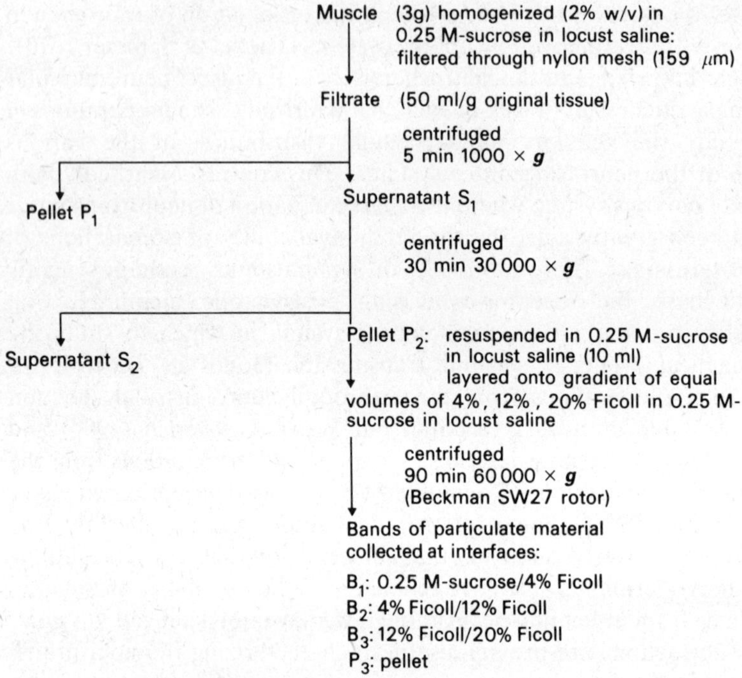

FIG. 1. Fractionation scheme for locust muscle.

particular organelle. P_2 was a mixture of mitochondria and a heterogeneous collection of vesiculated membrane fragments. Subfractionation of P_2 on the density gradient produced two major subfractions: P_3, which consisted almost totally of mitochondria, and B_3, which was made up almost entirely of large membrane-bound bodies (1–3 μm diameter) containing numerous small (0.05–0.08 μm) vesicles. Some mitochondria were also seen in most preparations of B_3.

An analysis of the distribution of glutamate and of GABA throughout the fractionation scheme showed that the bulk of the amino acids was recovered in the supernatant S_1. However, a small but significant amount of both amino acids was seen in B_3 (Table 1).

The possibility that the glutamate seen in B_3 was the result of non-specific trapping of soluble material by the membrane vesiculation presumed to occur during homogenization was discounted on the basis of control experiments in which radiolabelled glutamate and/or sucrose was added to the homogenization buffer. The radioactivity recovered in each fraction was measured in each fraction and it was found that although B_3 contained no radiolabelled sucrose, radiolabelled glutamate was present. The amount of radiolabelled glutamate

TABLE 1 Glutamate and GABA content of S_1 and B_3 fractions

	Amino acid content (μg/g original tissue)			
	Glutamate		*GABA*	
Fraction	*Free*	*Trapped*[a]	*Free*	*Trapped*
S_1	2277 ± 204		150.9 ± 15.6	—
B_3	—	32.6 ± 15.2	6.8 ± 0.5	

Results given as mean ± SEM of five experiments.
[a] Trapped amino acid is that which can be detected after dissolution of the membrane fraction with 0.1% Triton-X100.

accounted for about 10% of the total glutamate content of B_3 and suggested that perhaps glutamate was taken up into B_3 during the fractionation procedure. This possibility was investigated further and the uptake properties of B_3 are described below.

Amino acid uptake into fraction B_3

The uptake of amino acids into B_3 was measured by a rapid centrifugation assay that we have previously used for measuring glutamate binding (Filbin et al 1980). Uptake of L-glutamate was linear with respect to protein concentration. The uptake was temperature-sensitive and after 30 min at 0 °C only about 8% of the uptake seen at 25 °C was observed. The uptake of a number of amino acids was measured at 25 °C over a wide concentration range and the results are summarized in Table 2.

The uptake of amino acids was found to be markedly affected by sodium.

TABLE 2 Amino acid uptake into fraction B_3 of locust muscle

	High affinity site		*Low affinity site*	
Amino acid	K_m *(μm)*	V_{max} *(nmol min^{-1} mg^{-1} protein)*	K_m *(μm)*	V_{max} *(nmol min^{-1} mg^{-1} protein)*
L-Glutamate	19.4	0.20	664.0	0.95
D-Glutamate	No detectable uptake			
L-Aspartate	30.0	0.10	640.0	0.50
D-Aspartate	35.0	0.10	740.0	1.10
L-Glutamine	70.0	0.09	1200.0	0.80
L-Alanine	75.0	0.56	—	—
L-Leucine	No detectable uptake			

The results of these experiments are summarized in Table 3. As can be seen, the uptake associated with the high affinity site is particularly affected by the absence of sodium, except for L-alanine uptake.

TABLE 3 Sodium dependence of amino acid uptake into fraction B_3 of locust muscle

Amino acid	*Concentration (μm)*	*% of control uptake in absence of Na^+*
L-Glutamate	20	29
	500	54
L-Aspartate	20	20
	500	45
D-Aspartate	20	57
	500	88
L-Glutamine	20	66
	500	71
L-Alanine	20	66
	500	41

Muscle was fractionated in Na^+-free locust saline/sucrose and uptake experiments were done in the absence or presence of 0.15 M-NaCl. Uptake in the presence of Na^+ was taken as 100%.

In studies of this type it is often difficult to distinguish between true uptake into the membrane-bound compartment and binding of the ligand to the membrane surface. To some extent the observed sodium dependence suggests that the amino acids are indeed being taken up into the membranous vesicles. However, we investigated this aspect further in a series of experiments in which B_3 was disrupted. The uptake of both L-glutamate and L-aspartate was reduced by between 50% and 80% either by three cycles of freeze–thaw treatment, or by osmotically shocking the fraction by homogenizing it in 10 volumes of ice-cold distilled water and recovering the disrupted membranes by centrifugation at 100 000 × *g* for 20 min. Electron microscopy revealed that these treatments almost totally disrupted the large membranous profiles described earlier.

The kinetic experiments described above suggested that L-glutamate and L-aspartate showed similar initial rates of uptake. Therefore, we examined the effects of each amino acid on the uptake of the other. These experiments showed that the uptake of 20 mM-L-glutamate is inhibited by 50% by either L- or D-aspartate at a concentration of 20 μM. Increasing the concentration of either aspartate isomer further reduced the L-glutamate uptake. L-Alanine at 20 μm had no effect on the uptake of 20 μM-L-glutamate. D-Glutamate had no effect on L-glutamate uptake even at a concentration of 0.5 M. L-Aspartate inhibited the uptake of D-aspartate and vice versa. The uptake of both isomers was equally affected by L-glutamate.

These results strongly suggest a common uptake system for glutamate and aspartate, the system being stereospecific for L-glutamate but not discriminating between D- and L-aspartate. This is reminiscent of findings in mammalian brain.

Studies of glutamate-mediated synaptic transmission are to a considerable extent hampered by a lack of potent, well-characterized agonists and antagonists. Some agents have however been used in studies in mammalian brain and to some degree it is possible to distinguish between uptake systems and synaptic receptor sites (Roberts 1974, 1981).

Table 4 summarizes the results of a series of preliminary experiments in

TABLE 4 Effects of potential inhibitors on the uptake of glutamate and aspartate by B_3

	% inhibition of uptake	
Inhibitor (100 μm)	*L-Glutamate (20 μm)*	*L-Aspartate (20 μm)*
Glutamate dimethyl ester	0	5
Glutamate diethyl ester	12	5
γ-Hydroxylglutamate	28	12
4-Fluoroglutamate	19	—
2-Amino-4-phosphonobutyric acid	0	0
N-Methyl-D-aspartate	15	14
L-Cysteate	31	25
Cysteine sulphinate	20	23
p-Chloromercuribenzoic acid	29	29
Chlorpromazine	17	2

Inhibitors were added to B_3 samples 15 s before the addition of radiolabelled amino acid. Incubations were for 15 min at 25 °C.

which the effects of a number of agents on the uptake of both glutamate and aspartate have been examined. The results reveal that none of the compounds tested so far is a particularly potent inhibitor of the uptake system. However, the observed effects are in accordance with those seen in mammalian preparations (Roberts 1974, 1981).

Having established that fraction B_3 possesses specific high affinity uptake systems for L-glutamate and for aspartate, we wanted to know whether the accumulated amino acids were subsequently released. The induced release *in vitro* of neurotransmitters from a variety of mammalian brain preparations has proved to be a useful aid to the complete characterization of neurotransmitter systems. Release of neurotransmitters from nerve terminals can be provoked by depolarizing the terminals in the presence of calcium ions. The depolarization is frequently achieved by raising the K^+ concentration in the suspending medium. We adopted this procedure in a preliminary study. Samples of B_3 were incubated for 20 min at 25 °C with radiolabelled

L-glutamate or L-aspartate. The fractions were washed free of exogenous labelled amino acid by either centrifugation or rapid filtration. They were suspended for 5 min in either normal locust saline/sucrose solution or in the same buffer containing 55 mM-KCl. The release of accumulated amino acid was then measured. The results of these experiments were very variable but in every case 55 mM-KCl produced a 10- to 20-fold increase in the amount of either L-glutamate or L-aspartate released. However, this effect seems to be completely independent of calcium and we have been unable to demonstrate any dependence of the K^+-stimulated release on Ca^{2+}. In this important respect the fraction B_3 differs markedly from nerve terminal preparations from mammalian brain.

Origins of the B_3 fraction

Before we can begin to speculate about the possible involvement of the amino acid uptake systems of B_3 in neurotransmission at the locust neuromuscular junction, it is clearly necessary to determine the cellular and subcellular origins of the fraction. For subcellular fractions from tissues of the higher vertebrates there is a wide variety of so-called marker enzymes that permit the fractions to be identified. Marker enzymes are those enzymes whose subcellular distribution has been established, initially by histochemical procedures, and that are generally accepted to be characteristic of a particular organelle (Harrison & Lunt 1980). There are very few well-established marker enzymes for insect muscle and to a large extent we are forced to extrapolate from studies on vertebrate tissues. Donnellan et al (1974) in their study of fleshfly muscle looked at the distribution of a number of enzymes of glutamate metabolism and found that in general these were associated with the mitochondrial fraction. We have examined these and other enzyme activities in the fractions we obtain. The results of these studies are summarized in Table 5.

In general our findings agree with those of Donnellan et al (1974). Among the glutamate-metabolizing enzymes, GOT was enriched in the supernatants, consistent with its soluble nature, whereas GDH, GPT and GAD were associated with the mitochondrial fraction. GPox activity is characteristic of the inner mitochondrial membrane of insect flight muscle (Donnellan et al 1974) and its enrichment in B_3 suggests the presence of such membranes in this fraction. However, MDH, a marker for the mitochondrial matrix (Donnellan et al 1974), was not correspondingly enriched in B_3. AK activity was unexpectedly found to be enriched in the supernatant fraction, although it is well established as a mitochondrial enzyme located between the inner and outer mitochondrial membranes. Possibly the fractionation conditions that

TABLE 5 Distribution of enzyme markers in subcellular fractions of locust muscle

		Relative specific activity in fractions						
		F	S_1	S_2	P_1	P_2	P_3	B_3
GOT	Glutamate-metabolizing enzymes	1.0	1.1	1.3	0.6	0.8	0.7	0.2
GDH		1.0	1.1	0.5	1.8	2.0	2.4	0.6
GPT		1.0	0.9	0.6	1.9	2.4	2.9	0.8
GAD		1.0	1.4	0.3	1.8	3.4	10.6	4.0
GPox	Mitochondrial markers	1.0	0.8	0.2	1.0	2.0	1.6	1.6
MDH		1.0	0.9	0.7	1.2	1.6	1.9	1.0
AK		1.0	1.9	2.0	0.2	0.1	0.2	0.1
GPDH	Cytosolic markers	1.0	1.0	1.8	0.1	0.1	0.1	—
LDH		1.0	1.1	1.1	1.4	0.7	0.3	1.0
AChE	Plasma membrane markers	1.0	1.7	2.0	0.7	1.3	1.2	1.6
AC		1.0	1.5	0.6	0.8	1.3	0.2	3.1
5′N		1.0	1.4	1.8	0.3	0.3	0.1	0.9
ATPase(total)		1.0	—	1.3	—	1.7	2.6	4.0
ATPase (pNPP)		1.0	—	0.2	—	0.7	0.1	2.1

GOT, L-aspartate:2-oxoglutarate aminotransferase (EC 2.6.1.1); GDH, L-glutamate: NAD(P)$^+$ oxidoreductase (EC 1.4.1.3); GPT, L-alanine:2-oxoglutarate aminotransferase (EC 2.6.1.2); GAD, L-glutamate 1-carboxy-lyase (EC 4.1.1.15); GPox, *sn*-glycerol-3-phosphate: (acceptor) oxidoreductase (EC 1.1.99.5); MDH, L-malate:NAD$^+$ oxidoreductase (EC 1.1.1.37); GPDH, *sn*-glycerol-3-phosphate:NAD$^+$ 2-oxidoreductase (EC 1.1.1.8); AK, ATP:AMP phosphotransferase (EC 2.7.4.3); LDH, L-lactate:NAD$^+$ oxidoreductase (EC 1.1.1.27); AChE, acetylcholine acetylhydrolase (EC 3.1.1.7); AC, ATP pyrophosphate-lyase (cyclizing) (EC 4.6.1.1); 5′N,5′-ribonucleotide phosphohydrolase (EC 3.1.3.5); ATPase, ATP phosphohydrolase (EC 3.6.1.3); pNPP, K$^+$-stimulated, ouabain-sensitive *p*-nitrophenylphosphate phosphatase.

$$\text{Relative specific activity is defined as: } \frac{\%\ \text{enzyme activity recovered in fraction}}{\%\ \text{protein recovered in fraction}}$$

we use lead to some disruption of the mitochondria, allowing AK activity to leak out. A high GPDH activity is characteristic of insect muscle (Zebe & McShan 1957) and this enzyme was found to be enriched in S_2 with little activity in the pellets. The cytosolic marker, LDH, has been used to characterize synaptosomes from mammalian brain (Marchbanks 1967) but locust muscle contains very low activities (Crabtree & Newsholme 1975). Certainly there was no evidence of enrichment of these enzymes in B_3, which suggests that the fraction does not contain significant quantities of cytoplasm as might be expected for pinched-off nerve terminals.

The choice of plasma membrane markers is difficult. AChE has been assayed in locust muscle (Cleworth et al 1980) and is presumed to be associated with the plasma membrane of the muscle cells in spite of the fact that there is no cholinergic innervation of these cells. It is generally accepted that AC activity is exclusively associated with the plasma membrane in tissues of higher animals and 5′N is widely used as a plasma membrane marker in studies of mammalian cells. The distribution of AChE and particularly AC strongly suggests that B_3 is greatly enriched in plasma membrane. The 5′N activity, however, seems to be primarily soluble. It may be that the rather high ionic strength of the locust saline causes some detachment of this enzyme from the plasma membrane, but we have not investigated this possibility.

The distribution of total ATPase activity indicates considerable enrichment in B_3. This activity was not sensitive to ouabain. ATPase activity was stimulated by Ca^{2+} and inhibited by azide, features that are characteristic of myofibrillar and/or sarcoplasmic reticulum ATPase activity. Thus the enrichment in B_3 may indicate a considerable content of vesiculated sarcoplasmic reticulum in this fraction. The Na^+, K^+-ATPase activity usually associated with vertebrate plasma membranes is reflected in the pNPP activity. This was also enriched in B_3, lending further support to the suggestion that it is derived from plasma membrane.

In summary, the enzyme marker studies indicate that B_3 may consist of vesiculated plasma membrane associated with sarcoplasmic reticulum. We might speculate that the large membranous profiles are plasma membrane with enclosed vesicles of sarcoplasmic reticulum.

Glutamate and GABA binding studies

We have made numerous previous attempts to measure the binding of glutamate and of GABA to synaptic receptor sites that are presumed to exist on the surface membranes of insect muscle (Briley et al 1980, Filbin et al 1980). We are still unable to measure specific glutamate binding in a consistently reproducible manner. Certainly the methods used to measure glutamate binding in mammalian brain preparations (Roberts 1981, Briley et al 1981) are not satisfactory for measuring such activity in insect muscle preparations. At present, using a rapid centrifugation binding assay (Briley et al 1981), we can identify two binding sites in fraction B_3 for L-glutamate that are distinguishable from the sodium-dependent uptake sites described above. The sites have K_d values of 12.5 nM and 1.3 μM and B_{max} values of 0.57 pmol/mg protein and 10.9 pmol/mg protein respectively. The binding of GABA has been difficult to measure; as with the glutamate studies, methods that give consistently reproducible binding measurements with membrane fractions from mammalian brain do not seem appropriate for the insect muscle fractions. We have however identified two binding sites in B_3 that are independent of sodium and have K_d values of 500 nM and 0.25 nM. The binding of the compound muscimol gave similar values; in mammalian brain and crayfish muscle, muscimol is thought to bind specifically to GABA receptor sites.

We should emphasize that these binding studies are very preliminary and the binding sites have not been extensively characterized. Clearly, our methods are still not appropriate for the routine analysis of synaptic receptor sites for these two transmitter candidates, but it is hoped that a systematic analysis of the binding characteristics will allow the development of such assays.

Conclusions

We have obtained a vesiculated membrane fraction (B_3) from whole innervated locust muscle that shows specific, high affinity uptake systems for L-glutamate and for D- and L-aspartate. The fraction seems to be plasma membrane-derived and there is evidence also of a component derived from sarcoplasmic reticulum. We cannot at this stage conclude whether B_3 is exclusively from the muscle cells or whether sheath cell membranes contribute. We have no evidence that the nerve terminal membrane makes any substantial contribution to the fraction. Nevertheless, we feel that B_3 may be a useful preparation for further studies. It is generally agreed that glutamate is removed from the synaptic region by an uptake system and it seems probable that such a system could be associated with plasma membrane of both muscle and sheath cells. Thus the uptake system of B_3 may well be involved in the removal of synaptically released glutamate. Clearly, what we must now do is to investigate the pharmacological properties of the B_3 uptake system and attempt to correlate our findings with the known physiological effects of agents that are active at glutamate-mediated synapses.

The identification and characterization of the synaptic receptors is a more formidable task. The fact that B_3 shows binding of both glutamate and GABA lends support to the idea that it is derived from the muscle plasma membrane. However, if we are to make rapid progress in understanding the glutamate and GABA receptors on insect muscle we need to develop specific high affinity probes for those sites. The tremendous advances in our understanding of the nicotinic cholinergic receptor (see Barnard, this volume) are largely due to the availability of the highly specific, almost irreversible natural toxins that bind to the acetylcholine-binding site. There are venoms, particularly those of the braconid and sphecid wasps, whose site of action is the insect neuromuscular junction (see Piek, this volume). These or other venoms might provide the tools with which to isolate and characterize the synaptic receptors of insect muscle.

Acknowledgements

This work has been supported by grants from The Science and Engineering Research Council to G.G.L. M.T.F. is in receipt of a CAST postgraduate studentship from the Northern Ireland Department of Education in conjunction with Shell Research Ltd, Sittingbourne.

REFERENCES

Barnard EA 1982 Isolation of receptors from the central nervous system. This volume, p 109-117

Breer H 1982 Properties of synaptosomes from the central nervous system of insects. This volume, p 158-169

Breer H, Jeserich G 1980 A microscale flotation technique for the isolation of synaptosomes from nervous tissue of *Locusta migratoria*. Insect Biochem 10:457-463

Briley PA, Filbin MT, Lunt GG, Donnellan JF 1980 The biochemical characterisation of amino acid-mediated synaptic transmission at the locust neuromuscular junction. In: Insect neurobiology and pesticide action. Society of Chemical Industry, London, p 177-184

Briley PA, Filbin MT, Lunt GG, Turner PJ 1981 Glutamate receptor binding in mammals and insects. Mol Cell Biochem, in press

Cleworth JR, Robinson NL, Usherwood PNR 1980 Biochemical studies on glutamate receptors in insect muscle. In: Insect neurobiology and pesticide action. Soc Chem Ind (Lond) Monogr, p 279-280

Crabtree B, Newsholme EA 1975 Comparative aspects of fuel utilisation and metabolism by muscle. In: Usherwood PNR (ed) Insect muscle. Academic Press, London, p 405-500

Donnellan JF, Jenner DW, Ramsey A 1974 Subcellular fractionation of fleshfly flight muscle in attempts to isolate synaptosomes and to establish the localization of glutamate enzymes. Insect Biochem 4:243-265

Faeder IR, Salpeter MM 1970 Glutamate uptake by a stimulated insect nerve muscle preparation. J Cell Biol 46:300-307

Filbin MT, Lunt GG, Donnellan JF 1980 Glutamate receptor biochemistry. In: Sattelle DB et al (eds) Receptors for neurotransmitters, hormones and pheromones in insects. Elsevier/North-Holland Biomedical Press, Amsterdam, p 153-160

Gerschenfeld HM 1973 Chemical transmission in invertebrate central nervous systems and neuromuscular junctions. Physiol Rev 53:1-119

Harrison R, Lunt GG 1980 Biological membranes, 2nd edn. Blackie, Glasgow

Jones DG 1975 Synapses and synaptosomes. Chapman & Hall, London

Marchbanks RM 1967 The osmotically sensitive potassium and sodium components of synaptosomes. Biochem J 104:148-157

Piek T 1982 Solitary wasp venoms as tools for the study of neuromuscular transmission in insects. This volume, p 275-285

Roberts PJ 1974 Glutamate receptors in the rat central nervous system. Nature (Lond) 252:399-401

Roberts PJ 1981 Binding studies for the investigation of receptors for L-glutamate and other excitatory amino acids. In: Roberts PJ et al (eds) Glutamate: transmitter in the central nervous system. Wiley, Chichester, in press.

Usherwood PNR (ed) 1975 Insect muscle. Academic Press, London

Zebe EC, McShan WH 1957 Lactic and α-glycerophosphate dehydrogenases in insects. J Gen Physiol 40:779-790

DISCUSSION

A. T. Eldefrawi: With the *Torpedo* ACh receptor, an excellent way of making sure that the binding is to the receptor one is interested in is to take advantage of the allosteric interaction bewteen channel binding sites and receptor binding sites. If something like phencyclidine binds to the glutamate channel and this binding is activated by glutamate, that means that binding really is to the glutamate receptor.

Lunt: Yes. In addition, with respect to the question of protease activity in receptor preparations, as was discussed earlier, we have tested the whole range of anti-protease agents that we use routinely when preparing vertebrate muscle receptors. There is no systematic effect of these protease inhibitors on our glutamate binding results.

A. T. Eldefrawi: Have you looked at picrotoxin binding or diazepam binding? These compounds interact with other components of the GABA receptor channel system.

Lunt: We have looked at labelled picrotoxin and we saw specific binding to the same muscle membrane preparation, B_3. However, I am reluctant to follow up such findings until the basic question of whether one can detect a glutamate receptor, or a GABA receptor, or both, on locust muscles, has been answered.

Berridge: You spoke about binding sites and also about uptake mechanisms. Presumably the uptake mechanism also has a binding site. How do you distinguish between the two binding sites?

Lunt: You can't do this absolutely on the basis of, say, sodium dependency. In general, the uptake systems for these amino acids are dependent on sodium whereas binding to the receptor site is not sodium-dependent, but this doesn't enable you to distinguish the two kinds of binding sites.

Dudai: What about the temperature dependence of uptake and binding?

Lunt: Uptake of glutamate is temperature-dependent, whereas the four binding sites (p 184) don't show such a degree of temperature dependence.

Usherwood: Binding to the transport site might not be sodium-dependent, even though uptake is.

Berridge: How does one distinguish a transport site from the receptor binding site?

Lunt: The transport site could well be one of the sites we see in the specific binding in the disrupted vesicle preparations. To study that, one needs a range of compounds of known pharmacology and then one might distinguish between binding sites associated with a carrier, and the synaptic receptor binding site.

Barnard: Is it possible to calculate the maximum yield that should be expected for these receptors from this tissue, from the evidence from microscopy on the number of junctions, estimates of the density of receptors at a junction, and so on? Could one see whether the maximum order of magnitude of receptor content that one *could* in principle find is not less than what you do find? From that viewpoint, should the GABA-binding sites be less numerous than the glutamate sites, as is found?

Lunt: These calculations haven't been made, although Peter Usherwood (1980) has approached the question. Part of the problem is that to make these preparations we use the entire intrathoracic musculature of the locust. Not all of the muscles are well characterized electrophysiologically. I assume that they all have an excitatory innervation that is mediated by glutamate?

Usherwood: That is a reasonable assumption.

Lunt: They certainly don't all have a GABA inhibitory innervation. It is difficult to discover what differences would be expected in the degree of innervation of those muscles. Would one expect big differences in the number of terminals per muscle fibre between the various intrathoracic muscles?

Usherwood: I would imagine that the density of innervation varies from muscle to muscle in the thorax. In the big thoracic muscles the density of even excitatory (glutamatergic) innervation is probably low compared with that of leg muscle, for example. This is why it would be better to study leg muscle (e.g. extensor tibiae muscle). The femoral muscles of locusts have well-characterized GABA inhibitory innervations (Usherwood & Grundfest 1963, 1965, Brookes & Werman 1973, Clark et al 1979, Walther 1980).

Lunt: We have made the B_3 preparation from the extensor tibiae muscle. In fact, Richard Olsen tried to measure GABA binding on this membrane preparation unsuccessfully. The glutamate-binding characteristics are not very different from those seen in the membranes from the intrathoracic muscles.

Usherwood: Have you attempted to differentiate between binding to transport sites and binding to receptor sites by using ligands like quisqualate that aren't considered to bind to the uptake sites?

Lunt: Yes. The sodium-dependent binding/uptake component was not affected by quisqualate and the sodium-independent binding was (James et al 1977). However, we weren't aware then of the four separate binding sites which we have now found (two for GABA and two for glutamate).

Usherwood: Do the two glutamate binding sites have identical pharmacological specificities?

Lunt: We have not done that in that sort of detail yet.

Kelly: Is it possible that glutamate is taken up into glial cells? J. E. Treherne's group have shown glutamate uptake into glial cells. Also, your uptake hasn't saturated after 20 minutes, whereas glutamate uptake into synaptosomes from mammalian brain is saturated in three minutes (Logan & Snyder 1971, Balcar & Johnston 1972). How do you know that this isn't entry of glutamate into glial cell fragments rather than into nerve terminals?

Lunt: We don't. I am not suggesting that the B_3 fraction consists of nerve terminals. We don't know what its origins are. From the enzyme markers (Table 5, p 183) there is a significant amount of what appears to be myofibrillar ATPase.

Kelly: In glial cells one usually blocks glutamate uptake with SH inhibitors, so the idea that glial cells don't have ATPases isn't very substantial. You didn't test the classical range of mercurials and cardiac glycosides?

Lunt: No, we haven't yet.

Kelly: Even in vertebrate preparations I am not satisfied that the glutamate uptake into synaptosomes is really into nerve terminals (Storm-Mathisen & Iversen 1979). Granule cells of the cerebellum in tissue cultures, which most people regard as typical glutamate-secreting neurons, show no glutamate uptake. I don't know of a tissue culture preparation in which neurons take up glutamate; it is always the glial cells that take it up (Currie & Kelly 1981). This is also true of insects and crustaceans. Glutamate uptake is predominantly into the glial sheath which engulfs nerve terminals of the neuromuscular junction (see Faeder & Salpeter 1970).

Lunt: In those autoradiographic studies in insect muscle (Faeder & Salpeter 1970) the sheath cell showed the first accumulation of radiolabel. We don't know where a sheath cell membrane would end up in our fractionation scheme.

Kelly: DeFeudis (1980) thinks that the sodium-independent binding occurs only to postsynaptic receptors. How different were your methods from his?

Lunt: There is no direct evidence for postsynaptic membrane in the B_3 fraction; we only know that there is no glutamate or GABA binding in any other membranous fraction.

Kelly: What is the uptake if you simply take muscle tissue, chop it up with a McIlwain chopper and put it through a nylon sieve?

Lunt: I think that we would get glutamate uptake, as we do in our starting homogenate. We see an enrichment of glutamate uptake during the fractionation, per unit of protein. That is why we concentrated on B_3, because that fraction shows most enrichment of uptake and of binding.

Kelly: In mammalian brain tissue there is almost as much uptake of glutamate into a simple mince or chop as into purified synaptosomes.

Lunt: The same is true of glutamate *release* experiments in vertebrate brain. Homogenates release as much as do synaptosomes.

Cull-Candy: I wasn't too surprised at your problems with GABA, Dr Lunt, because in our ionophoretic studies on locust muscle fibres (Cull-Candy & Miledi 1981) in general GABA sensitivity is lower than glutamate sensitivity. In many fibres it is difficult to find anything other than a low level of extrajunctional GABA sensitivity.

Would comparing the K_d values with the K_d values for acetylcholine (ACh) to see whether they are in the same sort of physiological range tell one whether any of these binding sites is actually the receptor?

Barnard: I would think not, because we are still ignorant about the full significance of the K_d values measured for ACh: we can't reconcile the affinities found in the isolated nicotinic receptor for ACh with the apparent physiological affinities. There are evidently other steps which influence the latter, in the overall process of channel opening which is monitored physiologically. The isolated receptor has too high an affinity for ACh, compared with the *in vivo* apparent value.

M. E. Eldefrawi: That is true, but it depends on the assay used. We had a lot of difficulty in getting our study (Eldefrawi et al 1971a) on the binding of ACh to the cholinergic receptor of *Torpedo* electric organ accepted because the K_d of 50 or 60 nM seemed too high an affinity compared to a K_d of 50 μM calculated from electrophysiological data. However, later it was realized that high affinity binding occurred to the desensitized conformation of the receptor (Weber et al 1975). If one uses, say, the channel probe assay that we described in our paper here and looks at the concentrations of ACh that stimulate binding of $[^3H]H_{12}$-HTX to the channel sites in *Torpedo* membranes, a K_d of 50–60 μM is calculated (Aronstam et al 1981), which is in excellent agreement with the value obtained from the cellular response of the same tissue. Depending on the assay and how one is looking at the tissue, one may get different values.

I am intrigued by the two K_d values for glutamate and for GABA binding in Dr Lunt's preparation. The finding of multiple binding affinities is like what we saw with decamethonium binding to *Torpedo* membranes (Eldefrawi et al 1971b). One K_d may be responsible for binding of decamethonium to the receptor site, another K_d for its binding to acetylcholinesterase. A third K_d is for the binding of decamethonium to the ion channel site. It may turn out that these values obtained for glutamate binding are significant dissociation constants for various glutamate targets in the tissue, one or both of which may represent the receptor.

Narahashi: Another K_d may be for the voltage-dependent sodium channel, because decamethonium also blocks that (Yeh & Narahashi 1974)!

M. E. Eldefrawi: We have never looked at that!

A. T. Eldefrawi: There are also synaptic and extrasynaptic glutamate receptors in insect muscle, so that more than one K_d could still represent binding to glutamate receptors.

Lunt: That is right. One would certainly anticipate seeing more than one binding site in the binding assays.

Olsen: On the question of distinguishing receptors and uptake sites, it comes down to the pharmacological tools. One can measure uptake and one can measure electrophysiological responses. One needs to show the specificity for each subtype of receptor and each uptake mechanism, and then the types of binding can be distinguished. This is complicated by heterogeneity. If reproducible, muscimol-displaceable GABA binding were found in Dr Lunt's insect muscle fractions, one might conclude that muscle membrane is present, because the GABA receptor is probably a good marker for muscle membrane as compared to glia, though that is not established. For glutamate binding there are some analogues that block uptake and do not affect synapses. They could be used to rule out uptake for Dr Lunt's binding sites.

Kelly: You could raise antibodies against glial cell membranes and see whether B_3 includes glial cell membrane. I can't see any other way of being sure that you are not studying glial membrane rather than muscle or nerve membrane. Martin Raff has developed antibodies to vertebrate glial cell markers (Raff et al 1979). Similar techniques could be used to raise antibodies against different invertebrate cell types.

Berridge: Could you do a first-stage purification by using denervated muscle?

Lunt: My feeling is that this is too sophisticated for the stage we are at.

Berridge: Since you lack the drugs for a pharmacological analysis, it might be possible to abolish the uptake process by removing the nerves. There might be a problem in that the nerves might degenerate but the glial cells might not.

Usherwood: Such studies are possible but time-consuming because of the small size of locust muscle. When locust femoral (leg) muscle is denervated the motor axons degenerate but the glial cells remain. The results of biochemical studies on uptake and binding of L-glutamate to membrane fractions from denervated locust leg muscle look similar to those obtained from innervated muscle (J. Cleworth et al, in preparation).

M. E. Eldefrawi: It seems clear that progress on glutamate receptors is held up because of the lack of a high affinity probe. Perhaps we shall hear from Dr Piek about a wasp toxin that might help here. There are however several drugs that would interfere with glutamate conductance in a non-competitive manner, suggesting that there are channel blockers for the glutamate system. Various drugs or their derivatives (e.g. covalent acid derivatives) may be developed that would yield the high affinity labels that could be used here.

Barnard: Has the effect of benzodiazepines been studied electrophysiologically on these insect GABA receptors?

Cull-Candy: Not so far as I know in *locust* muscle, although other systems have been studied.

Dudai: We have investigated the binding of benzodiazepines to *Drosophila* homogenates, under conditions in which these drugs bind to vertebrate preparations. We failed to find binding in this insect.

Barnard: I just wondered if the much-discussed modulatory effect of benzodiazepines on mammalian GABA receptors had been excluded in the case of the insect peripheral GABA receptors? Given the problems that George Lunt described in looking at GABA binding itself in the locust muscle preparations, it might be difficult to detect benzodiazepine binding, *if* any such is associated with the low density population of sites in that preparation. Perhaps electrophysiologically one might be able more readily to test for benzodiazepine facilitation of the GABAergic transmission.

Olsen: We looked at the effect of benzodiazepines on chloride flux in crayfish abdominal muscle and found no effect on the response to GABA (Olsen et al 1978). We haven't done this in an insect.

Usherwood: Professor Barnard suggested in his paper that it is no longer necessary to use irreversibly binding ligands to isolate neuroreceptors from excitable tissues. Since we do not have an irreversibly binding ligand for locust glutamate receptors the ability to isolate these membrane proteins without irreversible probes would be a major step forward.

Barnard: I think both points of view are valid, in practice. On the one hand, it seems true to say that receptor neurochemistry has now reached the stage where one no longer *has* to use irreversible probes. It was important to have them when that field was first developed because nobody knew what type of component to look for or whether the reversible binding of ligands to isolated fractions could be significant. Now that all the experience with the nicotinic cholinergic receptor, gained by taking advantage of its abundance in the electric organ and the virtual irreversibility of its binding of α-BGTX, has been assimilated, it is seen that with such a receptor one can, in fact, use other specific ligands which bind reversibly to it. The condition for this is that the dissociation of the complex formed must be slow relative to the time occupied by a suitable assay, which generally means that the K_d of the complex should be not much greater than 10^{-8} M. It is also necessary, of course, that a medium should be found in which the receptor is extractable and active, but that applies to the use of both the reversible and the irreversible type of probe. The reversible specific ligand of the type I just mentioned can in favourable cases also be successfully immobilized in a gel for affinity chromatography, to purify the receptor in question.

One recent example of the use of reversible specific ligands of this type in studying or purifying other receptors is with the ACh receptor from the chick brain, as I mentioned in my presentation here, where the affinity of the toxin (used both for assay and for affinity chromatography) is such that the receptor complex in solution has a K_d of ~10^{-8} M. Another example is the assay of the mammalian brain enkephalin receptor, using reversibly bound opiates or enkephalins, or the benzodiazepine/GABA receptor using benzodiazepines (for review, see Olsen 1981). A third example is the recent complete purification of the β-adrenergic receptor (using its reversible binding to alprenolol) by Shorr et al (1981). The technology has thus improved since 1970, when the biochemical advances on the peripheral nicotinic ACh receptor began with the exploitation of the irreversible probe, α-BGTX: with a suitable specific *reversible* probe it should now be possible to deal with the very small amount of any receptor which is extractable in active form from the insect nervous system.

It seems to me, however, that the opposing point of view is still valid at present in this particular case. That is, the problem in isolating the glutamate receptor is that in the system described here you don't have *any* good ligand at present. Glutamate has too low an affinity to be applicable. You need a ligand that binds with a K_d of about 10^{-8} M or better. It need not be irreversible, but of course you need something with high specificity for the glutamate receptor. Probably in the future chemical ingenuity could provide an irreversible label based on glutamate, although there are greater limitations there than with the ACh receptor because the insect glutamate receptors seem to be less tolerant of change in the ligand. There may be some possibilities of chemical substitution in it, however: the ω-phosphonate of glutamate, for example, is active there (Cull-Candy et al 1976, Tinguy-Moreaud et al 1981). In such a case, one may need to introduce chemically an irreversibly reacting group of that kind in order to utilize a truly specific ligand which in its reversible form has too low an affinity. As Dr Eldefrawi mentioned, if a photo-affinity azido derivative of any specific ligand was developed, and if it is still bound, it should be applied in an attempt to isolate the glutamate receptor even from the small amounts occurring in insects.

M. E. Eldefrawi: Photo-affinity labels proved to be valuable probes for studies on receptors and other regulatory membrane proteins. One doesn't need to have an azide derivative to develop a photo-affinity label. The availability of the chemical in the vicinity of the binding site, combined with irradiation, is sufficient to produce an activated free radical that binds covalently to the protein (Oswald & Changeux 1981). This new development should be applied to the glutamate receptor, because electrophysiologists as well as chemists can take advantage of this tool.

Barnard: That is why I asked about the benzodiazepines. Bound flunitrazepam on irradiation with ultraviolet light covalently labels its receptor complex in mammalian brain membranes (Möhler et al 1980). If that reactivity is present in an insect system, it would provide a tool, but it is disappointing that benzodiazepines apparently don't bind to a GABA-receptor-associated site in the invertebrate nervous systems reported upon.

Cull-Candy: If the glutamate receptor has a low affinity, would that matter for experiments in which you determine whether there is a binding site? From patch clamp experiments made by Dr Parker, Professor Miledi and me (Cull-Candy et al 1981), the apparent affinity of the glutamate receptor is about an order of magnitude lower than the apparent affinity of the ACh receptor in the vertebrate end-plate.

Lunt: This would make it more difficult. In all these ligand-binding experiments with readily reversible ligands, where binding depends on being able to separate bound ligand from free ligand, with a readily dissociating ligand one is consistently underestimating the number of binding sites.

Cull-Candy: So that may be a problem as well?

Lunt: Yes. In our early binding studies, there was no question of picking up a site with a dissociation constant of 0.5 μM or 12 nM with our filtration assays.

Usherwood: The fact that the affinity of the receptor site for glutamate is low *in vitro* does not exclude the future discovery of probes with much higher affinities either for that site or for some other site on the receptor protein.

Kelly: Is there glutamate in your B_3 fraction?

Lunt: There is a small amount of glutamate, but not enough to inhibit the binding of radiolabelled glutamate.

Kelly: That assumes that the glutamate is all free and that the concentration is the same throughout the membrane.

Lunt: Yes, we assume that. But in the osmotically shocked B_3 fraction, when you re-pellet the membrane you would not have any free glutamate. But there is the idea, in mammalian brain, of an endogenous glutamate inhibitor.

Dudai: In mammalian tissues, when you use osmotic shock to remove GABA, in order to measure GABA or benzodiazepine binding without potential interference by the ligand, in some cases you still find either generated GABA or some retained GABA. So I am not sure whether the fact that you have binding of glutamate excludes the possibility that long-lived glutamate is still present and affects your results.

Lunt: Only repeated washing would remedy this.

Dudai: This may kill the receptor too!

Lunt: In effect, isn't that what happens with your GABA receptor, Dr Olsen?

Olsen: Yes, as you and Dr Dudai have pointed out, it is necessary to treat the membranes in such a way as to remove endogenous ligands (e.g. GABA; Napias et al 1980) without harming the binding activity. But you say that you measured glutamate in the fractions you assayed.

Lunt: We measured glutamate in the B_3 fraction before and after osmotic shock.

Olsen: Did you get the glutamate out with an organic solvent?

Lunt: No, with Triton, which just dissolved the membranes.

Dudai: Do you have a system there that generates glutamate?

Lunt: I don't think so.

Olsen: The answer to Dr Cull-Candy is that if the receptor in the homogenate has the affinity of the native receptor, which is measured by dose–response curves to be activated at 10–100 μM, we shall be unable to measure it by currently available binding assay techniques. For all the receptors that can be measured, either we use a higher affinity ligand than the natural transmitter; or the receptor has been altered by homogenization to give it a fortuitously higher affinity, presumably, than it has in the native form. So if the glutamate receptor and the GABA receptor retain their low native affinity, they won't be measurable *in vitro*. This might be related to Dr Lunt's difficulties.

Cull-Candy: Dr Parker, Professor Miledi and I estimate the apparent affinity of the receptor for glutamate to be in the order of 300–500 μM (Cull-Candy et al 1981).

M. E. Eldefrawi: When we eventually solve these problems and have a purified glutamate receptor, it will be interesting to know the effect of detergents on this receptor and on glutamate responses. Have you looked at the effect of any detergents on the glutamate responses of invertebrate muscle?

Cull-Candy: No, although it would be useful to do so.

Narahashi: To come back to the point about finding a specific ligand for glutamate receptors, we need an insect neurophysiologist to screen large numbers of compounds for possible agonistic or antagonistic actions on the glutamate receptor. I believe there is a kind of bee venom that blocks neuromuscular transmission in insects?

Piek: Yes, apamin, but its action has only been studied in vertebrates, where it seems to be very selective, in that the potassium permeability is blocked only in certain organs and not in others.

We find inhibition of the sodium-dependent glutamate uptake by *Philanthus* venom. This has been shown by autoradiography combined with electron microscopy (J. Van Marle et al, unpublished results). The sodium-dependent uptakes of glutamate by glial cells and by nerve terminals are both reduced to 30–40% of the control. We have evidence that the postsynaptic binding of glutamate to its receptor is not blocked by the venom, so maybe we can use such venoms to distinguish between uptake and postsynaptic membrane binding.

Kelly: In the mammalian work, Watkins et al (1981) showed that their best antagonists did not operate against the purest glutamate receptors, so the mammalian work hasn't given a very good lead here.

Dudai: Wasn't the affinity of Watkins' best antagonist about 10^{-6} M for 2-amino-5-phosphonovaleric acid?

Kelly: I think so, but this antagonist acts against the *N*-methylaspartate receptors.

Narahashi: Is there any way of developing a derivative of kainic acid?

Kelly: That would be a third receptor type. There are three receptor types—*N*-methylaspartate, kainate and quisqualate.

Barnard: Can the insect neuromuscular junction glutamate-sensitive receptor be identified with one of those?

Usherwood: Not kainic acid.

Kelly: It would be more like quisqualate.

Usherwood: There are some junctions that are depolarized by ibotenic acid, so the situation is complex. Subpopulations of receptors coexist on both junctional and extrajunctional membrane and these have not yet been fully characterized pharmacologically.

Kelly: Have you any evidence for presynaptic glutamate receptors in your insect preparations?

Usherwood: We have evidence for presynaptic glutamate receptors at glutamatergic junctions on locust leg muscle (Usherwood & Machili 1966, 1968). However, receptors for other putative transmitters are also present presynaptically at locust nerve-muscle junctions (Fulton & Usherwood 1977).

Lunt: May I mention the effect of the plant lectin concanavalin A (Con A) on the glutamate receptor? Michaelis (1979) has reported the isolation of what he refers to as a glutamate receptor from rat brain on the basis of affinity purification. He calls it a metalloglycoprotein. We have never really tried to repeat this. It seems remarkable to obtain such a purification simply with a Con A affinity column. There certainly are effects of Con A on the receptor state. Have you looked at Con A in any of your binding studies?

Usherwood: Not in any detail. We have never suggested that Con A binds to the glutamate receptor. Con A certainly influences the receptor-ionophore properties since it inhibits desensitization of some populations of glutamate receptors on locust leg muscle (Mathers & Usherwood 1976).

Evans: One problem in comparing physiological studies with binding studies is that you don't know what percentage of the receptors have to be filled for a maximum physiological response to be obtained. It could be a very small percentage. Thus the concept of 'spare receptors' needs to be considered when one tries to compare results obtained by these differing approaches.

REFERENCES

Aronstam RS, Eldefrawi AT, Pessah IN, Daly JW, Albuquerque EX, Eldefrawi ME 1981 Regulation of [^{3}H]perhydrohistrionicotoxin binding to *Torpedo ocellata* electroplax by effectors of the acetylcholine receptor. J Biol Chem 256:2843-2850

Balcar VJ, Johnston GAR 1972 The structural specificity of the high affinity uptake of L-glutamate and L-aspartate by rat brain slices. J Neurochem 19:2657-2666

Brookes N, Werman R 1973 The cooperativity of γ-aminobutyric acid action on the membrane of locust muscle fibers. Mol Pharmacol 9:571-579

Clark RB, Gration KAF, Usherwood PNR 1979 Relative 'trophic' influences of excitatory and inhibitory innervation of locust skeletal muscle fibres. Nature (Lond) 280:679-682

Cull-Candy SG, Miledi R 1981 Junctional and extrajunctional membrane channels activated by GABA in locust muscle fibres. Proc R Soc Lond B Biol Sci 21:527-535

Cull-Candy SG, Donnellan JF, James RW, Lunt GG 1976 2-Amino-4-phosphonobutyric acid as a glutamate antagonist in locust muscle. Nature (Lond) 262:408-409

Cull-Candy SG, Miledi R, Parker I 1981 Internal perfusion of patch-clamp electrodes for analysis of synaptic channels. J Physiol (Lond) 317:3P

Currie DN, Kelly JS 1981 Glial versus neuronal uptake of glutamate. J Exp Biol, in press

DeFeudis FV 1980 Binding studies with muscimol: relation to synaptic γ-aminobutyrate receptors. Neuroscience 5:675-688

Eldefrawi ME, Britten AG, Eldefrawi AT 1971a Acetylcholine binding to *Torpedo* electroplax: relationship to acetylcholine receptors. Science (Wash DC) 173:338-340

Eldefrawi ME, Eldefrawi AT, Gilmour LP, O'Brien RD 1971b Multiple affinities for binding of cholinergic ligands to a particulate fraction of *Torpedo* electroplax. Mol Pharmacol 7:420-428

Faeder I, Salpeter M 1970 Glutamate uptake by a stimulated insect nerve muscle preparation. J Cell Biol 46:300-307

Fulton BP, Usherwood PNR 1977 Presynaptic acetylcholine action at the locust neuromuscular junction. Neuropharmacology 16:877-880

James RW, Lunt GG, Donnellan JF 1977 Isolation of a glutaminergic receptor fraction from locust muscle. Biochem Soc Trans 5:170-172

Logan WJ, Snyder SH 1971 Unique high affinity uptake systems for glycine, glutamic and aspartic acids in central nervous tissue of the rat. Nature (Lond) 234:297-299

Mathers DA, Usherwood PNR 1976 Concanavalin A blocks desensitization of glutamate receptors on insect muscle fibres. Nature (Lond) 259:409-411

Michaelis EK 1979 The glutamate receptor-like protein of brain synaptic membranes is a metalloprotein. Biochem Biophys Res Commun 87:106-112

Möhler H, Battersby MK, Richards JG 1980 Benzodiazepine receptor protein identified and visualized in brain tissue by a photoaffinity label. Proc Natl Acad Sci USA 77:1666-1670

Napias C, Bergman MO, Van Ness PC, Greenlee DV, Olsen RW 1980 GABA binding in mammalian brain: inhibition by endogenous GABA. Life Sci 27:1001-1011

Olsen RW 1981 GABA-benzodiazepine-barbiturate receptor interactions. J Neurochem 37:1–13

Olsen RW, Ticku MK, Van Ness PC, Greenlee DV 1978 Effects of drugs on γ-aminobutyric acid receptors, uptake, release and synthesis *in vitro*. Brain Res 139:277-294

Oswald R, Changeux J-P 1981 Ultraviolet-induced labeling by noncompetitive blockers of the acetylcholine receptor from *Torpedo marmorata*. Proc Natl Acad Sci USA 78:3925-3929

Raff MC, Fields KL, Hakomori SI, Mirsky R, Pruss RM, Winter J 1979 Cell-type-specific markers for distinguishing and studying neurons and the major classes of glial cells in culture. Brain Res 174:283-308

Shorr RGL, Lefkowitz RK, Caron MG 1981 Purification of the β-adrenergic receptor. J Biol Chem 256:5820-5826

Storm-Mathisen J, Iversen LL 1979 Uptake of [^{3}H]glutamic acid in excitatory nerve endings: light and electronmicroscopic observations in the hippocampal formation of the rat. Neuroscience 4:1237-1253

Tinguy-Moreaud E De, Bioulac B, Neuzil E 1981 Phosphonic analogues of neuroactive biomolecules. Biochem Soc Trans 9:246-247

Usherwood PNR 1980 Amino-acid gated ion channels in locust muscle membrane. In: Rózsa KS (ed) Neurotransmitters in invertebrates. Pergamon, Oxford/Akadémiai Kiadó, Budapest (Adv Physiol Sci 22) p 191-203

Usherwood PNR, Grundfest H 1963 Inhibitory postsynaptic potentials in grasshopper muscle. Science (Wash DC) 143:817-818

Usherwood PNR, Grundfest J 1965 Peripheral inhibition in skeletal muscle of insects. J Neurophysiol (Bethesda) 28:497-518

Usherwood PNR, Machili P 1966 Chemical transmission at the insect excitatory neuromuscular synapse. Nature (Lond) 210:634-636

Usherwood PNR, Machili P 1968 Pharmacological properties of excitatory neuromuscular synapses in locust. J Exp Biol 49:341-361

Walther C 1980 Small motor axons in orthopteran insects: a re-investigation of the femoral retractor unguis muscle in a stick insect and two species of locust. J Exp Biol 87:99-120

Watkins JC, Davis J, Evans RH, Francis AA, Jones AW 1981 Pharmacology of receptors for excitatory amino acids. In: Di Chiara G, Gessa GL (eds) Glutamate as a neurotransmitter. Plenum Press, New York, p 263

Weber M, David-Pfeuty T, Changeux J-P 1975 Regulation of binding properties of the nicotinic receptor protein by cholinergic ligands in membrane fragments from *Torpedo marmorata*. Proc Natl Acad Sci USA 72:3443-3447

Yeh JZ, Narahashi T 1974 Noncholinergic mechanism of action of cholinergic drugs on squid axon membranes. J Pharmacol Exp Ther 189:697-707

Genetic approaches to insect neurochemistry

YADIN DUDAI

Department of Neurobiology, The Weizmann Institute of Science, Rehovot 76100, Israel

Abstract During the past few years genetic methods have been used to analyse neurochemical mechanisms in *Drosophila* and to determine the role of these mechanisms in development and behaviour. Three main experimental approaches can be adopted in such studies. Mutants isolated on the basis of abnormal development or behaviour, or both, can be screened for lesions in suspected metabolic events. A second approach is to isolate mutants on the basis of their susceptibility to neurotoxins and to screen these mutants for defects in macromolecules which are expected to interact with the drugs. A third and more straightforward approach is to locate the gene coding for a given macromolecule and to isolate mutations in it. The information gained on *Drosophila* neurochemistry, combined with the powerful genetic techniques available for this organism, make it possible to tackle problems which are difficult to approach by other methods.

Analysis of the effect of mutations on the structure of a neuronal apparatus and its operation may help to elucidate the nature of the macromolecular components of the apparatus and their role in normal development and behaviour. The fruitfly, *Drosophila melanogaster*, is especially suitable for such an approach. Indeed, during the past few years, several groups of investigators have used genetic techniques to dissect neurochemical mechanisms in *Drosophila*. It is possible to distinguish in such studies three main experimental approaches, which differ in the starting point of their research strategy, as described below.

Starting from the phenotype

In this approach mutants are selected on the basis of abnormal development or behaviour, or both, and the nature of the molecular lesion underlying the

1982 Neuropharmacology of insects. Pitman, London (Ciba Foundation symposium 88) p 199-206

abnormal phenotype is studied. The mutants can be selected either from the rich library of existing *Drosophila* mutants (Benzer 1973, Lindsley & Grell 1968), or by screening for a specific trait among the progeny of mutagenized flies.

The road from the isolation of an interesting mutant to the identification of the gene product responsible for the phenotypic aberration is long and tedious. Identification of the anatomical sites at which the gene exerts it effects (e.g. nerve, muscle or receptor cells), using either direct anatomical techniques or mosaics (Hotta & Benzer 1970), may be of value, although by itself it does not, of course, reveal the molecular lesion.

Another strategy is to assay electrophysiological or neurochemical parameters suspected of being involved in the defect. One may also screen for the ability of drugs to reverse the phenotype of the mutant, or attempt to induce a mutant-like phenotype in normal flies, and hence get some hint as to whether specific metabolic events are involved.

An additional approach is to map the mutation accurately and match the genetic data with information already available on the appropriate piece of chromosome, in the hope that something is already known about the gene product.

Many fascinating behavioural and neurological single-gene mutants of *Drosophila* are now available (reviewed by Hall & Greenspan 1979). They range from mutants defective in the embryonic development of their nervous system, through sluggish or non-phototactic flies, or temperature-sensitive paralytic, easily shocked or shaker flies, to love-song mutants, mutants with an abnormal biological clock, and even mutants that cannot learn and remember. Such a rich gallery of abnormalities contains bizarre phenotypes that may intrigue investigators in many branches of neurobiology. However, the temptation to select an odd mutant and use it to solve the riddles of life may be somewhat lessened by the fact that only in a few cases has our knowledge of the lesion been extended significantly beyond the phenotypic description. Nevertheless, the instances in which we have some idea about what might be going wrong at the molecular level not only illustrate the potential of the approach but have already added to our understanding of neurobiological processes.

For example, alleles of the *Shaker* locus (on the X chromosome) have been isolated on the basis of this mutant's abnormal leg-shaking under ether anaesthesia. Using the *Drosophila* larva neuromuscular preparation for electrophysiological analysis, Jan et al (1977) showed that the *Shaker* mutation affects synaptic transmission and suggested that potassium channels in the nerve terminal membrane are defective.

Electrophysiological studies have also been made of mutants that behave normally at a permissive temperature (e.g. 25 °C) but become paralysed at a restrictive temperature (e.g. above 30 °C). Two examples are *nap*^{ts} (*no action potential, temperature-sensitive*), a second-chromosome mutation, and *para*^{ts} (*paralysed, temperature-sensitive*), an X-linked mutation. Both these mutations affect axonal conduction. Wu & Ganetzky (1980) have provided evidence consistent with the idea that *nap*^{ts} alters the function of sodium channels and that in *para*^{ts} a polypeptide is affected whose function is related to the sodium channels. L. M. Hall and her colleagues (this volume) recently found that *nap*^{ts} contains a reduced number of binding sites for the sodium channel blocker, saxitoxin.

Many additional behavioural mutants were found to display electrophysiological abnormalities, including *comatose, bang-sensitive* and *shibire* (reviewed by Hall & Greenspan 1979). The last mutant, isolated as an X-linked paralytic mutant, has attracted many investigators because of its wide range of defects, including developmental and neuronal abnormalities. It has been suggested that this mutant has lesions in some membrane component that is common to many cells. It should be emphasized that although temperature-sensitive paralysis serves as a screening indicator for neuronal mutants, identification of a defect in neuronal function does not imply that the lesion is confined to the nervous system.

The story of *dunce* is an exciting example of a link that has been made, by a combination of the techniques mentioned above, between a complex behaviour and a neurochemical event. The *dunce* mutants were isolated in a screen for mutants that fail in an olfactory memory paradigm (Dudai et al 1976). The mutations were mapped to a small segment at the left tip of the X chromosome and were later identified by Byers et al (1981) as alleles of a previously identified locus of a putative structural gene coding for an isozyme of cyclic AMP phosphodiesterase. The direct relevance of the cyclic AMP phosphodiesterase defect to the early-memory defect is not yet known. The neurochemical defect in other learning and memory mutants (reviewed by Dudai & Quinn 1980) is still a mystery. The observation that *dunce* has a defect in cyclic nucleotide metabolism hints that additional learning and memory mutants might be screened for defects in other aspects of cyclic AMP metabolism, such as adenylate cyclase, protein kinase, or aminergic neurotransmission (e.g., Uzzan & Dudai 1981).

Starting from pharmacology

Drugs that interact with various components of the nervous system, including enzymes, receptors, ion pumps or ion channels, are often lethal when administered at the appropriate doses. One can select for drug-resistant mutants and then test whether the resistance is due to an alteration in the presumed macromolecular target.

This pharmacological approach has been used in *Drosophila* in an attempt to isolate mutations affecting several components of the nervous system. The search for mutations in the gene(s) coding for the α-bungarotoxin (α-BGTX)-binding component is one example. The nervous system of *Drosophila* is a rich source of α-BGTX-binding sites (reviewed by Dudai 1980, Sattelle 1980). Nicotine is a potent inhibitor of these sites and is readily absorbed into the nervous system. Hall and her colleagues (L.M. Hall 1980) decided to screen for nicotine-resistant flies in an attempt to isolate receptor mutants. Several resistant strains were found.

The structure of the receptor from a nicotine-resistant strain, as revealed by isoelectric focusing of an ^{125}I-labelled α-BGTX–receptor complex, differed from that of a non-resistant strain (L. M. Hall et al 1978). Reciprocal crosses between the strains indicated that the gene controlling the isoelectric-point shift segregates with the X chromosome. The major nicotine-resistant factor in the resistant stock was also reported to be X-linked. This suggests that gene(s) on the X chromosome affect receptor structure. It is worth noting that a nicotine-resistant gene was also identified on the third chromosome but was later lost (Lindsley & Grell 1968).

Several other biochemical systems which play a role in neuronal function have been screened for mutations by pharmacological methods. Wright et al (1976) isolated mutants resistant and hypersensitive to α-methyldopa, an inhibitor of dopa decarboxylase, the enzyme which synthesizes the putative neurotransmitter, dopamine. However, this method failed to yield structural gene mutations for dopa decarboxylase and more straightforward genetic methods had to be used (see below). *Drosophila* strains resistant to various neuropharmacological agents were isolated by Howard et al (1975), but no correlations with possible alterations in neurotransmitter or receptor systems were established. In the latter study, strains resistant to inhibitors of acetylcholinesterase were not found to have altered acetylcholinesterase activity *in vitro*. In another study, Gitschier et al (1980) tested a mutant, *ttx*^{s}, isolated through its high sensitivity to dietary tetrodotoxin (a potent sodium channel blocker), for the binding *in vitro* of [^{3}H]saxitoxin (another channel blocker, as mentioned above); no abnormal binding was detected. These studies emphasize that one should not *a priori* expect a direct correlation between resistance (or high sensitivity) to a drug and a modification in the macromolecule considered to be the pharmacological target of the drug.

Starting from the gene

Methods exist for *Drosophila* which enable one to identify directly the gene(s) coding for a given macromolecule and to isolate specific mutations which may be used to further investigate the effect of the defective gene product on development and behaviour.

If a protein exists in several electrophoretic variants, one may localize the gene(s) coding for that protein by mapping the variants. Several genes coding for enzymes in *Drosophila* have been localized by that method (reviewed by Courtright 1976). However, in many cases no electrophoretic variants could be detected. A different method, segmental aneuploidy (Lindsley et al 1972), may be tried. The strategy is based on the assumption that the amount of a macromolecule in an organism is proportional to the number of copies of the structural gene for that macromolecule. One thus constructs duplications and deficiencies of the chromosomes and screens for gene-dosage effects—that is, for regions of the genome which lead in three doses, or copies, to increased biochemical activity, and in one dose to decreased activity, when compared to the normal diploid number. Once the approximate location of the gene in the chromosome has been identified, the locus can be more accurately mapped by examining the effect of small deficiencies in the suspected region on the measured biochemical activity. A mutation that eliminates a key component of the organism is expected to be lethal when homozygous. The heterozygote, which contains only one normal dose of the same gene, is expected to have a significantly reduced level of the gene product. One may therefore isolate point mutations in the gene, including conditioned alleles, by screening the chemically mutagenized chromosome for recessive lethality in the region spanned by the appropriate deficiency, and later assay the heterozygous lethals for reduced activity of the gene product.

The structural loci for several enzymes in *Drosophila* have been identified by segmental aneuploidy. Among them are some of special interest to neurochemists. Two have already been mentioned: cyclic AMP phosphodiesterase and dopa decarboxylase. Another example is provided by studies of enzymes that metabolize the neurotransmitter, acetylcholine (ACh).

The structural gene for the enzyme that synthesizes ACh, choline acetyltransferase (CAT; Greenspan 1980), and the structural gene for the enzyme that hydrolyses ACh, acetylcholinesterase (AChE; Hall & Kankel 1976), have been mapped by segmental aneuploidy. The results indicate that in *Drosophila* there is only one gene for each of these enzymes. Mutants, including temperature-sensitive mutants, have been isolated. The latter are especially valuable, since they permit gene function to be experimentally switched on and off during development.

In an elegant set of studies J. C. Hall et al (1979) used mutants to analyse the role of both AChE and CAT in development and behaviour. The mosaic technique was used to determine the crucial foci in which lack of enzyme causes lethality, and the effects, during development, of lack of enzymic activity on various parts of the nervous system. One of the most striking observations was that the ganglionic neuropil adjacent to patches of AChE-deficient cortex displayed abnormal organization. AChE (Greenspan et al 1979) and CAT (Greenspan 1980) mosaics were also used to reveal the role of cholinergic transmission in vision and in courtship.

CAT activity in homogenates prepared from temperature-sensitive mutants (raised to adulthood at a permissive temperature) was inactivated faster than normal at high temperature. These results suggest, but do not prove, that the CAT protein molecule in the temperature-sensitive mutants is abnormally thermolabile (Greenspan 1980). AChE extracts prepared from several temperature-sensitive mutants raised to maturity at a permissive temperature also displayed heat lability or low substrate affinity when assayed at a restrictive temperature. However, in the AChE mutants viability did not correlate with enzyme activity *in vivo* or *in vitro*. It was therefore suggested that these mutants have enzyme molecules that function abnormally at the restrictive temperature, at least in part because of defective interactions of the catalytic subunits with other polypeptides *in situ* (J. C. Hall et al 1980). Such mutants may prove of value in studying the quaternary structure and assembly of AChE which, in vertebrate tissue, has been found to be a complex, multi-subunit entity whose exact mode of association with the cell membrane is not yet clear.

Both the AChE locus and the CAT locus were located on chromosome 3, but at least 230 gene bands apart (J. C. Hall et al 1980). Genes for the putative nicotinic receptor and the putative muscarinic receptor were shown not to be adjacent to the AChE gene (reviewed by Dudai 1980, Sattelle 1980). Thus, mutations also make it possible to determine whether genes coding for various components of a transmitter system are grouped together on the chromosome. For the cholinergic system in *Drosophila*, the answer is apparently no.

It is not yet clear whether localization methods based on the gene-dosage effect would work for other molecules, such as receptors or ionophores. The latter may, for example, be composed of several different polypeptides, the genes for which may be scattered throughout the genome, and in such a case establishing gene-dosage effects might be difficult. Segmental aneuploidy is therefore not a magic tool: the search by this method for the gene(s) coding for the α-BGTX-binding nicotinic receptor did not prove successful, and an alternative approach had to be followed (see above; L. M. Hall 1980).

New genetic engineering methods have recently been added to the arsenal of techniques which could be used in *Drosophila* neurogenetics. Thus, when one has a localized gene, one can, in principle, clone it in a bacterial plasmid. The cloning permits the production of large quantities of a macromolecule which is difficult to purify by conventional methods. Studies of ionic channels or receptors for neurotransmitters may benefit immensely from such an approach.

When surveying the field, one cannot escape the feeling that the state of neurochemical knowledge, rather than the genetic techniques, has been the limiting factor in expanding the use of genetic dissection methods in studies of the neurochemistry of the fruitfly. This situation is rapidly changing. Much information has been accumulating on *Drosophila* neurochemistry. Information is now available on various types of receptors for ACh (reviewed by Dudai 1980, Sattelle 1980), on cholinergic enzymes (J. C. Hall et al 1979), on glutamatergic transmission (Jan & Jan 1976), on aminergic receptors and neurotransmitter-sensitive adenylate cyclase (Uzzan & Dudai 1981), on catecholamine metabolism (Dewhurst et al 1972) and on ion channels (Gitschier et al 1980). It now seems possible to use genetics to tackle questions that are difficult to approach by other methods, and to acquire information complementary to that acquired in other systems. For example, what are the roles of specific neuronal components (e.g. receptors, channels) in the development of the nervous system, and are there crucial periods for their expression? Do different physiological and behavioural phenomena (e.g. various steps in the release of a transmitter, various phases of memory) share common components of a neuronal machinery, and could each of the observed events be dissected into more elementary steps by the isolation of specific mutations? Are subclasses of receptors for neurotransmitters, or different gates for different ions, coded by different genes, or are these macromolecules differentiated only by their microenvironment or by metabolic modifications? And do various components of a neuronal membrane or of a neurotransmitter system share common genetic regulatory mechanisms? The following papers by Hall and by Jan may thus be regarded as examples of steps along a promising road.

Acknowledgement

The work at the author's laboratory is supported by the United States–Israel Binational Science Foundation, Jerusalem.

REFERENCES

Benzer S 1973 Genetic dissection of behavior. Sci Am 229(6):24-37

Byers D, Davis RL, Kiger JA 1981 Defect in cyclic AMP phosphodiesterase due to the *dunce* mutation of learning in *Drosophila melanogaster*. Nature (Lond) 289:79-81

Courtright JB 1976 *Drosophila* gene-enzyme systems. Adv Genet 18:249-314

Dewhurst SA, Croker SG, Ikeda K, McCaman RE 1972 Metabolism of biogenic amines in *Drosophila* nervous tissue. Comp Biochem Physiol 43B:975-981

Dudai Y 1980 Cholinergic receptors of *Drosophila*. In: Sattelle DB et al (eds) Receptors for neurotransmitters, hormones and pheromones in insects. Elsevier/North-Holland Biomedical Press, Amsterdam, p 93-110

Dudai Y, Quinn WG 1980 Genes and learning in *Drosophila*. Trends Neurosci 3:28-30

Dudai Y, Jan YN, Byers D, Quinn WG, Benzer S 1976 *dunce*, a mutant of *Drosophila* deficient in learning. Proc Natl Acad Sci USA 73:1684-1688

Gitschier J, Strichartz GR, Hall LM 1980 Saxitoxin binding to sodium channels in head extracts from wild-type and tetrodotoxin-sensitive strains of *Drosophila melanogaster*. Biochim Biophys Acta 595:291-303.

Greenspan RJ 1980 Mutations of choline acetyltransferase and associated neural defects in *Drosophila melanogaster*. J Comp Physiol 137:83-92

Greenspan RJ, Finn JA, Hall JC 1979 Acetylcholinesterase mutants in *Drosophila* and their effects on the structure and function of the central nervous system. J Comp Neurol 189:741-774

Hall JC, Greenspan RJ 1979 Genetic analysis of *Drosophila* neurobiology. Annu Rev Genet 13:127-195

Hall JC, Kankel DR 1976 Genetics of acetylcholinesterase in *Drosophila melanogaster*. Genetics 83:517-535

Hall JC, Greenspan RJ, Kankel DR 1979 Neural defects induced by genetic manipulation of acetylcholine metabolism in *Drosophila*. In: Cowan WM et al (eds) Society for Neuroscience Symposia vol 4. Society for Neuroscience, Bethesda, Md, p 1-42

Hall JC, Alahiotis S, Strumpf DA, White K 1980 Behavioral and biochemical defects in temperature-sensitive acetylcholinesterase mutants of *Drosophila melanogaster*. Genetics 96:939-965

Hall LM 1980 Biochemical and genetic analysis of an α-bungarotoxin-binding receptor from *Drosophila melanogaster*. In: Sattelle DB et al (eds) Receptors for neurotransmitters, hormones and pheromones in insects. Elsevier/North-Holland Biomedical Press, Amsterdam, p 111-124

Hall LM, Von Borstel RW, Osmond BC, Hoeltzli SD, Hudson TH 1978 Genetic variants in an acetylcholine receptor from *Drosophila melanogaster*. FEBS Lett 95:243-246

Hall LM, Wilson SD, Gitschier J, Martinez N, Strichartz GR 1982 Identification of a *Drosophila melanogaster* mutant that affects the saxitoxin receptor of the voltage-sensitive sodium channel. This volume, p 207-217

Hotta Y, Benzer S 1970 Genetic dissection of the *Drosophila* nervous system by means of mosaics. Proc Natl Acad Sci USA 67:1156-1163

Howard BD, Merriam JR, Meshul C 1975 Effects of neurotropic drugs on *Drosophila melanogaster*. J Insect Physiol 21:1397-1405

Jan LY, Jan YN 1976 L-Glutamate as an excitatory transmitter at the *Drosophila* larval neuromuscular junction. J Physiol (Lond) 262:215-236

Jan YN, Jan LY, Dennis MJ 1977 Two mutations of synaptic transmission in *Drosophila*. Proc R Soc Lond B Biol Sci 198:87-108

Lindsley DL, Grell EH 1968 Genetic variations of *Drosophila melanogaster*. Carnegie Institution of Washington, DC

Lindsley DL, Sandler L, Baker BS et al 1972 Segmental aneuploidy and the genetic gross structure of the *Drosophila* genome. Genetics 71:157-184

Sattelle DB 1980 Acetylcholine receptors of insects. Adv Insect Physiol 15:215-315

Uzzan A, Dudai Y 1981 Aminergic receptors in *Drosophila melanogaster:* responsiveness of adenylate cyclase to putative neurotransmitters. J Neurochem, in press

Wright TRT, Bewley GC, Sherald AF 1976 The genetics of DOPA decarboxylase in *Drosophila melanogaster*. II. Isolation and characterization of DOPA-decarboxylase-deficient mutants and their relationship to the α-methyl-DOPA-hypersensitive mutants. Genetics 84:287-310

Wu CF, Ganetzky B 1980 Genetic alteration of nerve membrane excitability in temperature-sensitive paralytic mutants of *Drosophila melanogaster*. Nature (Lond) 286:814-816

Identification of a *Drosophila melanogaster* mutant that affects the saxitoxin receptor of the voltage-sensitive sodium channel

LINDA M. HALL, SUSAN D. WILSON, JANE GITSCHIER, NANCY MARTINEZ and GARY R. STRICHARTZ†

Albert Einstein College of Medicine, Department of Genetics, 1300 Morris Park Avenue, Bronx, NY 10461 and †Department of Physiology and Biophysics, State University of New York at Stony Brook, Stony Brook, NY 11790, USA

Abstract There is a *no action potential, temperature-sensitive (nap^{ts})* mutant of *Drosophila melanogaster* in which nerve conduction is blocked at increased temperatures which do not block conduction in the wild-type. This blockade is reversible and the conduction returns rapidly when the temperature is lowered. In this mutant strain both larvae and adults show a reversible, temperature-induced paralysis. Electrophysiological recordings at the permissive temperature have shown that this mutant is abnormally sensitive to tetrodotoxin and that the refractory period of the action potential is longer in the mutant than in the wild-type. These studies suggest that this mutation affects the voltage-sensitive sodium channel.

The molecular defect in the *nap^{ts}* mutant was analysed using [^{3}H]saxitoxin as a biochemical probe to study the sodium channel. The mutant has fewer saxitoxin-binding sites than the wild-type controls. There are no differences between *nap^{ts}* and the wild type in K_d, pH sensitivity, and temperature effects on saxitoxin-binding parameters. There appear to be no structural differences between sodium channels in the mutant and wild-type: the mutant characteristics can all be explained by the reduced numbers of sodium channels.

The voltage-sensitive sodium channel plays a key role in the propagation of action potentials along the axons of nerves in both invertebrates and vertebrates. Functionally, this membrane component includes a cation selectivity filter, a transmembrane pore, and voltage- and time-dependent activation and inactivation mechanisms. Through the use of a variety of neurotoxins this

1982 Neuropharmacology of insects. Pitman, London (Ciba Foundation symposium 88) p 207-220

channel has been shown to be composed of at least three pharmacologically distinct sites. (See Catterall 1980 for a review.) Tetrodotoxin and saxitoxin bind to the channel at the outside of the membrane and block the flux of sodium ions, thereby abolishing the action potential. The lipid-soluble toxins, such as veratridine and batrachotoxin, alter the activation and inactivation processes. The polypeptide toxins from scorpion and sea anemone act synergistically with the lipid-soluble toxins and block the inactivation process.

We are interested in characterizing this complex channel in insects and have begun by analysing the properties of the saxitoxin-binding receptor in the fruitfly *Drosophila melanogaster* (Gitschier et al 1980). In addition to characterizing the biochemical and pharmacological properties of the sodium channel in an insect, we hoped that the use of *Drosophila melanogaster* would allow us to begin genetic analysis of this important but complex membrane macromolecule. As a first step towards genetic analysis of the sodium channel we have studied [^{3}H]saxitoxin binding in a variety of mutants which display a temperature-induced paralysis and also show temperature-induced changes in nerve conduction. These mutants were felt to be likely candidates for sodium channel defects. In this paper we describe one such mutant, named *no action potential, temperature-sensitive* or *nap^{ts}*, and we show that this mutant has reduced numbers of sodium channels in flies homozygous for the mutant gene.

Characterization of the saxitoxin receptor in *Drosophila melanogaster*

It is well established that the neurotoxins, saxitoxin and tetrodotoxin, both specifically block the sodium current of the action potential at low concentrations and thereby abolish the action potential (Narahashi 1972). Radioactively labelled toxins have been used as probes with which to study the biochemical properties of the sodium channel in a variety of organisms. The good agreement between the biochemically and electrophysiologically measured binding parameters provides convincing evidence that the toxin receptor is associated with the voltage-sensitive sodium channel. (See recent reviews by Ritchie & Rogart 1977, Ritchie 1979.) As a first step in our analysis of the insect sodium channel, we wished to determine whether [^{3}H]saxitoxin binding in *Drosophila* extracts displayed the properties expected of specific binding to the sodium channel.

Our first criterion for specific binding was that action potentials in *Drosophila* neurons should be blocked by low concentrations of tetrodotoxin and saxitoxin. To show this we used a dissected third-instar larval preparation in which nerves and muscles could be bathed directly in solutions containing the toxins (Gitschier et al 1980). We monitored nerve conduction by intracellular recording of the excitatory junctional potential in a muscle fibre in response to electrical stimulation of the corresponding motor nerve. A loss in evoked response of the muscle was indicative of a block in nerve conduction by the toxins, since the muscle itself was insensitive to tetrodotoxin and saxitoxin. Control experiments showed that muscles in which the evoked response was blocked responded normally when L-glutamate (the putative excitatory neuromuscular transmitter) was applied iontophoretically. Nerve conduction was found to be completely blocked after 20 minutes in a solution containing 200 nM-tetrodotoxin or saxitoxin and conduction was impaired at even lower concentrations. No difference was observed between the sensitivity to tetrodotoxin and to saxitoxin. We concluded that action potentials in *Drosophila* nerves resemble those of other organisms in that conduction is blocked by tetrodotoxin and saxitoxin.

Having established the potency of saxitoxin electrophysiologically, we prepared a crude membrane fraction from *Drosophila* heads and used that to see whether the binding of [^{3}H]saxitoxin was saturable. As described by Gitschier et al (1980), a rapid filtration assay was used to separate the membrane-bound [^{3}H]saxitoxin, which is retained on the filter, from the free toxin which passes through. The net binding to wild-type extracts as described by Gitschier et al (1980) and illustrated by the closed symbols in Fig. 1 was

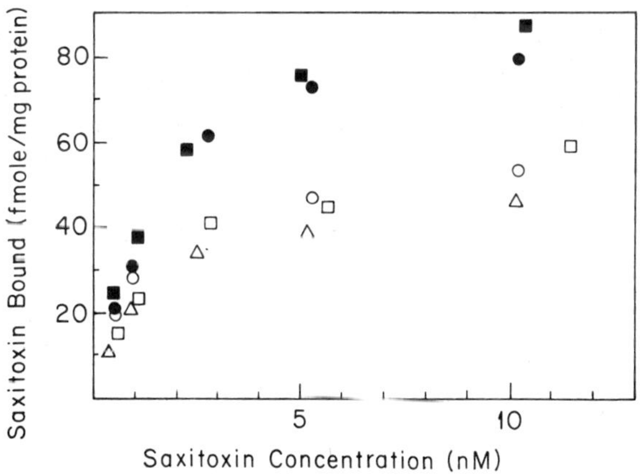

FIG. 1. Saturable saxitoxin binding to *Drosophila* head extracts. Closed symbols, Canton-S wild-type; open symbols, *nap*^{ts} mutant. The different symbols refer to independent experiments.

shown to be saturable. When the data were replotted as described by Scatchard (1949), the equilibrium dissociation constant (K_d) was calculated to be 1.9 nM and the total number of saturable binding site(s) was estimated to be 97 fmoles of binding sites/mg protein. (These values represent the averages from four experiments done on the Canton-S wild-type at 4 °C and were taken from Gitschier et al 1980.)

If saxitoxin is binding to and blocking the selectivity filter of the sodium channel, this binding should show the pharmacological specificity predicted from electrophysiological studies. Tetrodotoxin, which is functionally identical to saxitoxin in its effect on the action potential, would be expected to be an effective inhibitor of [^{3}H]saxitoxin binding. Similarly, inorganic cations might be expected to inhibit toxin binding with an order of effectiveness that parallels the permeability of the channel to these ions. Our early experiments (Gitschier et al 1980) showed that, as expected, tetrodotoxin was an effective inhibitor of [^{3}H]saxitoxin binding to *Drosophila* head extracts. The K_i for tetrodotoxin inhibition was shown to be 0.5 nM. As summarized in Table 1,

TABLE 1 Ability of inorganic monovalent cations to inhibit [^{3}H]saxitoxin binding to *Drosophila* head extracts (from Gitschier 1981)

Ion	*K_i (nM)* *(mean ± SD)*
Li^+	40.7 ± 3.5
Na^+	46.3 ± 4.5
K^+	195 ± 43
Cs^+	216 ± 13

we have recently demonstrated that the order of effectiveness of monovalent cations in inhibiting [^{3}H]saxitoxin binding to *Drosophila* extracts (Li^+ is more effective than Na^+, etc.) paralleled the order of channel permeability for these ions ($Li^+ > Na^+ > K^+ > Cs^+$) (Chandler & Meves 1965).

Thus, by all the criteria used, including functional blockade, binding saturability and pharmacological specificity, [^{3}H]saxitoxin appears to be binding specifically and with high affinity to the ion selectivity filter of the voltage-sensitive sodium channel and thus would be a suitable probe for analysing mutants suspected to have defects in this ion channel.

Mutant analysis

The *no action potential, temperature-sensitive (nap^{ts})* was found to be a single-gene, recessive mutation that was mapped to position 54.5 on the second chromosome and had many properties that suggested it might affect the voltage-sensitive sodium channel (Wu et al 1978). This mutant showed normal locomotor behaviour at the permissive temperature (21 °C), but both the larvae and adults became rapidly paralysed at 37 °C, while the wild-type still showed normal activity at that temperature. The paralysis was rapidly reversed if the mutants were shifted back to the permissive temperature. Electrophysiological studies showed that axonal conduction failed at high temperature in this mutant but returned again at lower temperatures (Wu et al 1978).

More recent electrophysiological studies (Wu & Ganetzky 1980) have demonstrated that even at permissive temperatures axonal conduction in nap^{ts} larvae was abnormally sensitive to tetrodotoxin. The same workers also found that the refractory period between action potentials was prolonged in the mutant relative to the wild-type, again at permissive temperatures. These characteristics were those expected for a mutant affecting the sodium channel. Therefore, we decided to investigate [^{3}H]saxitoxin binding to extracts from nap^{ts} as a way of investigating the molecular nature of the defect in this mutant stock.

As shown in Fig. 1, when saturable saxitoxin binding in nap^{ts} extracts (open symbols) was compared with that in wild-type extracts (closed symbols) the mutant had many fewer saturable binding sites than the wild-type when binding was measured at 4 °C. When results from a number of experiments were pooled, we calculated that the number of saturable binding sites in the nap^{ts} mutant was 57.2 ± 6.4 fmol/mg protein while that for the wild-type was 97.2 ± 4.0. There was, however, no discernible difference between the K_d of the mutant (1.3 ± 0.3 nM) and that of the wild-type (1.5 ± 0.4 nM) at 4 °C.

Since the nap^{ts} mutant exhibits its most pronounced effects at higher temperatures, we were interested to see whether the binding parameters showed marked changes with temperature. Such changes might be expected if the mutant receptor contained an amino acid substitution that rendered the channel protein complex more susceptible to heat denaturation. As summarized in Table 2, there was no difference in the effects of temperature on the

TABLE 2 Effects of temperature on saxitoxin-binding parameters of head extracts from the nap^{ts} mutant and the Canton-S wild-type (from Gitschier 1981)

Drosophila stock	*K_d at 4 °C (nM)*	*K_d at 30 °C (nM)*	*M[a] at 4 °C (fmol/mg protein)*	*M[a] at 30 °C (fmol/mg protein)*
Canton-S wild-type	1.7 ± 0.3	3.2 ± 0.9	114.8 ± 24.3	112.0 ± 21.0
nap^{ts}	1.8 ± 0.4	3.0 ± 0.6	69.7 ± 3.2	72.3 ± 8.1

[a] M, total number of saturable binding sites.

K_d and M (total number of saturable binding sites) of the nap^{ts} mutant and the wild-type when binding at 4 °C was compared with that at 30 °C. It should be noted that the difference in the number of saturable binding sites in the mutant and wild-type was found at both temperatures tested. Binding studies could not be done at the higher temperature of 37 °C because even a short

(5-min) incubation at this temperature caused a significant loss of binding activity in wild-type extracts. These experiments demonstrate that the *nap*^{ts} mutation affects the numbers of saxitoxin-binding sites in the sodium channel but do *not* provide any indication that the mutation causes a change in channel structure. Such structural changes might be expected to affect the K_d, or the thermal sensitivity of the binding parameters, or both.

To investigate further the possibility of structural changes induced in the sodium channel by the *nap*^{ts} mutant we compared the effects of pH on saxitoxin-binding activity in mutant and wild-type extracts. As summarized in Fig. 2 the mutant *nap*^{ts} and wild-type Canton-S extracts show exactly the same

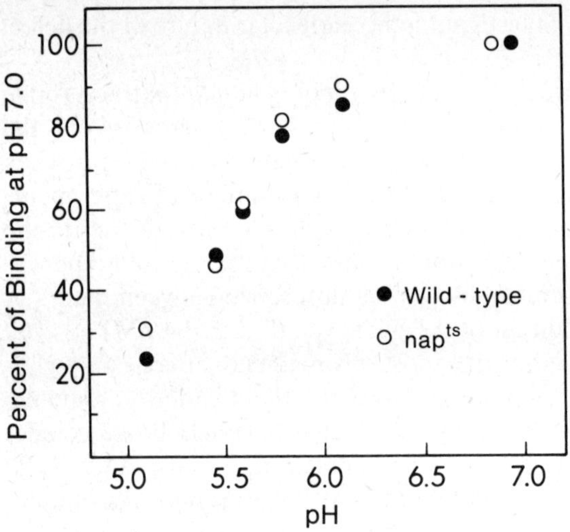

FIG. 2. Effects of hydrogen ions on saxitoxin binding to *Drosophila* head extracts from the *nap*^{ts} mutant and the Canton-S wild-type.

pH sensitivity profile, again providing no evidence for effects on sodium channel structure by the *nap*^{ts} mutant.

Since the *nap*^{ts} mutant has been consistently shown to reduce the numbers of saxitoxin receptors, we were interested to see whether this effect was specific for the sodium channel or whether the mutant affected other neuronal membrane components as well. To investigate this, we compared the levels of saturable α-bungarotoxin-binding activity in *nap*^{ts} mutant and wild-type head extracts. ^{125}I-labelled α-bungarotoxin has been shown to bind to a *Drosophila* membrane component with the properties expected of a nicotinic acetylcholine receptor (Schmidt-Nielsen et al 1977, Dudai 1979). Using a filter binding assay similar to that described by Schmidt-Nielsen et al (1977) we were able to show that the mutant *nap*^{ts} and the wild-type Canton-S have exactly the same number of saturable α-bungarotoxin binding sites (Fig. 3). We found no evidence that the *nap^{ts}* mutant is causing a *general* reduction in *Drosophila* neural tissue. It will be interesting to determine whether *nap^{ts}* affects levels of the other pharmacologically distinct sites that are associated with the sodium channel.

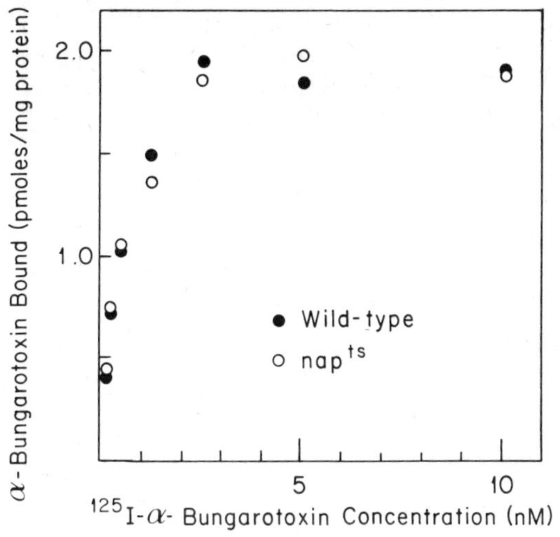

FIG. 3. Saturable ^{125}I-labelled α-bungarotoxin binding to *Drosophila* head extracts from the *nap^{ts}* mutant and the Canton-S wild-type.

Explanation of the *nap^{ts}* phenotypes

Since we can find no evidence for structural changes in the sodium channel caused by the *nap^{ts}* mutation, we have considered the possibility that all the phenotypic features of the mutant may be due to a reduction in the number of functional sodium channels. In brief, the mutant characteristics which must be accounted for include: heat-induced paralysis, heat-induced loss of axonal conduction, increased sensitivity to tetrodotoxin blockade of the action potential at permissive temperatures, and increased length of the refractory period at permissive temperatures. If all these alterations in the mutant are caused by a decrease in the number of functional sodium channels, it should be possible to make a *phenocopy* of these mutant characteristics by pharmacologically reducing the number of functional sodium channels in the wild-type strain. In the experiment summarized in Table 3 we applied tetrodotoxin solutions of varying concentrations to filter paper disks and allowed wild-type adults to feed from these disks for 24 hours at 21 °C. At the end of this time we scored the number of survivors, transferred them to 37 °C

TABLE 3 Effects of sublethal doses of tetrodotoxin on behaviour of the Canton-S wild-type *Drosophila* at 37 °C

Tetrodotoxin concentration (μg/ml)	*% lethality (24 h, 21 °C)*	*Behaviour after 1 min at 37 °C*
0	0	Normal locomotion
10	47	Convulsive motions
20	77	Fall over; some leg shaking
40	92	All paralysed
60	98	All paralysed
80	99	All paralysed

for 1 minute, and then observed their behaviour. Flies fed a toxin-free control solution showed normal locomotor activity at 37 °C while the flies which had ingested tetrodotoxin showed debilitation that varied in severity depending on the concentration of tetrodotoxin which they were fed. This is consistent with the suggestion that the temperature-induced paralysis in the *nap^ts^* mutant is due to a reduced number of functional sodium channels.

In the light of these observations, we suggest that the heat-induced failure of axonal conduction and the resulting paralysis could be caused by an increase in the leakiness of the neuronal membranes at higher temperatures coupled with the mutation-induced reduction in numbers of sodium channels. It is known that the cable input resistance of neurons decreases with increasing temperature (Lagerspetz 1974). This means that at higher temperatures a signal will be passively conducted over a shorter distance before it dies out. This increased leakage at higher temperatures could result in failure of the propagation of a signal between the more sparsely distributed sodium channels in the *nap^ts^* neurons whereas, with the same leakiness, propagation could continue in wild-type neurons with higher channel density.

The lengthening of the refractory period can be explained on the basis of reduced numbers of sodium channels in the *nap^ts^* mutant. If the channels that *are* present in this mutant have the normal kinetics of activation and inactivation, a greater percentage of the sodium channels will have to be reactivated in the mutant for the same number of channels to be active at the end of the refractory period as in the wild-type. Since the kinetics are unchanged, this reactivation will take longer for the mutant than for the wild-type. (It is assumed that in the wild-type there are 'spare' sodium channels so that some of the channels can still be inactive at the end of the observable refractory period.)

The enhanced sensitivity to tetrodotoxin observed in electrophysiological studies of the *nap^ts^* mutant at permissive temperatures can also be explained by reduced numbers of sodium channels. Although the affinity for tetrodotoxin is apparently unchanged in the *nap^ts^* mutant, the presence of fewer channels means that a smaller percentage of channels need to bind tetrodotoxin in the mutant for the same blockade to be achieved as in the wild-type. Thus, equivalent blockade in the mutant will occur at a lower concentration of tetrodotoxin than required for wild-type.

Discussion and future prospects

Through the use of ligand-binding studies and a neuropharmacological approach, we have developed a hypothesis to explain one of the temperature-sensitive neurological mutants of *Drosophila melanogaster*. Future studies will be directed at testing this hypothesis further. We shall determine whether phenocopies of other characteristics of the mutant, such as a lengthened refractory period, can be made by pharmacologically reducing the number of sodium channels in the wild-type. We shall investigate in more detail the relationship between temperature-induced paralysis and the number of functional sodium channels. By screening for other alleles at the *nap* locus we should be able to isolate other mutants with varying numbers of functional sodium channels. These mutations should allow us to define the minimal number of sodium channels required for survival. Single-gene mutations affecting the sodium channels act like very specific pharmacological agents and should ultimately allow us to dissect structure–function relationships in the channel. Temperature-sensitive mutants, such as nap^{ts}, allow us to turn neuron excitability on and off simply by manipulating the temperature. Thus, this mutant, used in conjunction with the powerful mosaic analysis tools (Hall et al 1976), should allow us to define the role of cell excitability in various developmental and behavioural processes.

As mentioned at the beginning of this paper, the sodium channel is composed of at least three pharmacologically distinct sites. It is of interest to know whether the different toxin-binding sites and channel functions are found in a single gene product or in different gene products. Genetic analysis of a variety of sodium channel mutants would be an ideal approach to this problem. A possible strategy for such an approach is suggested by the recent observation of Wu & Ganetzky (1980) that double mutants consisting of the $para^{ts}$ mutant (on the X chromosome) and nap^{ts} (on the second chromosome) fail to survive even at normally permissive temperatures. One interpretation of these results is that $para^{ts}$ and nap^{ts} affect the same physiological pathway. Since nap^{ts} has been shown to affect the number of sodium channels, perhaps $para^{ts}$ also affects some aspect of channel function. New mutations affecting sodium channel function might be isolated by looking for mutants which die when they also carry the nap^{ts} mutation.

Finally, our studies with the nap^{ts} mutant have provided some information about the consequences of a reduction in sodium channels under temperature-stress conditions. The saxitoxin that we use for these studies is the toxic agent in the shellfish poisoning that affects seafood-eaters. Our results suggest that organisms are better able to survive a pharmacological reduction in their sodium channel numbers if they are maintained at lower temperatures. This would suggest that a reduction in body temperature might aid the survival of victims of shellfish-poisoning. Saxitoxin binds reversibly to sodium channels even at low temperatures (Gitschier et al 1980), so reduction in body temperature may help neurons to keep functioning until the toxin dissociates from its receptor and can be removed. Such therapy could be readily tested in mammalian models.

Acknowledgements

This work was supported by National Institutes of Health grants NS 16204 to L.M.H. and NS 12828 to G.R.S. We thank Dr T. R. Venkatesh for help in preparing the manuscript.

REFERENCES

Catterall WA 1980 Neurotoxins that act on voltage-sensitive sodium channels in excitable membranes. Annu Rev Pharmacol Toxicol 20:15-43

Chandler WK, Meves H 1965 Voltage clamp experiments on internally perfused giant axons. J Physiol (Lond) 180:788-820

Dudai Y 1979 Cholinergic receptors in insects. Trends Biochem Sci 4:40-44

Gitschier JM 1981 Tritiated saxitoxin as a probe for voltage-sensitive sodium channels in wild-type, mutant, and aneuploid *Drosophila melanogaster*. PhD thesis, Massachusetts Institute of Technology, Cambridge, Mass.

Gitschier J, Strichartz GR, Hall LM 1980 Saxitoxin binding to sodium channels in head extracts from wild-type and tetrodotoxin-sensitive strains of *Drosophila melanogaster*. Biochim Biophys Acta 595:291-303

Hall JC, Gelbart WM, Kankel DR 1976 Mosaic systems. In: Ashburner M, Novitski E (eds) The genetics and biology of *Drosophila*, vol 1a. Academic Press, New York, p 265-314

Lagerspetz KYH 1974 Temperature acclimation and the nervous system. Biol Rev 49:477-514

Narahashi T 1972 Mechanism of action of tetrodotoxin and saxitoxin on excitable membranes. Fed Proc 31:1124-1132

Ritchie JM 1979 A pharmacological approach to the structure of sodium channels in myelinated axons. Annu Rev Neurosci 2:341-362

Ritchie JM, Rogart RB 1977 The binding of saxitoxin and tetrodotoxin to excitable tissue. Rev Physiol Biochem Pharmacol 79:1-50

Scatchard G 1949 The attractions of proteins for small molecules and ions. Ann NY Acad Sci 51:660-672

Schmidt-Nielsen BK, Gepner JI, Teng NNH, Hall LM 1977 Characterization of an α-bungarotoxin binding component from *Drosophila melanogaster*. J Neurochem 29:1013-1029
Wu CF, Ganetzky B 1980 Genetic alteration of nerve membrane excitability in temperature-sensitive paralytic mutants of *Drosophila melanogaster*. Nature (Lond) 286:814-816
Wu CF, Ganetzky B, Jan LY, Jan YN, Benzer S 1978 A *Drosophila* mutant with a temperature-sensitive block in nerve conduction. Proc Natl Acad Sci USA 75:4047-4051

DISCUSSION

Barnard: Could the reduction in the number of saxitoxin binding sites be a gene-dosage effect? Are there two active copies of a channel-determining gene in the normal genome?

Hall: We tried to identify the structural gene for the sodium channel using the segmental aneuploidy technique that Yadin Dudai discussed. We put in duplications and expected to find increased levels of saxitoxin binding. A couple of segmental duplications looked promising. One was extremely exciting because it covered the *nap^ts* gene and we felt we had found a structural gene for the sodium channel. When we broke this particular region down into smaller segments, the gene-dosage effect disappeared. We don't think we are dealing simply with a gene-dosage effect here. I think we were unable to find a consistent gene dosage because the sodium channel, like the acetylcholine receptor, is probably a complex macromolecule. If several different polypeptides are needed to make a functional channel, the genes coding for these polypeptides might be scattered throughout the genome. Unless all the genes are duplicated at once, that is, unless they are clustered, we wouldn't see an enhanced binding effect. We tried to use the segmental aneuploidy technique to map the gene(s) for the α-bungarotoxin-binding receptor and were also unable to find a gene-dosage effect. Here we are certain the receptor consists of several different polypeptides and failure to find a gene-dosage effect is taken as evidence that the different genes are not clustered together. In all cases to date where the segmental aneuploidy technique has been successfully applied, the proteins studied have been either monomers or homopolymers consisting of perhaps a dimer of the same polypeptide. In complicated systems like the heteropolymers that form membrane-bound receptors this technique apparently doesn't work.

Barnard: What other mechanisms, then, could be expected to explain a genetically determined reduction in the amount of a gene product in *Drosophila* to one half?

Hall: I don't know. There may be alterations in the membranes, or specific sites in the membranes where these things can plug in, and this gene may affect those sites somehow. Alternatively, the *nap* mutant may represent a malfunctioning of a gene which regulates channel levels in response to physiological cues. We have no evidence that we have altered the structure of the channel so I find it hard to invoke mechanisms such as increased degradation which would accompany a mutant receptor, for example.

Dudai: The relation between an apparent defect in a gene and the product of the gene is often not simple. J. C. Hall and his colleagues isolated temperature-sensitive acetylcholinesterase mutants (Hall et al 1980). The flies are paralysed at a non-permissive temperature, at which wild-type flies are not affected. It was shown by Hall et al (1980) that if you extract the enzyme from the mutants and assay it, in some mutations the temperature sensitivity of the enzyme *in vitro* does not parallel the sensitivity of the organism. This suggests that you may be looking not only at the enzyme but also at the relationship of the enzyme to something else, for example a membrane in which it has to be integrated. An interesting example is a cold-sensitive mutant of acetylcholinesterase in *Drosophila* in which the enzyme activity *in vitro* is not abolished at a low, restrictive temperature (Hall et al 1980). It is possible that it is the incorporation of the enzyme into the membrane that is affected by this mutation. So not all the properties encountered in the phenotype will be displayed by the gene product *in vitro*.

Barnard: Does the acetylcholinesterase activity go down to one half in a heterozygote?

Dudai: It goes down to 60% in the heterozygote (which contains one dose of the defective gene and one of the normal gene). There is usually some dosage compensation, perhaps 10–15%.

Usherwood: Are other membrane components also modified in this system?

Hall: The only one that we have looked at is the α-bungarotoxin-binding component and that is not affected. We know nothing about calcium channels, potassium channels or other components.

Usherwood: So it is possible that there is some general defect within the membrane which influences channel activity in a variety of different ways?

Hall: It is possible, and along these lines I would like to see whether the other pharmacological sites in the sodium channel are affected.

Usherwood: Dr Jan, have you any evidence for changes in the properties of the skeletal muscles in this mutant?

Jan: There are no obvious changes. In this mutation, Dr Hall, can you really rule out the possibility that the properties of the sodium channels have remained the same as in the normal fly, since you couldn't do the study at the non-permissive temperature?

Hall: That is a possibility. We showed that the pH profiles were the same and that there was no difference in the K_d values of mutant and wild-type at 4 °C and 30 °C.

Jan: But nerve conduction is not affected in the mutant at 30 °C.

Hall: That is right. We have been concerned about this issue and so we also looked at thermal inactivation profiles of the mutant and wild-type at higher temperatures (>40 °C). These experiments are not as reproducible as I would like them to be but we saw no difference between the mutant and the wild-type. We are still pursuing experiments that would allow us to look for alterations in the properties of the saxitoxin receptor in the mutant, but I don't think we are going to find them.

Kelly: I am a bit worried about the neurophysiology. As you raised the temperature the time constant of the neurons decreased; in other words, either the resistance or the capacitance of the membrane changed. But in the mammalian system action potentials rise faster as the temperature is raised. Even up to 41 °C in brain slices the spikes become slightly higher and shorter in duration, which is the opposite of what you are saying.

Hall: My evidence for that is indirect. It comes from published work which suggests that as the temperature is increased the membranes become leakier, so the length constant would decrease (Lagerspetz 1974).

Kelly: But the experiment has to be done. There could be a depolarization, with the sodium pumps becoming less efficient. Or there may be a difference in the time constants. My impression is that halving the number of sodium channels wouldn't have a dramatic effect. There would have to be a really large change in the time constant to block conduction completely.

Hall: Is it known how many sodium channels are required for conductance at various temperatures? This mutation might allow us to look at that.

Kelly: The one thing that fits in here is that people with disseminated sclerosis say they are much better off when it is cold. In the acute phase the nerves demyelinate and the underlying neuronal membrane is devoid of sodium channels. Later they grow out from the nodal tissue to cover the exposed membrane (Bostock & Sears 1978).

Hall: I was concerned about whether a reduction in the number of sodium channels would cause a temperature-induced paralysis and that is why I set up the tetrodotoxin feeding experiment which allowed us to reduce the number of sodium channels in the wild-type pharmacologically. When that was done, we saw a temperature-induced paralysis in the wild-type reminiscent of that seen with the mutant. When flies are fed a dose of tetrodotoxin that kills 50% of them, the wild-type survivors are certainly more sensitive than untreated flies to temperature-induced paralysis. There are a number of possible reasons for that, including what you have just suggested. The fact that I can pharmacologically reduce the number of channels and mimic mutant paralysis argues that the reduction in channel number is sufficient to cause the effect. The actual temperature sensitivity may have a variety of explanations involving temperature effects on things such as membrane resistance, potassium channels, or the sodium pump.

Narahashi: You could also do *in vitro* experiments in support of your hypothesis. You would have to choose the preparation carefully for appropriate membrane properties and channel density. With a fraction of sodium channels blocked by tetrodotoxin, you can repeat the experiment to see the temperature sensitivity.

Hall: There may be species differences and we have to worry about that.

Narahashi: I think the species differences have something to do with the density of the sodium channels and the diameter of the fibres. Can you translate your figure of 114 fmol/mg protein into the channel density?

Hall: I can't do that off the top of my head.

REFERENCES

Bostock H, Sears TA 1978 The internodal axon membrane: electrical excitability and continuous conduction in segmental demyelination. J Physiol (Lond) 280:273-301

Hall JC, Alabiotis SN, Strumpf DA, White K 1980 Behavioral and biochemical defects in temperature-sensitive acetylcholinesterase mutants of *Drosophila melanogaster*. Genetics 96:939-965

Lagerspetz KYH 1974 Temperature acclimation and the nervous system. Biol Rev 49:477-514

Genetic and immunological studies of the nervous system of *Drosophila melanogaster*

YUH NUNG JAN and LILY YEH JAN

Department of Physiology, University of California, San Francisco, School of Medicine, San Francisco, California 94143, USA

Abstract The fruitfly *Drosophila melanogaster* has been a favourite organism in classical genetic studies and in studies of gene structure and regulation using recombinant DNA techniques. The information accumulated from these studies is potentially useful in studying neurobiology. It is therefore worth characterizing the nervous system in *Drosophila* in some detail. Electrophysiological and immunological studies are described:

(1) The neuromuscular junctions in *Drosophila* larvae approach those of the frog in their accessibility to electrophysiological analysis. Single-gene mutations affecting synaptic transmission can be isolated. Three such mutations have been studied in some detail.

(2) So far, substance P-like, ACTH-like and enkephalin-like immunoreactivities have been found in the nervous system and reproductive system of *Drosophila*. Monoclonal antibodies that recognize specific subclasses of nerve fibres have also been obtained. Techniques unique to *Drosophila* genetics allow the structural genes to be precisely localized.

One approach to studying the nervous system is to use genetics as a tool to augment electrophysiological, immunological and biochemical studies. The fruitfly *Drosophila melanogaster* is often used in this approach because its genetics has been analysed extensively (Ashburner & Novitski 1976). Single-gene mutations that perturb neural function or development have been isolated and analysed by many investigators hoping to gain insight into the molecular mechanisms underlying sensory transduction, neurospecificity, or complex behaviour (for review, see Benzer 1973, Hall et al 1981). Our own interests have been various aspects of the function and development of the synapse. Specifically, we are interested in identifying the elements necessary for synaptic transmission or changes in synaptic efficacy, and the various

1982 Neuropharmacology of insects. Pitman, London (Ciba Foundation symposium 88) p 221-239

transmitter systems in the nervous system. Both electrophysiological and immunological methods have been used and will be described here.

Neuromuscular transmission in *Drosophila* larvae

For genetic analysis of the synapse one needs a functional assay. The fruitfly is so small that most of its synapses are not readily accessible to electrophysiological experimentation but fortunately there are a few exceptions. For example, the neuromuscular junctions of full-grown larvae proved to be a favourable preparation (Jan & Jan 1976a,b). Although a larva is only 3–4 mm in length, the fibres of its body wall musculature are rather large: a typical fibre has a diameter of 60–80 μm, comparable to the diameter of frog muscle fibres. Therefore, impaling a fibre with two microelectrodes for intracellular recording and current passing is relatively easy. Further, these muscle fibres are only 400 μm long. Hence, each fibre is essentially isopotential; excitatory junctional potentials (EJPs) and miniature excitatory junctional potentials (MEJPs) can be recorded with little attenuation everywhere in the fibre. The nerve–muscle preparation is thin, so that synaptic boutons in living preparations are visible with Nomarski optics. This allows one to monitor synaptic currents with focal extracellular recording and to apply chemicals iontophoretically near the synapse.

Muscles in the larval body wall receive excitatory innervation from the central nervous system. The neuromuscular junction in *Drosophila* larvae resembles other chemical synapses in that the spontaneous and evoked release of transmitter is quantal and follows a Poisson distribution, and transmitter release varies with approximately the fourth power of the external calcium concentration (Jan & Jan 1976a). The transmitter released at this neuromuscular junction is probably L-glutamate, as in other insect neuromuscular junctions (Usherwood 1980). Iontophoretic applications of L-glutamate at synaptic regions depolarize the muscle and induce permeability changes similar to those occurring during an EJP, even in solutions containing so little calcium that transmitter release is blocked. The reversal potentials of the EJP and the L-glutamate potential are identical; in both cases the ionic dependence of the reversal potentials can be predicted from an extended constant-field equation using a ratio of sodium to potassium permeabilities of $P_{Na}/P_K = 1.3$, and a ratio of magnesium to potassium permeabilities of $P_{Mg}/P_K = 4.7$ (Jan & Jan 1976b). The involvement of magnesium ions in the generation of EJPs is unexpected. However, the haemolymph in *Drosophila* larvae is rich in magnesium and calcium, and an inward current carried by divalent ions is probably an important component of the excitatory junctional current. Glutamate receptors on *Drosophila* muscles, unlike those in locust

(Cull-Candy 1976), are localized only in subsynaptic regions. Glutamate iontophoresis experiments showed that the glutamate sensitivity dropped from 200 mV/nC at the neuromuscular junction to less than 0.1 mV/nC as the glutamate pipette was moved away from the synapse.

Mutations affecting neuromuscular transmission

Having established the basic properties of neuromuscular transmission in normal larvae we screened behavioural mutants for mutations affecting synaptic transmission. Several mutations were found to alter neuromuscular transmission. Three of these have been studied in some detail (Table 1). We shall briefly describe these mutants and discuss their potential use.

TABLE 1 Mutations affecting neuromuscular transmission

Mutations	*Localization*	*Behavioural abnormality*	*Abnormality at neuromuscular junction*	*Possible molecular defect*
bang-sensitive^{MW1} (*bas*^{MW1})	Between *g*(1–44.4) and *sd* (1–51.5)	Paralysed by sudden jolt	Abnormal long-term facilitation	Unknown[a]
Shaker (Sh)	1–58	Shaking under ether anaesthesia	Prolonged nerve action potential	Abnormal function of potassium channels
no action potential, temperature-sensitive (nap^{ts}*)*	Between *pr* (2–54.5) and *cn* (2–57.5)	Paralysed at high temperature	Nerve conduction blocked at high temperature	Abnormal function of sodium channels

[a] The defect in *bang-sensitive*^{MW1} could be either an abnormally rapid accumulation of sodium inside nerve terminals during nervous activities, or an abnormal sodium sensitivity of the machineries responsible for long-term facilitation.

Dissection of synaptic facilitation by a genetic lesion: bang-sensitive (Jan & Jan 1978)

At many neuromuscular junctions the amount of transmitter released in response to a nerve stimulus is increased by previous impulses. This is called facilitation and involves at least two processes, each with a characteristic time course of development and of decay. Facilitation also takes place at the

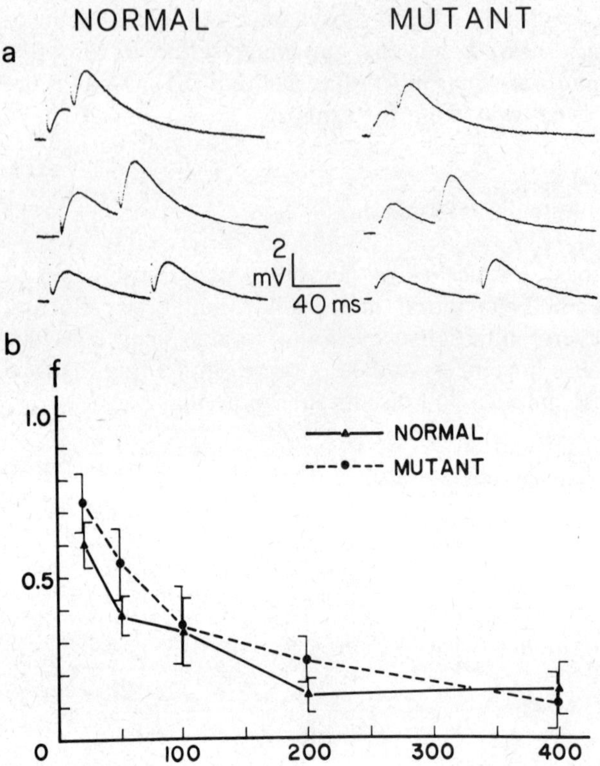

FIG. 1. Short-term facilitation in normal (C-S) (▲) and mutant (*bas*^{MW1}) (●) larvae in 0.14 mM-Ca^{2+}.

(a) Two pulses at 20-, 50- or 100-ms intervals were given once every 10 s; 32 responses were averaged to smooth out the quantal fluctuations. The amount of facilitation is similar for normal and mutant larvae.

(b) Amount of short-term facilitation (f) as a function of the interval between the two pulses; f is the fractional increase of the second response above the first. Error bars indicate SD; results from 4–8 experiments on different larvae are averaged for each point. (From Jan & Jan 1978.)

Drosophila larval neuromuscular junction. When two nerve stimuli were given within a few hundred milliseconds, the second excitatory junctional potential (EJP) was on the average larger than the first (Fig. 1). This is called short-term facilitation and is due to an increase in the average number of packets of transmitter released per impulse. Repetitive stimulation causes the EJP amplitude to rise to a plateau within one second, through summation of the short-term facilitatory effects. As stimulation is maintained for tens of seconds, however, there is a further growth of the EJP (Fig. 2). This process is called long-term facilitation and is also presynaptic in nature. The first clue

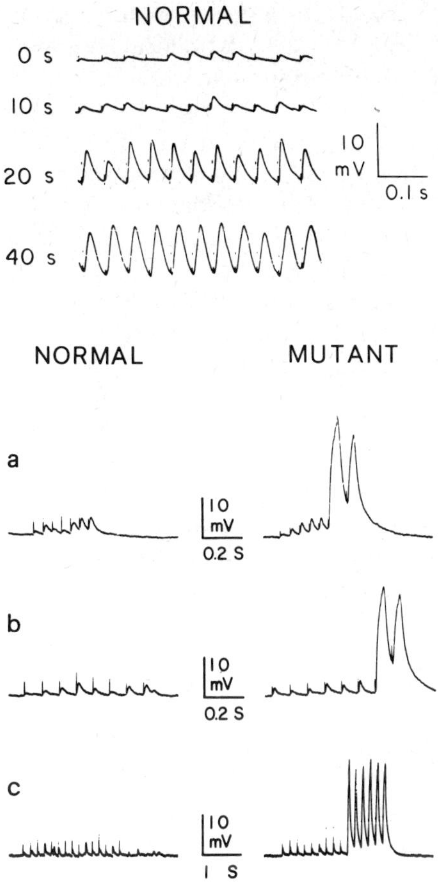

FIG. 2. Long-term facilitation in normal (upper half) and *bas*^{MW1} (lower right) larvae. *Upper half:* EJPs in normal larvae during a train of stimuli at 10 Hz. From top to bottom: four segments of records at different times (0, 10, 20, 40 s) after the start of the train of stimuli. *Lower half:* EJPs in normal (left) and mutant (right) larvae during trains of stimuli at 20 Hz(a), 10 Hz(b) and 5 Hz(c). External Ca^{2+} concentration was 0.14 mM. (From Jan & Jan 1978.)

that short-term and long-term facilitation in *Drosophila* may involve different underlying mechanisms was derived from the isolation of a mutant which had normal short-term facilitation but abnormal long-term facilitation.

The facilitation mutant was isolated as a behavioural mutant and named *bang-sensitive*^{MW1}. When the culture vial is given a sudden jolt these flies are immobilized for one to two minutes, whereas normal flies are not paralysed by this treatment. At the neuromuscular junction *bang-sensitive*^{MW1} larvae show normal short-term facilitation (Fig. 1). Long-term facilitation in this mutant, on the other hand, is greatly enhanced. For instance, with repetitive stimulation at 10/s, ten nerve stimuli were enough to induce long-term facilitation in the mutant, while over a hundred stimuli were needed in normal larvae (Fig. 2).

The fact that a mutation altered long-term but not short-term facilitation suggests that these two processes have different underlying mechanisms. This was confirmed by physiological and pharmacological experiments. Tetrodotoxin (TTX) or removal of external sodium abolishes nerve conduction along larval nerves. However, short-term facilitation can still be induced by repetitively stimulating the nerve at a small distance from the neuromuscular junction, so that the nerve terminal is depolarized passively, through electrotonic spreading of currents. Long-term facilitation, on the other hand, is totally abolished by blocking nerve conduction. In fact, experimental manipulations designed to accelerate the Na^+,K^+-ATPase activity or to decrease the amount of sodium influx during an action potential always depress long-term facilitation, while manipulations that decrease the sodium pump activity enhance it. Therefore, unlike short-term facilitation, long-term facilitation is probably caused by the accumulation of sodium inside the nerve terminal. Similar sodium dependence of the long-term facilitatory effects has been found in crustacean and frog neuromuscular junctions (Atwood et al 1975, Rahamimoff et al 1978).

In summary, facilitation of synaptic transmission in *Drosophila* involves at least two processes with different underlying mechanisms. This was first suggested by the existence of a mutation that affects long-term but not short-term facilitation and was later confirmed by pharmacological studies. Thus, a single-gene mutation has served as a genetic lesion for the dissection of processes involving changes of synaptic efficacy.

Mutations causing defective potassium channels: Shaker (Jan et al 1977)

Shaker flies shake their legs vigorously under ether anaesthesia (Kaplan & Trout 1969) whereas normal flies are immobile or shake their legs only slightly. The excitatory junctional potentials in *Shaker* larvae are abnormally large and prolonged, because nerve-evoked transmitter release is greatly prolonged in the mutant. This abnormality is due to an abnormally prolonged nerve action potential, as indicated by calcium iontophoresis and extracellular recording experiments (Jan et al 1977, Jan & Jan 1979).

Prolonged nerve action potentials have also been demonstrated by intracellular recordings from giant axons of *Shaker* flies (Tanouye et al 1981). Pharmacological experiments on both larval and adult preparations showed that nerve action potentials in *Shaker* remained prolonged after sodium channels were blocked by TTX or after calcium channels were blocked by cobalt. Further, the same phenotype of *Shaker* was mimicked by treating normal flies or larvae with a potassium channel-blocking agent, 4-aminopyridine (Jan et al 1977, Tanouye et al 1981). These studies indicated that the *Shaker* mutation altered the functioning of potassium channels and caused the action potentials to be prolonged.

Conditional mutations: no action potential, temperature-sensitive (nap^{ts}) (Wu et al 1978)

It is conceivable that a mutation causing severe defects in normal function or development will be lethal to the animal. One way to get round this problem is to look for conditional mutants, such as temperature-sensitive paralytic mutants (Grigliatti et al 1973, Siddiqi & Benzer 1976). One such mutant, *nap*^{ts}, causes a rapid and reversible paralysis of larvae and adult flies at restrictive temperatures (above 34 °C). At the larval neuromuscular junction axonal conduction in *nap*^{ts} fails at high temperature, while synaptic transmission is unimpaired, since transmitter release can still be induced by passively depolarizing the nerve terminals (Wu et al 1978). At the permissive temperature nerve conduction in *nap*^{ts} is abnormally sensitive to TTX, a specific blocker of sodium channels. The refractory period of nerve action potentials in *nap*^{ts} is also longer than normal (Wu & Ganetzky 1980). Moreover, [^{3}H]saxitoxin binding studies have revealed that this mutant has reduced numbers of sodium channels (Hall et al, this volume). Taken together, these results suggest that the *nap*^{ts} mutation affects the voltage-dependent sodium channels.

In addition to *Shaker* and *nap*, which may be the structural genes of potassium and sodium channels respectively, several other genes coding for neuronal components have been identified recently. These include genes for acetylcholinesterase (Hall & Kankel 1976), choline acetyltransferase (Greenspan 1980) and dopa decarboxylase (Sherald & Wright 1974). Once a large number of genes crucial for neural functions have been localized, one can begin to ask the following questions. Are there organizations of these genes at the genomic level? Do they form clusters on the chromosome? Are they under common control for expression? How are these genes regulated during development of the nervous system? Some of these questions can be tackled at the molecular level. A number of genes in *Drosophila* have been cloned in bacteria. With the small genome of *Drosophila* it will soon be possible to clone any gene of interest, so long as its location on the chromosome is known and chromosomal rearrangements or mutations of that gene exist. This will make it possible to study in detail the expression and regulation of genes crucial for neural functions. These techniques will also open up the possibility of isolating the molecular entities constituting the neuronal components of interest, despite their scarcity *in vivo*, by cultivating and transcribing their genes.

For the reasons described above it seems worthwhile to identify as many genes crucial for neural function or development as possible. For this purpose, we have used immunological methods to augment the electrophysiological studies.

Immunological studies

In the past decade immunological techniques have had tremendous impact on neurobiology. We have recently adopted two approaches that have been widely used.

(1) Purified, well-defined, neuronal antigens have been used to raise antibodies. These antibodies were then used in radioimmunoassays or immunohistochemical experiments to detect or to localize minute amounts of the neuronal antigen in tissues. Recent studies of neuropeptides represent a good example of this approach. Antibodies to bioactive peptides have been used to determine the distribution of these peptides throughout the nervous system, so providing insights to their functions. Many of the neuropeptides discovered in the vertebrate nervous system are thought to function as neurotransmitters (e.g. Nicoll et al 1980). If so, these molecules might be highly conserved during evolution, since many 'classical' transmitters such as acetylcholine, monoamines, glutamate and γ-aminobutyric acid serve similar functions in both vertebrates and invertebrates. With this in mind, we used antibodies specific for mammalian peptides to look for immunoreactive materials in *Drosophila*.

First, we did indirect immunofluorescence staining on cryostat sections of flies or larvae using antibodies against luteinizing hormone-releasing hormone, thyrotropin-releasing hormone, substance P, angiotensin I, angiotensin II, leucine-enkephalin (Leu-enkephalin) or adrenocorticotropic hormone (ACTH). Antibodies to substance P, Leu-enkephalin or ACTH showed specific binding to tissues in *Drosophila* (Table 2); preadsorption of the primary antisera with synthetic peptides totally eliminated the staining. The substance P-like immunoreactivity was found in accessory glands and ejaculation bulbs of the male reproductive system (Fig. 3). The ACTH-like immunoreactivity and enkephalin-like immunoreactivity were both found in fibres and varicosities in the brain, optic ganglia, and cervical connectives (Fig. 4), as well as in the reproductive system. We are now trying to establish

TABLE 2 Localization of materials cross-reactive with mammalian peptides

Antibodies specific for:	*Structures stained in immunohistochemical experiments*
Substance P	Male reproductive system (accessory glands, ejaculation bulbs)
ACTH	Fibres in central nervous system, male and female reproductive system
Leu-enkephalin	Fibres in central nervous system, male and female reproductive system

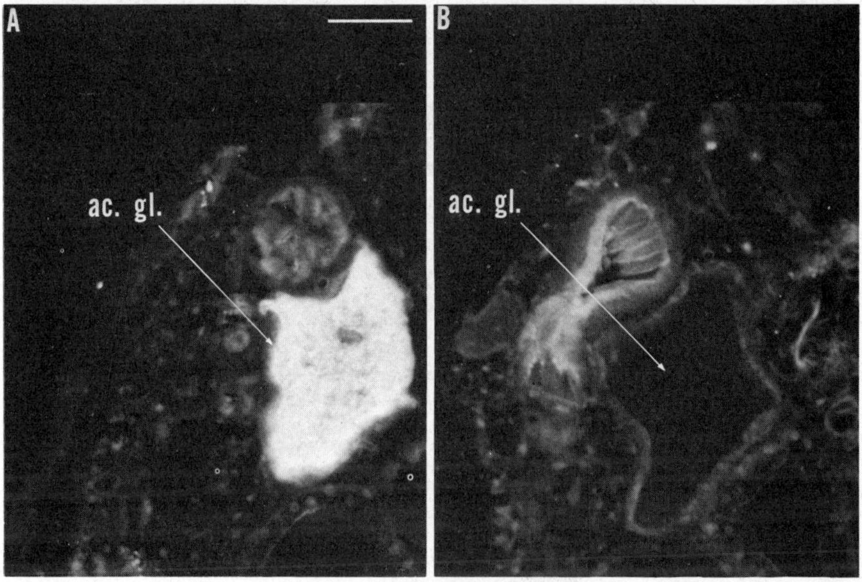

FIG. 3. (a) Immunofluorescent staining of substance P-like immunoreactivity in the accessory gland (ac.gl.) in the abdomen of a male fly. Horizontal sections of the fly were cut on a cryostat, then treated sequentially with rabbit antisera specific for substance P and rhodamine-labelled goat antibodies that were specific for rabbit antibodies. All antibody solutions contained 0.3% Triton.

(b) Preadsorption control in which the rabbit antisera were incubated with 70 μg/ml substance P before they were applied to the sections. Calibration bar indicates 50 μm.

whether structures containing these immunoreactive materials in the reproductive system are neural in nature. Radioimmunoassays coupled with biochemical separation techniques are being used to determine whether these materials are indeed peptides, and whether they are identical to mammalian peptides. These experiments might also lead to the identification of structural genes coding for substance P-like, ACTH-like and enkephalin-like materials, since the amount of a gene product in the fly usually varies in proportion to the number of gene copies contained in that fly's genome. Since there are

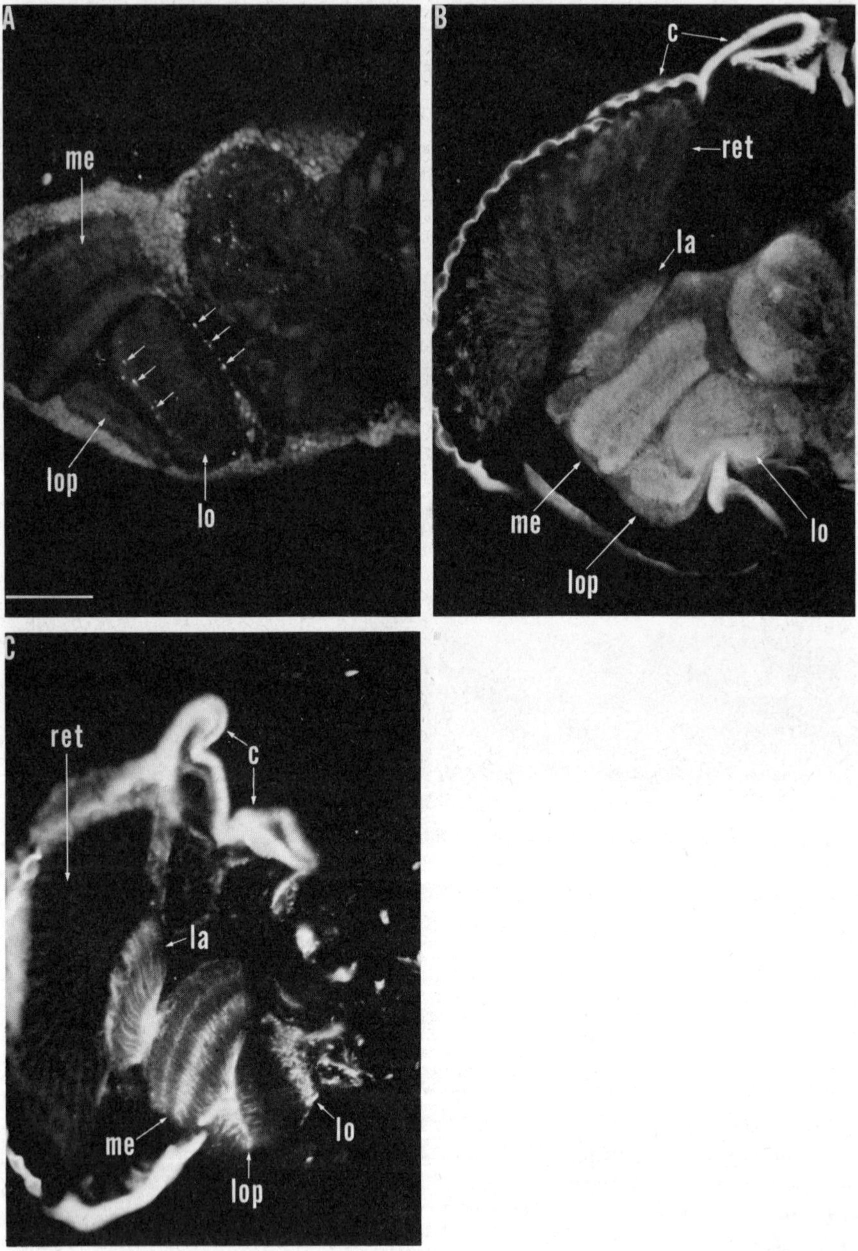

FIG. 4. A. Immunofluorescent staining of the enkephalin-like immunoreactivity in the central nervous system (CNS) of a fruitfly. 10-μm cryostat sections of a fruitfly were treated first with

stocks of flies carrying duplications or small deletions covering the entire genome of *Drosophila* (segmental aneuploidy: see Lindsley et al 1972), one may be able to identify the structural gene(s) of, say, the substance P-like materials by measuring substance P-like immunoreactivity in flies carrying duplications or deletions of various segments of the genome. For ACTH (and endorphin), clones of cDNA complementary to its messenger RNA already exist (Nakanishi et al 1979). Thus, if the ACTH-like material in *Drosophila* is sufficiently similar to mammalian ACTH, its structural gene could be localized by *in situ* hybridization of those cDNAs to salivary chromosomes. Localizing these structural genes would allow mutations of these genes to be isolated. The physiological modifications induced by these mutations would then provide clues to the functions of these substances.

(2) The second immunological approach to neurobiology is the use of monoclonal antibodies. The invention of hybridoma techniques for making monospecific antibodies (Köhler & Milstein 1975) was a major breakthrough, allowing investigators to obtain specific monoclonal antibodies from heterogeneous antigens. This technique has been applied to *Drosophila* in several laboratories. A number of useful clones have emerged from our preliminary studies. Some clones of hybridomas make antibodies which recognize only a subset of nerve fibres in embryos, larvae and adult flies (Figs. 4 and 5). Thus, hybridoma techniques may be used to identify neuronal specific molecules. For those molecules that are found to play an important physiological or developmental role, one could set up radioimmunoassays using monoclonal antibodies and use segmental aneuploidy to localize their structural genes. As mentioned earlier, genetic mapping of these genes facilitates the isolation of mutations affecting these particular genes and may lead to molecular studies of their expression and regulation using recombinant DNA techniques.

Conclusions

Electrophysiological and immunological methods can be used to identify molecules crucial for neural function or development in *Drosophila*. The

---

rabbit antisera specific for Leu-enkephalin, then with rhodamine-labelled second antibodies. The general brightness associated with neuronal cell bodies was due to non-specific binding of rabbit antibodies; this background of staining persisted after preabsorbing the rabbit antisera with Leu-enkephalin. Cross-sections of fibres which appear as bright dots (arrows) are specific staining. Similar patterns of staining were found in successive serial sections, but are absent in sections treated with preabsorbed antisera.

B. Immunofluorescent staining of the fruitfly's CNS using an antisera specific for neuronal membranes in general.

C. Immunofluorescent staining of the fruitfly's CNS using a monoclonal antibody which recognizes only a subset of nerve fibres. ret, retina; la, lamina; me, medulla; lo, lobula; lop, lobula plate; c, cuticle (which has yellow autofluorescence). Calibration bar represents 50 μm.

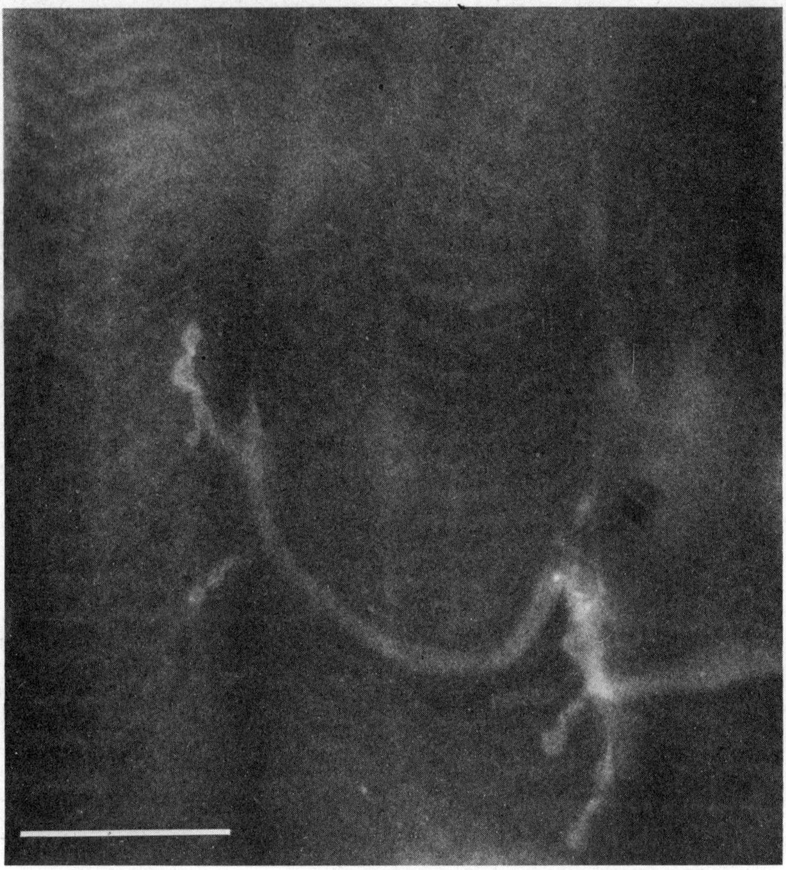

FIG. 5. Immunofluorescent staining of nerves and synaptic boutons on the surface of larval body-wall muscles. A whole mount of a larval nerve–muscle preparation was treated with antibodies specific for nerve membranes. Calibration bar represents 50 μm.

wealth of information on *Drosophila* genetics allows the genetic mapping of their structural genes and, in the long run, will make it possible to clone these genes or generate mutations of these genes for further study.

Acknowledgements

We thank Drs M. Brown, A. C. Cuello and M. S. Brownfield for antibodies and Ms Louise Evans for photography. The electrophysiological studies were initiated when we were in Dr Seymour Benzer's laboratory at Caltech. This work is supported by NIH grant NS 15757 to L.Y.J. and an MDA grant to Y.N.J.

REFERENCES

Ashburner M, Novitski E 1976 The genetics and biology of *Drosophila*, vol. 1. Academic Press, London

Atwood HL, Swenarchuk LE, Gruenwald CR 1975 Long-term synaptic facilitation during sodium accumulation in nerve terminals. Brain Res 100:198-204

Benzer S 1973 Genetic dissection of behavior. Sci Am 229(6):24-37

Cull-Candy SG 1976 Two types of extrajunctional L-glutamate receptors in locust muscle fibres. J Physiol (Lond) 255:449-464

Greenspan RJ 1980 Mutations of choline acetyltransferase and associated neural defects in *Drosophila melanogaster*. J Comp Physiol 137:83-92

Grigliatti T, Hall L, Rosenbluth R, Suzuki DT 1973 Temperature-sensitive mutations in *Drosophila melanogaster*. XIV. A selection of immobile adults. Mol Gen Genet 120:107-114

Hall JC, Kankel DR 1976 Genetics of acetylcholinesterase in *Drosophila melanogaster*. Genetics 83:517-535

Hall JC, Greenspan RJ, Harris WA 1981 Genetic neurobiology. Neurosci Res Bull (in press)

Hall LM, Wilson SD, Gitschier J, Martinez N, Strichartz GR 1982 Identification of a *Drosophila melanogaster* mutant that affects the saxitoxin receptor of the voltage-sensitive sodium channel. This volume p 207-217

Jan LY, Jan YN 1976a Properties of the larval neuromuscular junction in *Drosophila melanogaster*. J Physiol (Lond) 262:189-214

Jan LY, Jan YN 1976b L-Glutamate as an excitatory transmitter at the *Drosophila* larval neuromuscular junction. J Physiol (Lond) 262:215-236

Jan YN, Jan LY 1978 Genetic dissection of short-term and long-term facilitation at the *Drosophila* neuromuscular junction. Proc Natl Acad Sci USA 75:515-519

Jan YN, Jan LY 1979 Genetic dissection of synaptic transmission in *Drosophila melanogaster*. In: Insect neurobiology and pesticide action. Soc Chem Ind (Lond) Monogr, p 161-168

Jan YN, Jan LY, Dennis MJ 1977 Two mutations of synaptic transmission in *Drosophila*. Proc R Soc Lond B Biol Sci 198:87-108

Kaplan WD, Trout WE III 1969 The behavior of four neurological mutants of *Drosophila*. Genetics 61:399-409

Köhler G, Milstein C 1975 Continuous cultures of fused cells secreting antibody of predefined specificity. Nature (Lond) 256:495-497

Lindsley DL, Sandler L, Baker BS et al 1972 Segmental aneuploidy and the genetic gross structure of the *Drosophila* genome. Genetics 71:157-184

Nakanishi S, Inoue A, Kita T, Nakamura M, Chang ACY, Cohen SN, Numa S 1979 Nucleotide sequence of cloned cDNA for bovine corticotropin-β-lipotropin precursor. Nature (Lond) 278:423-427

Nicoll RA, Schenker C, Leeman SE 1980 Substance P as a transmitter candidate. Annu Rev Neurosci 3:227-268

Rahamimoff R, Erulkar SD, Lev-Tov A, Meiri H 1978 Intracellular and extracellular calcium ions in transmitter release at the neuromuscular synapse. Ann NY Acad Sci 367:583-598

Sherald AF, Wright TRF 1974 The analogue inhibitor α-methyl dopa as a screening agent for mutants elevating levels of dopa decarboxylase in *Drosophila melanogaster*. Mol Gen Genet 133:25-36

Siddiqi O, Benzer S 1976 Neurological defects in temperature-sensitive paralytic mutants of *Drosophila melanogaster*. Proc Natl Acad Sci USA 73:3253-3257

Tanouye MA, Ferrus A, Fujita SC 1981 Abnormal action potential associated with the *Shaker* complex locus of *Drosophila*. Proc Natl Acad Sci USA 78:6548-6552

Usherwood PNR 1980 Peripheral glutamatergic synapses in insects. In: Taxi J (ed) Ontogenesis and functional mechanisms of peripheral synapses. Elsevier, Amsterdam, p 367-384

Wu CF, Ganetzky B 1980 Genetic alteration of nerve membrane excitability in temperature-sensitive paralytic mutants of *Drosophila melanogaster*. Nature (Lond) 286:814-816

Wu CF, Ganetzky B, Jan LY, Jan YN, Benzer S 1978 A *Drosophila* mutant with a temperature-sensitive block in nerve conduction. Proc Natl Acad Sci USA 75:4047-4051

DISCUSSION

Narahashi: In your electrophysiological experiments the normal wild-type *Drosophila* mimicked the prolonged action potentials of the *Shaker* mutant when you blocked the potassium channels with 4-aminopyridine. You deduced that when potassium channels are inhibited you can mimic the mutants. But the prolongation of the action potential could be due to a slowing of the sodium inactivation.

Jan: Even if we blocked the sodium channels or the calcium channels in the mutant or 4-aminopyridine-treated normal fly, we still got prolonged depolarization. In *Drosophila* the nerve action potentials have sodium and calcium components but if one of them is removed you still get the action potential.

Narahashi: How can you come to that conclusion about the potassium channel on the basis of your 4-aminopyridine experiments?

Jan: All we are saying is that we couldn't find any evidence of an abnormal sodium channel or an abnormal calcium channel, so we were left with the possibility that the potassium channel did not function properly. In addition, the mutant phenotype could be mimicked by treating the normal animal with 4-aminopyridine. All the evidence is consistent with there being an abnormal potassium channel in *Shaker* mutants, but we don't have direct evidence.

Narahashi: What happens if you apply 4-aminopyridine to *Shaker* mutants?

Jan: It intensifies the behaviour somewhat. We can mimic the shaking behaviour by feeding the animal with 4-aminopyridine as well.

Barnard: Is there any other way of determining whether the shaking is due to a structural gene mutation or a regulatory genetic change?

Jan: If one could obtain a temperature-sensitive allele of the *Shaker* mutant, one would have some indication that it is a structural gene. Better yet, one could do gene-dosage studies. From such a study, Tanouye et al (1981) obtained evidence suggesting that the *Shaker* gene is a structural gene.

Hall: I would also question, from our results, whether a temperature-sensitive mutation actually gives evidence for a structural gene. There could be other explanations.

Jan: For instance, you could have a mutation affecting membrane lipid composition rather than the ionic channels. Such mutants could exhibit temperature sensitivity in channel activity.

M. E. Eldefrawi: I am fascinated by the antibody selectivity against muscle and nerve and so forth. Is it possible that a concentrated effort may produce an antibody that is specific for end-plates? Maybe you could produce a very selective probe for the glutamate receptor.

Jan: Our screening is based on immunocytochemistry. In *Drosophila* the glutamate receptors are found only in the junctional region. There is no evidence of extrajunctional glutamate receptors. If we obtain an antibody which stains only the postsynaptic membrane at the neuromuscular junction and blocks the glutamate response, we may have an antibody that is specific for the glutamate receptor.

Hildebrand: To what cells in the optic ganglia did your antibody against Leu-enkephalin bind?

Jan: We don't know yet.

Hildebrand: Were the binding sites found mainly in the medulla?

Jan: Yes.

Hall: Was there any staining in the lobules?

Jan: Most of the staining was found in the medulla lobula and the fibre tracts connecting the brain and the thoracic ganglion.

Hall: Candice Pert and I looked for opiate-receptor-like activity in *Drosophila* head extracts. Using alkaloids as probes, we couldn't find activity but using polypeptides we found receptors. It is interesting that you find Leu-enkephalin-like activity because that is consistent with our findings that there are opiate receptors in the insect.

Dudai: But there are several classes of opiate receptors.

Usherwood: I must admit to some sense of disappointment over progress made using the genetic approach to understanding insect neuropharmacology. Am I being unfair in concluding that the euphoria which accompanied the first results that were obtained using this approach has been tempered by the growing realization of its limitations?

Jan: The strength of the genetic approach lies in the potential of recombinant DNA techniques, which should prove to be a very powerful method.

M. E. Eldefrawi: I have wondered whether there is any proof that the electrogenic sodium and potassium channels on excitable membranes are separate independent entities or whether they are components of a single molecular entity. Electrophysiologically independent currents can be measured, but that doesn't mean they are independent or separate molecular entities. From the results presented here, however, I can see that the *nap*[ts] appears on a second chromosome associated with the sodium channel while the *Shaker* mutation associated with a change in the potassium channel is on the first chromosome. That genetic approach to neuropharmacology may provide us with the answer. If that turns out to be the case, we can say that we have two genes on different loci in the chromosomes that control the two electrogenic channels.

Dudai: Let me give an example of what could be done now. Let us take a relatively simple question that has indeed already been answered by other methods, namely whether acetylcholinesterase and the acetylcholine receptor are the same molecule. Since mutations exist in the gene for acetylcholinesterase in *Drosophila* (Hall & Kankel 1976) and since such mutations do not affect either the nicotinic or the muscarinic binding sites (Dudai 1978, Haim et al 1979), the receptors and the enzyme are not the same molecules. We got the same results with deletions that cover the gene for acetylcholinesterase from both sides. Moreover, these deletions were used to prove that the gene for acetylcholinesterase (Hall & Kankel 1976) and the unidentified genes for the receptors do not lie next to one another on the chromosome (Dudai 1978). This means that, for the cholinergic system in *Drosophila*, genes coding for different parts of the cholinergic apparatus are not located on the same small piece of chromosome. A similar conclusion was drawn by Hall and Greenspan on the lication of the genes for acetylcholinesterase and choline acetyltransferase (Hall & Greenspan 1979). In other words, there are interesting questions in neurochemistry which can be answered even if we have only a limited amount of genetic information.

Usherwood: Let me present you with a problem and see how you would solve it using the genetic approach. Let us imagine that there are GABA receptors on *Drosophila* muscle fibres. How would you attempt to identify such receptors and elucidate their structure?

Dudai: I would not use genetics alone to study the receptors. First of all, I would try to identify a biochemical probe that would label GABAergic receptors in *Drosophila*, such as muscimol. I would then try to find mutations which could be suspected of affecting inhibitory processes, e.g. flies which are very excited, and test whether they have GABA-binding sites with abnormal properties. I would also search, by segmental aneuploidy, for chromosomal segments that affect the level of the binding sites when deleted or duplicated. One should however remember that the latter method has been useful, so far, only for enzymes that are coded by a single gene. When talking about genetics and neuropharmacology, one should remember that we are not practising genetics alone; we usually combine the genetics with studies of behaviour, and/or electrophysiology, and/or pharmacology, and by such combinations we can, sometimes, successfully tackle complex phenomena and not, as you suggested, only questions related to a relatively well-defined neuronal component.

Consider short-term memory. Using the *dunce* mutation in *Drosophila*, an isozyme of cyclic AMP phosphodiesterase was implicated in the process (Byers et al 1981, and this volume, p 201). In this case information pointing to the involvement of cyclic AMP was also provided by Kandel and his colleagues (Kandel 1979), using a direct electrophysiological and biochemical approach in *Aplysia*. So we may argue whether genetics is the only method of obtaining such information. I would say it is usually not, but it is a very powerful way of obtaining additional, complementary information and, in some cases, the only information available so far.

Hall: I agree. Genetics can't stand by itself but has to be used with other methods. If I were going to look for a GABA receptor, I would not use the segmental aneuploidy approach, since my laboratory has used this approach twice for membrane receptors (saxitoxin receptor and α-bungarotoxin receptor) without finding a gene-dosage effect. A more rational approach would be to develop a photoactivatable or cross-linking ligand that would label the receptor; then one could look for an electrophoretic variant. That is a reasonable approach to identifying the structural genes.

Narahashi: I am not particularly disappointed with the genetic approach because I feel it will be useful not only for characterizing sodium, calcium and other channels but also for studying the insecticide resistance mechanism which has been studied for many years by a number of people. We and others have been able to identify a gene that controls DDT dehydrochlorinase or a gene that is responsible for low nerve sensitivity to DDT. In a DDT-resistant housefly strain, we identified a recessive gene on the second chromosome (the chromosome number has now been changed) that controls the low nerve sensitivity to DDT (Tsukamoto et al 1965). From our early work, DDT exerts its nerve-stimulating action through a modification of sodium channels (Narahashi & Haas 1968). I think that the genetic approach is a very useful one.

Kelly: Some of the explanations still seem more simple-minded than I would like. A reduction in tetrodotoxin receptors could be regulated in all sorts of ways. There are no sodium channels under the glial cells on myelinated nerve fibres (Bostock & Sears 1978), so you could argue that the genes control the amount of nerve membrane covered by the glial cells, or something like that.

Hall: I don't disagree with that. All we have done so far is to say that we have a mutant where the sodium channel is implicated and we have no evidence that it is a structural gene.

Kelly: I am not satisfied with the argument that, as the sodium channel and the potassium channel proteins are on different genes, they must make up separate components of the macromolecular structure of the membrane.

M. E. Eldefrawi: But doesn't one usually select a very simple model to start with and then discover the complexity as one goes along?

Kelly: Yes, but I am already dissatisfied with the simple model!

Jan: One of the main difficulties in deciding whether the mutations affect the sodium channel or the potassium channel directly is the small size of the animal. Had the mutation affected one of the channels present in the muscle, we would have been able to obtain direct evidence. The neurons are small but there are ways of overcoming the problem. For instance, people working with cell cultures use agents which fuse neurons. You can start with a number of small neurons and fuse them into one large one. This method has been tried in *Drosophila* with some success by C. F. Wu (unpublished). If we can develop methods like this to overcome some of those problems, we shall perhaps reach definitive answers.

Dudai: Let me stress again that genetics is not a universal answer. One of its most important uses, to my mind, is to probe something that is really unknown. If you have a receptor for acetylcholine it is nice to have mutants and to learn many things about the molecule by studying the effect of lesions on its structure and function. However, I would rather go into something totally unknown, such as steps in a physiological process or components of a neuronal apparatus which I can't identify by other methods—for example, elementary steps in the operation of the visual process or in sensitization or memory.

A word of caution is needed here on our expectations from genetics. In many cases mutations that affect the nervous system in a way that seems to be specific are not in fact mutations in molecules that are specific to the nervous system. You may find that the mutation affects an enzyme which looks trivial. It would have been much nicer if instead of identifying cyclic AMP phosphodiesterase as the lesioned gene product in the *dunce* mutant in *Drosophila*, we had found a new, unknown polypeptide, a 'dihydromemorase' enzyme or an ionophore for uranium. But nature is less innovative, and here we meet again an old acquaintance, an enzyme which seems to participate in many metabolic pathways. However, even such a semi-trivial answer enables us to form a link between a discrete metabolic event and a mysterious behavioural event and it sheds light on discrete steps in a complex behaviour. Genetics is in fact a delicate, accurate and unique microsurgical tool. And last, but of course not least, let us not forget that genetic engineering, as Dr Jan mentioned, may provide us not only with tools but also, by cloning the appropriate genes, with the biochemical material to work with.

REFERENCES

Bostock H, Sears TA 1978 The internodal axon membrane: Electrical excitability and continuous conduction in segmental demyelination. J Physiol (Lond) 280:273-301

Byers D, Davis RL, Kiger JA 1981 Defect in cyclic AMP phosphodiesterase due to the *dunce* mutation of learning in *Drosophila melanogaster*. Nature (Lond) 289:79-81

Dudai Y 1978 Properties of an α-bungarotoxin-binding cholinergic nicotinic receptor from *Drosophila melanogaster*. Biochim Biophys Acta 539:505-517

Haim N, Nahum S, Dudai Y 1979 Properties of a putative muscarinic cholinergic receptor from *Drosophila melanogaster*. J Neurochem 32:543-552

Hall JC, Greenspan RJ 1979 Genetic analysis of *Drosophila* neurobiology. Annu Rev Genet 13:127-195

Hall JC, Kankel DR 1976 Genetics of acetylcholinesterase in *Drosophila melanogaster*. Genetics 83:517-535

Kandel E 1979 Cellular insights into behavior and learning. Harvey Lect 73:19-92

Narahashi T, Haas HG 1968 Interaction of DDT with the components of lobster nerve membrane conductance, J Gen Physiol 51:177-198

Tanouye MA, Ferrus A, Fujita SC 1981 Abnormal action potential associated with the *Shaker* complex locus of *Drosophila*. Proc Natl Acad Sci USA 78:6548-6552

Tsukamoto M, Narahashi T, Yamasaki T 1965 Genetic control of low nerve sensitivity to DDT in insecticide-resistant houseflies. Sci Pest Control (Botyu-Kagaku) 30:128-132

Activation of ion channels in locust muscle by amino acids

K. A. F. GRATION

Department of Zoology, University of Nottingham, University Park, Nottingham, NG7 2RD, UK

Abstract L-Glutamate is the putative transmitter at excitatory synapses on insect muscle. Activation of locust muscle postjunctional membrane with glutamate increases the permeability of this membrane mainly to Na^+ ions. Glutamate receptors are also found on the extrajunctional membrane of locust muscle as two mixed but pharmacologically distinct populations. One population gates an increase in Cl^- permeability (H-receptors); the other population (D-receptors) gates an increase in cation permeability.

The open–close kinetics of extrajunctional D-receptor–channel complexes (glutamate channels) are investigated using the patch clamp technique. It is shown that ligand concentration influences glutamate channel life-times and that the life-time of the glutamate channel is governed by two exponential time-dependent processes, one of which has the effect of delaying channel closure. It is also shown that the potency of glutamate agonists is determined by (1) the life-times of agonist-activated channels and (2) the ability of the agonist to activate the channel. The results of recent experiments on the temperature dependence of channel opening and closing, and on the effects of collagenase on channel kinetics, are discussed.

There is good evidence that L-glutamate is the transmitter at excitatory synapses on skeletal muscles of insects (Beránek & Miller 1968, Usherwood & Machili 1968; see also reviews by Usherwood 1978, 1979). The application of glutamate to excitatory synapses on locust muscle activates glutamate-sensitive receptors on the postsynaptic membrane, causing an increase in membrane conductance. This increase results from the opening of cationic channels which are associated with, and whose function is governed (i.e. 'gated') by, the glutamate receptors. In the open state, the channels increase membrane permeability to Na^+ and K^+, with Ca^{2+} making a minor contribution to the current through the channel (Anwyl 1977). The increase in

---

Present address: Pfizer Ltd, Sandwich, Kent, CT13 9NJ, UK

1982 Neuropharmacology of insects. Pitman, London (Ciba Foundation symposium 88) p 240–259

membrane conductance is accompanied by fluctuations in membrane current ('noise') which may be attributed to statistical variations in the number of ion channels in the open state (see, for example, Katz & Miledi 1970, 1972). The properties of postsynaptic glutamate channels have been investigated by analysing voltage-clamped, glutamate-induced 'noise' (Anderson et al 1978). A mean single-channel conductance of about 120 pS was estimated from power spectra of the 'noise', with a mean channel life-time of about 2.5 ms. These studies showed that whereas channel conductance is independent of the agonist used, channel life-time is agonist-dependent.

Recently, the resolution of electrophysiological recording apparatus has improved to the point where the current generated by the opening, and subsequent closing, of single channels can be directly observed. In 1976 Neher & Sakmann introduced the patch clamp technique for recording currents through individual open channels, gated by acetylcholine (ACh), in denervated frog muscle membrane. These currents, like those passing through individual open channels in artificial membranes, have a square-pulse waveform; the transitions from the closed (non-conducting) state to the open (conducting) state, and vice versa, are rapid relative to the average life-time of the open channel. The patch clamp technique has since proved to be a powerful tool for the detailed analysis of channel kinetics. Neher & Steinbach (1978) showed that ACh channel openings in the presence of channel-blocking drugs may be grouped in bursts, as had been inferred indirectly from noise and relaxation measurements (see, for example, Ruff 1977). Recently, Sakmann et al (1980) measured the concentration dependence of the opening of single ACh channels.

In a previous study from this laboratory, Patlak et al (1979) showed that the patch clamp technique can be successfully applied to locust muscle preparations. These preparations have the advantage of being more or less devoid of connective tissue and are suitable, therefore, for patch clamping without enzymic pretreatment. Furthermore, the conductance (about 120 pS) of the open glutamate channel, which is almost five times greater than that of ACh channels in vertebrate muscle, is among the largest conductances yet found in biological membranes. This large conductance gives clear signal-to-noise resolution, while the use of the plant lectin, concanavalin A, to inhibit receptor desensitization (Mathers & Usherwood 1976) allows single-channel currents to be recorded over a wide range of concentrations of glutamate or agonist. Our research in this laboratory is now being directed towards understanding the kinetic properties of membrane ionic channels gated by the interaction of amino acids with glutamate receptors in locust muscle. Glutamate receptors on the extrajunctional membrane are of two kinds: H-receptors, which gate an increase in Cl^- permeability, and D-receptors, which gate an increase in cation permeability (Lea & Usherwood 1973, Cull-Candy & Usherwood 1973). We have investigated the activity of single, extrajunctional, D-receptor-gated channels in these muscle fibres.

Currents through single channels associated with extrajunctional D-receptors were recorded from innervated metathoracic extensor tibiae muscle fibres in adult locusts (*Locusta migratoria*) using the extracellular patch clamp technique (Neher & Sakmann 1976, Neher et al 1978) to apply agonists and to record membrane currents. Single-channel currents were recorded after pretreating the muscle preparations with 1–2 μM-concanavalin A. Most experiments were done at the muscle resting potential (−55 mV to −60 mV), except for those designed to measure channel conductance and to determine the temperature dependence of channel operation when membrane potential was controlled with a standard two-point voltage-clamp technique. Only those recorded sites that showed no evidence of overlapping events were used in the studies to be described. This suggests that only one receptor was active within the area of membrane under the patch electrode (Neher et al 1978). Single-channel currents were usually recorded through a low-pass filter with a 3 dB cut-off frequency of 1 kHz. However, some preliminary experiments at a cut-off frequency of 3 kHz will be described. We are now building a patch clamp with a cut-off frequency of 10 kHz to obtain better signal resolution.

Distribution of channel-open times

With concentrations of glutamate $\leqslant 5 \times 10^{-5}$ M, distributions of channel life-times, constructed from data recorded through a low-pass filter with a 3 dB cut-off frequency of 1 kHz, can be fitted by a single exponential. Mean channel life-times estimated from these exponential fits agree well with values obtained from 'noise' analysis at locust neuromuscular junctions (Anderson et al 1978) and with mean values calculated directly by averaging the life-times of single events. However, since brief transient openings (<300 μs) cannot be resolved under these recording conditions, the distributions are deficient in the region 0–1 ms (see also Sakmann et al 1980). To obtain a more complete description of the distribution of channel-open times and, consequently, of the mechanisms governing channel closure, we have recorded channels at a cut-off frequency of 3 kHz. This is possible with locust muscle because at hyperpolarized membrane potentials (about −120 mV) single-channel currents often exceed 18 pA (Patlak et al 1979) and signal-to-noise ratios ⩾4:1 can be obtained. At this increased bandwidth it is possible to record single-channel currents with durations ⩾100 μs, and from these data we can construct frequency histograms of channel-open times using bin widths of 50 μs.

Typical distributions of channel life-times, constructed from the same recording sample but presented in 0.5 ms and 50 μs time bins, respectively, are compared in Fig. 1. The distribution of channel life-times for the data grouped in 0.5 ms time bins appears to be reasonably well fitted by a single exponential. However, at very short life-times (<1 ms) the number of observed channel openings falls below that predicted from a single exponential. When these same data are presented in 50 μs time bins (Fig. 1B) a peak is evident in the life-time distribution and, as a first approximation, the distribution can be described by two exponentials. This suggests that channel closure is governed by two rate-limiting processes, one of which promotes channel closure and a second that prevents, or at least delays, closure.

All previous reports have suggested that closure of ligand-activated ionic channels is governed by a single rate-limiting step—a conclusion drawn from the exponential distribution of channel life-times (Nelson & Sachs 1979, Sakmann et al 1980). However, high frequency resolution data obtained by the patch clamp method are not yet available for channels other than the glutamate channel. The possibility remains that the distributions of life-times for channels gated by ligands other than glutamate, such as ACh or γ-aminobutyric acid (GABA), may prove to be non-exponential.

Concentration dependence of glutamate channel function

Preliminary results (Patlak et al 1979) with a single concentration of glutamate (10^{-4} M) suggested that glutamate channels can suddenly change their kinetic properties and switch from a short life-time mode to a long life-time mode, and vice versa. These preliminary experiments have since been extended to include investigations of channel open–close kinetics at a range (5×10^{-6} M to 5×10^{-3} M) of glutamate concentrations (Gration et al 1981a,b). These new studies show that the life-time of the glutamate channel, and the interval between successive open events, depend on the concentration of glutamate in the patch electrode. At 5×10^{-6} M-glutamate, channel life-times are rarely >12 ms (mean ±SD, 2.96 ± 0.26 ms) and open events are separated by long (mean ±SD, 1174 ± 94 ms) closed intervals. At the other end of the concentration range (5×10^{-3} M) channel life-times often exceed 1 s and the mean channel life-time is 273 ± 182 ms (±SD).

Histograms of frequency distributions of channel life-times at various glutamate concentrations are illustrated in Fig. 2. These show that the range of channel life-times increases with glutamate concentration, while the proportion of channels of <10 ms duration is reduced. These trends were apparent over the entire range of concentrations studied but their impact on channel life-time was not significant until the concentration exceeded 5×10^{-5} M (see Fig. 3A). Although channel life-time distributions recorded

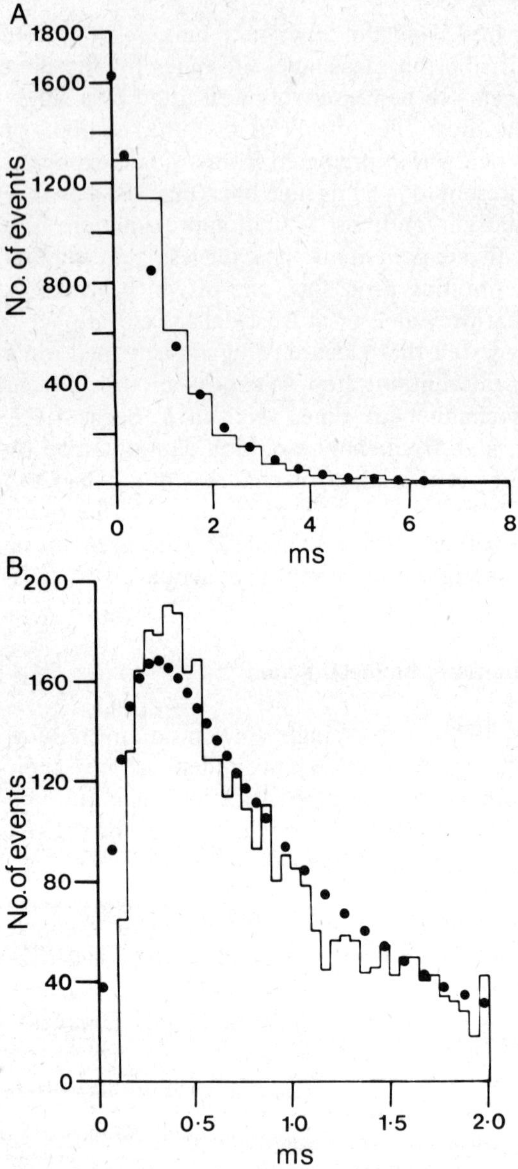

FIG. 1. Analysis of the frequency distribution of glutamate channel life-times. Glutamate concentration, 5×10^{-5} M; muscle fibre membrane potential, −125 mV; temperature, about 20 °C; channels recorded at a cut-off frequency of 3 kHz.

A. Histogram of life-times (4000 events) plotted in 0.5 ms bins. The dotted line is an exponential fitted to the distribution by the method of least squares, with $\tau = 1.15$ ms.

B. The data of (A) plotted in 50 μs bins. The dotted line is given by the relationship: $A[(e^{-t/\tau_1}) - (e^{-t/\tau_2})]$, where $\tau_1 = 0.9$ ms, $\tau_2 = 0.14$ ms. The values were selected to give the best fit by eye. A is the intercept.

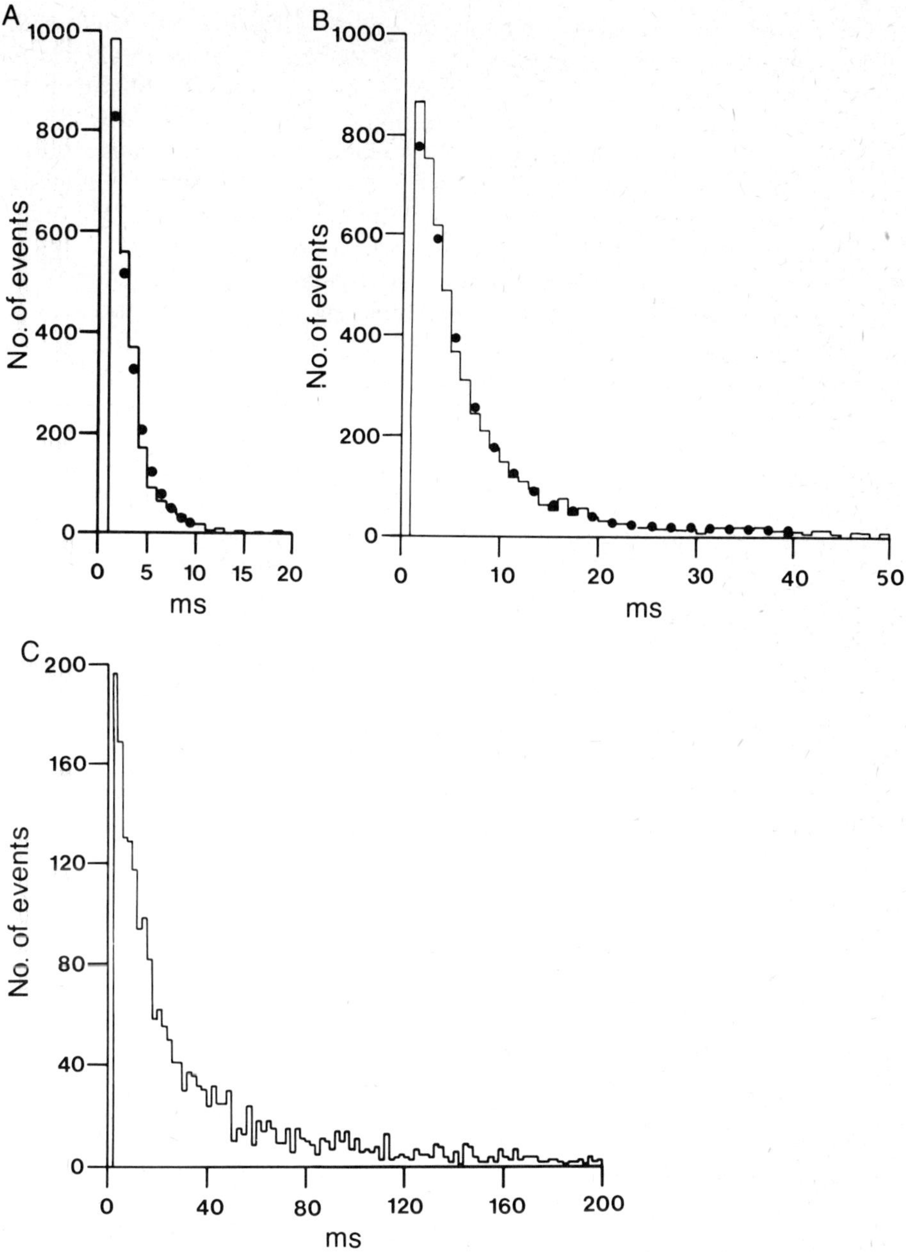

FIG. 2. Effect of glutamate concentration on the distribution of channel life-times. Histograms of channel life-times recorded at (A) 5×10^{-5} M, (B) 5×10^{-4} M and (C) 5×10^{-3} M-glutamate. The dotted line fitted to (A) is a single exponential, $\tau = 2.1$ ms; the dotted line fitted to (B) a double exponential $\tau_f = 2.8$ ms, $\tau_s = 7.5$ ms. Recording cut-off frequency, 1 kHz; temperature, about 20 °C.

with low concentrations of glutamate show a peak, life-time histograms with time bins of 1 ms can be fitted with a single exponential (see earlier). A similar presentation of channel life-times recorded at 5×10^{-4} M-glutamate produces histograms (Fig. 2B) that can only be adequately fitted with two exponentials (see also Patlak et al 1979); values for higher concentrations appear non-exponential (Fig. 2C). Because of the impossibility of fitting exponentials to many of the life-time distributions, mean channel life-times were calculated directly by averaging from measured data.

The relationship between the reciprocal of mean closed time (Mc^{-1}) and glutamate concentration, plotted on double-log axes, can be fitted by a straight line with a slope of approximately one (see Fig. 3B). This suggests that the binding of at least one molecule of glutamate is required to activate D-receptors. An estimate of the apparent dissociation constant for extrajunctional D-receptors can be obtained from the glutamate concentration at which the channel remains in the open (conducting) state during 50% of the recording period. In our recording conditions this concentration occurs at about 5×10^{-4} M. This compares to apparent dissociation constants of 2.2×10^{-6} M at -90 mV and 1.7×10^{-6} M at -130 mV, respectively, for ACh receptors in the extrajunctional membrane of denervated frog muscle (Sakmann et al 1980).

The interaction of ACh with receptors at the vertebrate muscle end-plate (del Castillo & Katz 1957, Katz & Miledi 1972) and of glutamate with postsynaptic glutamate receptors in locust muscle (Anderson et al 1978) has been described using the two-state scheme:

$$\text{closed} \underset{\alpha}{\overset{\beta'}{\rightleftharpoons}} \text{open}$$

In this scheme β' and α are the rate constants of channel opening and closing respectively. Whereas α should be independent of ligand concentration, β' should increase with concentration since it is an apparent rate constant, whose value is assumed to depend on agonist binding steps preceding channel opening.

Within the range of glutamate concentrations, 5×10^{-6} M to 5×10^{-5} M, the kinetics of single glutamate channels are seemingly consistent with this hypothesis. However, at concentrations $\geqslant 10^{-4}$ M, glutamate channel life-time is concentration-dependent. We (Gration et al 1981a) have proposed a receptor model incorporating two glutamate-binding sites to explain the concentration-dependent increase in glutamate channel life-time. Within the assumptions of this model, receptors occupied by one or two ligand molecules can gate channel openings, with the constraint that both binding sites must be vacated before the channel can close. The increasing frequency of glutamate binding which would be the consequence of increased glutamate concentrations raises the probability of overlapping occupancy of the two postulated binding sites, resulting in an increase in mean channel life-time. This model, with a provision for the non-random opening of glutamate channels (Gration et al 1981a,b), will also account for the abrupt changes in channel kinetics observed with 10^{-4} M-glutamate.

The concentration-dependent increase in the life-time of the glutamate channels seems to contrast with the behaviour of ACh channels. Within the range of ACh concentrations, 5×10^{-6} to 5×10^{-5} M, the open times of these channels are independent of ligand concentration. However, ACh channels have not been recorded at concentrations of $>5 \times 10^{-5}$ M and, in the absence of an agent that blocks ACh receptor desensitization (cf. the effect of concanavalin A on glutamate receptors in locust muscle), it seems unlikely that the open–close kinetics of these channels can be studied over as wide a concentration range as is possible with the glutamate channel.

Agonist potency determinations

In this laboratory we are now investigating structure–activity relationships for different glutamate agonists in terms of their ability to activate single channels and to maintain the activated channel in the open state. A range of glutamate agonists, including L-quisqualate, L-4-methylene glutamate, L-hydroxyglutamate, L-fluoroglutamate, L-cysteate and L-2-amino-3-sulphinopropionate, have been used in these studies, but I shall concentrate here on results obtained with glutamate, L-quisqualate and L-2-amino-3-sulphinopropionate. The relative potencies of these agonists, as determined by conventional techniques (e.g. measurement of depolarization, current, input conductance or twitch tension), vary as follows: quisqualate > glutamate > 2-amino-3-sulphinopropionate. By investigating the open–close kinetics of channels activated by these agonists and by measuring the conductance of the open agonist-activated channel we can gain information about the factors determining agonist potency.

Fig. 3A compares the relationships between mean channel life-time and ligand concentration for the three agonists. In each case, increasing the concentration of agonist above a given concentration (5×10^{-5} M for glutamate, 3×10^{-5} M for quisqualate and 3×10^{-3} M for 2-amino-3-sulphinopropionate) results in an increase in channel life-time. At or below these concentrations the mean channel life-time is about 6.0 ms for quisqualate, 2.5 ms for glutamate and 1.0 ms for 2-amino-3-sulphinopropionate. The observation that the life-time of channels in the extrajunctional membrane of locust muscle is agonist-dependent is directly analogous to the

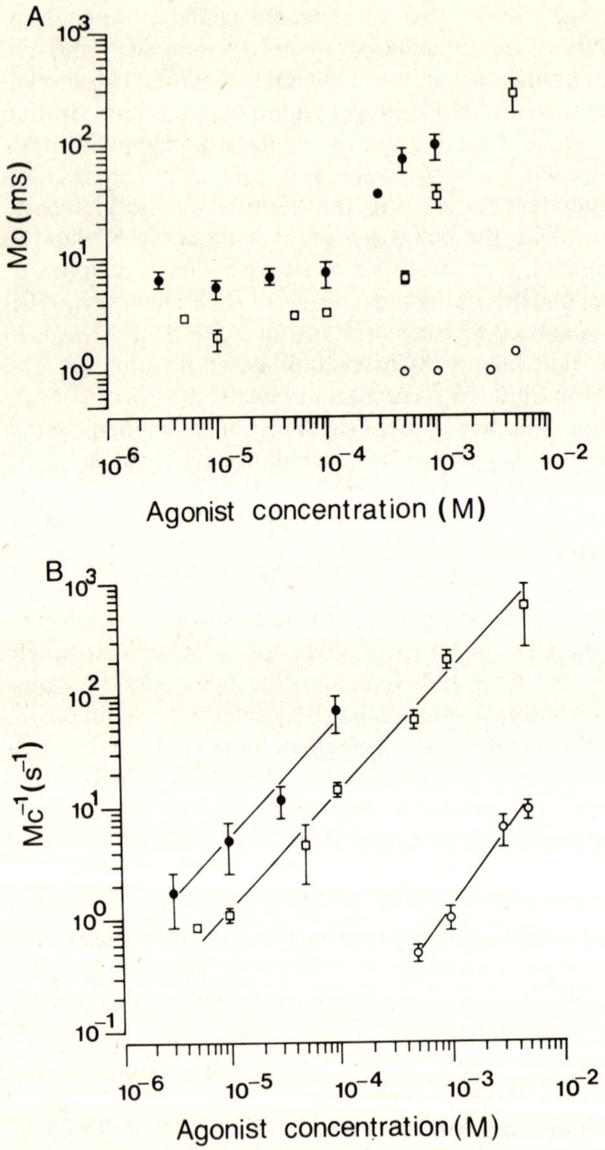

FIG. 3. Factors affecting the potency of different agonists.

A. Log–log plots of mean channel life-time (M_o) against agonist concentration for L-quisqualate (●), L-glutamate (□) and L-2-amino-3-sulphinopropionate (○).

B. Log–log plots of the reciprocal of channel closed time (Mc^{-1}) against agonist concentration for the same three agonists. The solid lines, which have slopes of 1.02 for quisqualate, 0.99 for glutamate and 1.32 for L-2-amino-3-sulphinopropionate, were fitted by the method of least squares.

situation previously described for the action of cholinergic agonists at ACh receptors on vertebrate muscle (Katz & Miledi 1972, Colquhoun et al 1975) and for glutamate agonists at excitatory synapses on insect (Anderson et al 1978) and crustacean (Crawford & McBurney 1976) muscle.

The relationships between reciprocal mean closed times (Mc^{-1}) and agonist concentration (Fig. 3B), which presumably reflect the ability of different agonists to activate the channel, suggest that differences in agonist potency may result, in part, from differences in the probability of channel activation. The plots have parallel slopes, suggesting that the agonists interact with the same site(s) on the receptor. However, it is clear from Fig. 3B that for a given concentration of agonist the probability of channel activation is as follows: quisqualate > glutamate > 2-amino-3-sulphinopropionate.

There was no significant difference between the conductances of the agonist-activated channels; therefore, in the absence of desensitization, a direct index of potency can be determined by measuring the total time that a single channel remains open during the recording period (Gration et al 1981a). Using this as an index for quisqualate:glutamate:2-amino-3-sulphinopropionate, relative potencies of 2185:560:1 are obtained.

These results indicate that the relative potencies of glutamate agonists at extrajunctional D-receptors on locust muscle are determined by the life-times of the channels that they gate and by their probability of activating the glutamate receptor–channel complex. Furthermore, the agonists that most readily activate the channel will gate channels with longer life-times than agonists which are less efficient in this respect.

Collagenase treatment

Electron micrographs of locust muscle show that the muscle fibres are sheathed in a thin (0.1–0.25 μm) layer of connective tissue, which contains collagen fibres within a matrix of amorphous material (Clements & May 1974). This sheath may act as a diffusion barrier and reduce the rate at which glutamate in the patch electrode can diffuse to the muscle fibre membrane. It seemed of interest, therefore, to determine the effects of collagenase treatment on the open–close kinetics of single glutamate channels.

In control experiments on non-treated muscle preparations the mean closed time for channels recorded at 7.5×10^{-5} M-glutamate was 535 ± 358 ms (mean $\pm$SD; M. Ashford, personal communication 1981). Electron micrographs of collagenase-treated (Collagenase type I, Sigma; 0.001–0.01% for 10 to 20 min) locust muscle show a widespread disruption of the connective tissue with dispersal of the collagen fibre matrix (I. R. Duce, personal communication 1981). After this treatment the frequency of single-channel events recorded at 7.5×10^{-5} M-glutamate increases, with a concomitant decrease in closed time to 79 ± 40 ms. The mean life-time of the glutamate channel was unaffected by collagenase, being 2.2 ± 0.7 ms before collagenase treatment compared to 2.3 ± 0.4 ms after exposure to the enzyme. The decrease in closed time may result from increased access of glutamate to the receptor sites when the connective tissue had been disrupted.

After collagenase treatment, the amplitude of the single-channel currents increases. In control experiments at the resting potential (about −60 mV) the distribution of channel amplitudes approximates to a normal distribution with a modal value of about 8 pA. After collagenase treatment the mode increases to about 15 pA. We have yet to see whether this increase results from a change in channel conductance. We are investigating this now.

Effects of temperature on single-channel currents

Arrhenius plots of both channel conductance and mean channel life-time for postsynaptic glutamate channels in locust muscle show a 'break point' at about 6 °C (Anderson et al 1978). In their studies Anderson et al used 'noise analysis' which, in effect, averages the simultaneous activity of many channels, raising the important question: to what extent do individual channels show the same (or different) break points as a function of temperature? In this section I discuss results on the temperature dependence of single glutamate channels.

A decrease in temperature reduces the conductance of the open glutamate channel. However, the zero current potential, determined under voltage clamp, is unaffected by changes in temperature and the relationship between channel current amplitude and membrane potential remains linear. Fig. 4A compares the relationship between single-channel conductance and temperature (in the form of a scattergram) with the Arrhenius plot derived by Anderson et al (1978) for postsynaptic glutamate channels. A comparison of these relationships shows that whereas the conductance of postsynaptic channels falls to about 10 pS at 0.8 °C, the conductance of channels that can be recorded using the patch clamp technique is at least 2.5 times this value, with single-channel conductances ranging from 25 pS to 60 pS with temperatures of 0.6 °C to 1.5 °C (P. Rand et al, unpublished results 1981). Although use of the patch clamp probably 'selects' for channel conductances of >25 pS, the presence of high conductance states at low temperature (<1.5 °C) suggests that at least some extrajunctional glutamate receptor–channel complexes do not experience a marked temperature transition at 6 °C. This view

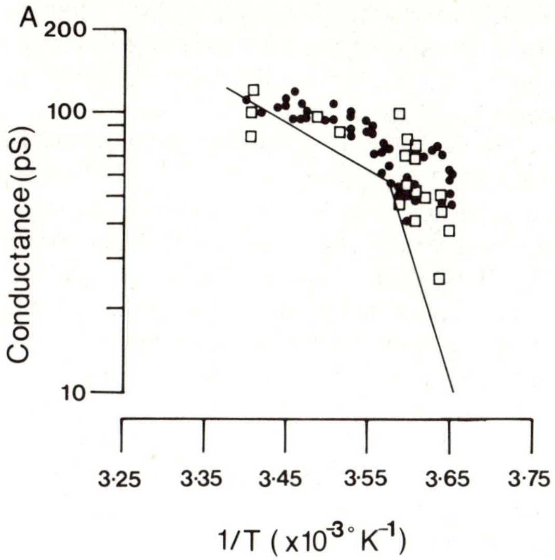

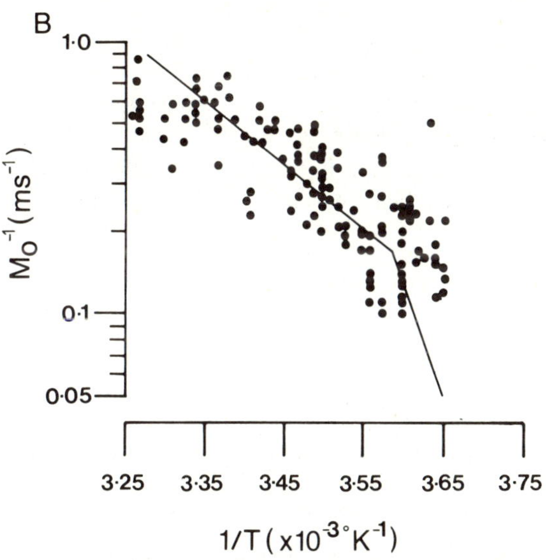

FIG. 4. Temperature-dependence of single channel conductance and life-time.

A. Semilog plot of channel conductance against T^{-1} in °K. Open squares (□), channel conductance measured from the slope of the best-fit line to the relationship between channel amplitude and membrane potential. Filled circles (●), conductances estimated from mean channel amplitude at a fixed membrane potential.

B. Semilog plot of mean channel life-time against T^{-1} in °K.

The solid lines are redrawn from Fig. 9 of Anderson et al (1978).

is confirmed by the data of Fig. 4B, which shows the relationship between mean channel life and temperature. Within the temperature range of 0.6 °C to 5 °C mean life-times for single channels are usually greater than values predicted by the Arrhenius plot of Anderson et al (1978).

Two possible explanations for the temperature transition observed in the 'noise' studies have been proposed: (1) a temperature-induced conformation change in the protein molecules associated with the receptor–channel complex, and (2) a phase transition in the lipid component of the membrane associated with the channel (Anderson et al 1978). There are several explanations for the existence of high conductance, short life-time extrajunctional channels at low temperatures. Possibly, the lipid environment of the extrajunctional membrane differs from that of the postsynaptic membrane. Alternatively, individual channels in the extrajunctional membrane may experience different membrane environments, such as variations in lipid content or receptor density, or both, which might affect the temperature dependence of channel operation.

Acknowledgements

I thank Professor P. N. R. Usherwood for helpful comments during the preparation of this manuscript. Many thanks also go to Drs J. J. Lambert, R. Ramsey and M. Ashford and Professor P. Rand. This work was supported by a Science and Engineering Research Council grant to P.N.R.U. M.A. is supported by a Medical Research Council training fellowship.

REFERENCES

Anderson CR, Cull-Candy SG, Miledi R 1978 Glutamate current noise: post-synaptic channel kinetics investigated under voltage clamp. J Physiol (Lond) 282:219-242

Anwyl R 1977 Permeability of the post-synaptic membrane of an excitatory glutamate synapse to sodium and potassium. J Physiol (Lond) 273:367-388

Beránek R, Miller PL 1968 The action of iontophoretically applied glutamate on insect muscle fibres. J Exp Biol 49:83-93

Clements AN, May TE 1974 Studies on locust neuromuscular physiology in relation to glutamic acid. J Exp Biol 60:673-705

Crawford AC, McBurney RN 1976 The post-synaptic action of some putative excitatory transmitter substances. Proc R Soc Lond B Biol Sci 192:481-489

Colquhoun D, Dionne VE, Steinbach JH, Stevens CF 1975 Conductance of channels opened by acetylcholine-like drugs in muscle end-plate. Nature (Lond) 253:204-206

Cull-Candy SG, Usherwood PNR 1973 Two populations of L-glutamate receptors on locust muscle fibres. Nature (Lond) 246:62-64 del Castillo J, Katz B 1957 Interaction at end-plate receptors between different choline derivatives. Proc Roy Soc Lond B Biol Sci 146:369-381

Gration KAF, Lambert JJ, Ramsey R, Usherwood PNR 1981a Non-random openings and concentration-dependent lifetimes of glutamate-gated channels in muscle membrane. Nature (Lond) 291:423-425

Gration KAF, Lambert JJ, Usherwood PNR 1981b A comparison of glutamate single-channel activity at desensitizing and non-desensitizing sites. J Physiol (Lond) 310:49P

Gration KAF, Lambert JJ, Usherwood PNR 1981c Potency of glutamate and agonists determined by patch clamp analysis of individual receptors in locust muscle. Proceedings of the Physiological Society, J Physiol (Lond), in press

Katz B, Miledi R 1970 Membrane noise produced by acetylcholine. Nature (Lond) 226:962-963

Katz B, Miledi R 1972 The statistical nature of the acetylcholine potential and its molecular components. J Physiol (Lond) 224:665-700

Lea TJ, Usherwood PNR 1973 The site of action of ibotenic acid and the identification of two populations of glutamate receptors on insect muscle fibres. Comp Gen Pharmacol 4:333-350

Mathers DA, Usherwood PNR 1976 Concanavalin A blocks desensitisation of glutamate receptors on insect muscle fibres. Nature (Lond) 259:409-411

Neher E, Sakmann B 1976 Single-channel currents recorded from membrane of denervated frog muscle fibres. Nature (Lond) 260:799-802

Neher E, Sakmann B, Steinbach JH 1978 The extracellular patch clamp: a method for resolving currents through individual open channels in biological membranes. Pflügers Arch Eur J Physiol 375:219-228

Neher E, Steinbach JH 1978 Local anaesthetics transiently block currents through single acetylcholine-receptor channels. J Physiol (Lond) 277:153-176

Nelson DJ, Sacks F 1979 Single ionic channels observed in tissue-cultured muscle. Nature (Lond) 282:861-863

Patlak JB, Gration KAF, Usherwood PNR 1979 Single glutamate-activated channels in locust muscle. Nature (Lond) 278:643-645

Ruff RL 1977 A quantitative analysis of local anaesthetic alteration of miniature end-plate currents and end-plate current fluctuations. J Physiol (Lond) 264:89-124

Sakmann B, Patlak J, Neher E 1980 Single acetylcholine-activated channels show burst-kinetics in presence of desensitizing concentrations of agonists. Nature (Lond) 286:71-73

Usherwood PNR 1978 Amino acids and synaptic transmission. Adv Comp Physiol Biochem 7:227-309

Usherwood PNR 1979 Glutamate receptors in Eucaryotes. Adv Pharmacol Chemother 1:107-116

Usherwood PNR, Machili P 1968 Pharmacological properties of excitatory neuromuscular synapses in the locust. J Exp Biol 49:341-361

DISCUSSION

Sattelle: Could you clarify the temperature effects on single-channel currents? In Fig. 4 (p 251) you compared your conductance figures with the Arrhenius plots from Stuart Cull-Candy's work (Anderson et al 1978). Have you actually fitted your own data?

Gration: Yes. For temperatures above 6 °C the activation energies obtained from single-channel conductance measurements and open times agree with those derived by Anderson et al (1978). However, Arrhenius plots of our single-channel data do not show temperature transitions.

Usherwood: Our impression was that success in finding channels did not significantly decrease when we lowered the temperature of the muscle. At some sites channel life-time was not significantly reduced when the temperature was lowered, whereas channel conductance was considerably reduced. At other sites the channel life-time was reduced when muscle temperature was lowered but channel conductance at low temperatures was not significantly different from that seen at room temperature. It looks as if, at least for extrajunctional receptors on locust muscles, channel life-time and channel conductance are either unrelated properties or can become dissociated as the temperature of the muscle is changed. Obviously, these results suggest a difference between glutamate receptor-ionophores in extrajunctional and junctional membrane of locust leg muscle. Perhaps this is due to differences in the packing density of the receptors at the two locations, the packing density of the extrajunctional receptors being much lower than that of their junctional counterparts. At low temperatures receptors in the extrajunctional membrane might, because of their low packing density, be able sometimes to seek an appropriate environment which did not cause a change in channel properties.

Dudai: Is the collagenase effect on channel conductance an all-or-none process? Do you convert one state to another, or does it depend on the time of incubation, or the concentration of collagenase?

Gration: Gradual transitions in channel conductance were not seen. Presumably a transition does not occur until the connective tissue sheath overlying the muscle fibre has been disrupted.

Dudai: Most collagenase preparations contain proteases. When a tissue is treated with collagenase a long incubation is needed during which even a small amount of protease may function efficiently. So are you sure you are dealing with a collagenase effect?

Gration: In preliminary experiments, protease (Sigma, type V) didn't seem to have any effect on the properties of the glutamate channel, but I can't rule out a contribution from these enzymes.

Dudai: Your results indicated that you don't have an in-between stage.

Gration: That is correct. Although we sampled sites on control muscles, then treated the muscle with collagenase and re-sampled during the treatment, my impression is that the transition in conductance is all-or-none.

Cull-Candy: Would it be possible to put collagenase in the patch electrode and see if there is a gradual change when the sheath is presumably being digested, while you are recording?

Gration: That is a good suggestion.

Usherwood: We have studied locust muscle fibres ultrastructurally before and after collagenase treatment. We found it difficult to control the collagenase digestion. Sometimes collagenase had no effect on the muscle. At other times it was too effective and the muscle deteriorated rapidly. When we were able to control digestion it was clear that the collagenase removed collagen fibres from the basal lamina. This takes on a more diffuse appearance after collagenase treatment, giving the impression that access to the surface membrane of the muscle fibre has improved. This could conceivably account for the increased conductance of the glutamate channel, although there are other possible explanations for this phenomenon, such as removal of surface charge.

Dudai: Is glutamate removed by an uptake process?

Gration: No. Although the nerve terminal is surrounded by glial cells, which actively sequester L-glutamate, these are absent at extrajunctional sites. Presumably, removal of glutamate from the extrajunctional membrane is diffusional. Clark et al (1980) have suggested that diffusion is primarily responsible for the termination of transmitter action at excitatory junctions on locust muscle.

Cull-Candy: Did you measure the temperature sensitivity in muscles that were not treated with Con A? Con A may bind in all sorts of places as well as on the glutamate receptor.

Gration: I envisage that Con A stabilizes the muscles, since it cross-links membrane components (Nicolson 1974). Whether that would eliminate temperature dependence, I do not know. Experiments on preparations not treated with Con A are a difficult proposition because of receptor desensitization.

Usherwood: The answer might be to look at glutamate noise rather than at individual channels. However, patch clamp analysis of isolated patches of muscle membrane rather than sites on intact muscle fibres would greatly simplify studies of the effects of temperature on the properties of individual glutamate-gated channels.

Cull-Candy: You could treat a junction with Con A and measure glutamate noise at the junctional site.

Usherwood: Yes. Or you could do that with an extrajunctional site, if you could get a suitable preparation.

Cull-Candy: The easiest way would be to treat a junctional site with Con A, then measure noise at the junction and see if Con A abolishes some of the temperature effects we see.

A. T. Eldefrawi: How does Con A prevent desensitization?

Gration: We do not know the mechanism. Desensitization may result from a conformation change in the receptor–channel complex. If this change in conformation was prevented by Con A treatment, possibly by membrane stabilization, then desensitization would be blocked.

Usherwood: There is an interesting relationship between receptor packing density and recovery from desensitization after application of glutamate or a glutamate agonist. If the receptor packing density is high, recovery from desensitization is more rapid than if it is low (Clark et al 1979, Gration et al 1979). Extrajunctional receptors therefore recover from desensitization more slowly than their junctional counterparts. After denervation the population density of extrajunctional receptors increases and this is accompanied by an increase in their rate of recovery from desensitization. This observation suggests that receptor mobility could play a role in desensitization.

Narahashi: What is the internal diameter of the capillary?

Gration: About 1 μm.

Narahashi: Do you have any idea what the channel density is?

Gration: It would seem that the density is low, since in at least 50% of our patch recordings we obtain single-channel currents with no superimpositions (multiple events). This indicates that there is only one receptor–channel complex within the area of the electrode. I cannot give a quantitative estimate of channel density.

Narahashi: I am interested in your statement about the partial block caused by quisqualate. Is this glutamate agonist what one might call a partial agonist?

Gration: In those terms I think it is, but the kinetics of the channels evoked by quisqualate are complex. Not all receptor–channel complexes appear to be susceptible to block by quisqualate; for example, with a patch electrode filled with 10^{-3} M-quisqualate mean channel life-times usually exceed 50 ms ($n = 10$ sites) but we have observed some sites ($n = 3$) at which mean life-time was less than 3 ms. In many respects the activity recorded at these sites resembles the effect of local anaesthetics on ACh channels.

Narahashi: Have you studied the effects of any ion substitutes?

Gration: No.

M. E. Eldefrawi: The quisqualate effect is reminiscent of what decamethonium does to the conductance of the end-plate (Adams & Sakmann 1978). Again, decamethonium is an agonist up to a certain concentration and above that it blocks ionic conductances and alters the time course of end-plate currents. Could you have stumbled on the first depolarizing blocker of the glutamate channel?

Gration: This would seem to be the case, but we have yet to determine why only a small proportion of the channel population is susceptible to quisqualate block.

Olsen: Have you investigated the effects of barbiturates or any other anaesthetics?

Gration: No.

House: Have you seen channel opening from activation of the H-receptor?

Gration: We have not looked at that yet, and there are at least two problems to overcome. Firstly, the resting potential of locust muscle membranes is close to the equilibrium potential of Cl^-, but this can be overcome by voltage clamp. Secondly, the desensitization of H-receptors is not blocked by Con A. Recent studies (see Gration et al 1981) have shown that desensitization of extrajunctional D-receptors is long-lasting. Concentrations of glutamate $>10^{-4}$ M in preparations not treated with Con A produce D-receptor desensitization, thereby eliminating most single-channel activity within 30–40 s of the patch electrode being placed against the muscle membrane. Subsequent recording from such sites yields long periods (30 s–10 min) of inactivity with occasional bursts of channel opening. Presumably, H-receptor–channel complexes behave in a similar manner.

House: You said that the H- and D-receptors were mixed. What are the chances of getting an H- and a D-receptor on the same patch?

Gration: Very high, I would say. I would be surprised if one didn't get an H-receptor in the patch when looking at a D-receptor.

Cull-Candy: We haven't had any evidence for H-receptor activity, although H-receptors are known to be widely distributed in the extrajunctional membrane (Cully-Candy 1976). There is no reason why the H-receptor should have such a large conductance as the D-receptor. The excitatory channel associated with the D-receptor has an unusually large conductance compared with other systems. As you mentioned, there is also the problem of the H-receptor equilibrium potential, which is close to the resting potential. The conductance of the H-channel may be intrinsically small and it desensitizes, so the chances of recording the single-channel current accidentally would be rather remote. If you contrived to get round all those problems, it should be possible to detect H-receptors.

Gration: The perfusion could help there. If you washed on ibotenic acid you might see transient channel activity from the H-receptor.

M. E. Eldefrawi: You mentioned that streptomycin affects the glutamate channel. Have you tested that in collagenase-treated as well as untreated muscle? And what does Con A treatment do to the response to streptomycin?

Gration: We tested streptomycin only on muscle *not* treated with collagenase. We have looked at the effects of streptomycin on the amplitude and decay time of the excitatory postsynaptic current in Con A-treated and non-treated muscle and it has identical effects in the two preparations.

Barnard: Is there evidence for mobility—that is, a significant degree of lateral diffusion—of these extrajunctional receptors?

Gration: No.

Barnard: Do you mean it hasn't been looked for, or it doesn't exist?

Gration: I don't know of any studies of the lateral mobility of extrajunctional receptors in locust muscle membrane.

Usherwood: One of the problems of recording from a single area of extrajunctional membrane is that the receptors sometimes disappear!

Barnard: That certainly suggests a nice image—that the receptors float away in the membrane, out of range. To determine whether there *is* such lateral mobility, one would need to use photo-bleaching experiments of the type used with the extrajunctional ACh receptor (Axelrod et al 1978) or—better—measurements of receptor mobility in an electric field across an isolated cell (Poo & Robinson 1977), both requiring a tightly bound fluorescent probe. These, however, would require a technology not yet available for the glutamate receptor. The simplest reason for expecting a much less sharp temperature effect on the extrajunctional receptors is their much lower population density in the membrane. One can imagine that in the densely populated junctional regions each receptor has a tightly bound domain of lipid around it, because the distances from the neighbouring receptors are so small that the intervening lipid is in a more or less fixed position. Therefore, changes in the state of the lipid with temperature would be expected to cause a much greater effect than extrajunctionally, where presumably this packing effect is not exerted on the lipid. Another possibility is that a different lipid composition, giving greater rigidity, characterizes the postsynaptic region. At least your evidence is consistent with these ideas.

Cull-Candy: Have you been able to record at different temperatures, keeping one channel under the pipette all the time? You appear to be pooling your data and it would be nice to see what happens to an individual channel.

Gration: We have not yet done a long temperature run at a single site to our satisfaction. We would like to go from 18 °C down to 1 °C; so far, our experiments have been over a more restricted range, from 0 °C up to 12 °C.

Usherwood: It was easier to 'hold' a single channel when raising the temperature of the muscle than when lowering its temperature. On occasions we were able to study single channels over the temperature range of 0 °C to 12°C.

Gration: In these experiments it was interesting that we often observed a change in channel conductance with no concomitant change in open time, and vice versa. It would be advantageous to do a dozen or so long-term temperature runs.

Usherwood: It would be so much easier to take off a patch of membrane with a receptor in it and do the temperature studies on that piece of isolated membrane. I think that is the best goal to aim for, rather than attempting to record the activity of single channels from sites on intact muscle preparations.

REFERENCES

Adams PR, Sakmann B 1978 Decamethonium both opens and blocks endplate channels. Proc Natl Acad Sci USA 75:2994-2998

Anderson CR, Cull-Candy SG, Miledi R 1978 Glutamate current noise: post-synaptic channel kinetics investigated under voltage clamp. J Physiol (Lond) 282:219-242

Axelrod D, Ravdin P, Podleski TR 1978 Control of acetylcholine receptor mobility and distribution in cultured muscle membranes. Biochim Biophys Acta 511:23-38

Clark RB, Gration KAF, Usherwood PNR 1979 Desensitization of glutamate receptors on innervated and denervated locust muscle fibres. J Physiol (Lond) 290:551-568

Clark RB, Gration KAF, Usherwood PNR 1980 Influence of glutamate and aspartate on time course of decay of excitatory synaptic currents at locust neuromuscular junctions. Brain Res 192:205-216

Cull-Candy SG 1976 Two types of extrajunctional L-glutamate receptors in locust muscle fibres. J Physiol (Lond) 255:449-464

Gration KAF, Clark RB, Usherwood PNR 1979 Denervation of insect muscle: a comparative study of changes in L-glutamate sensitivity on locust retractor unguis and extensor tibiae muscle. Neuropharmacology 18:201-208

Gration KAF, Lambert JJ, Usherwood PNR 1981 Glutamate-activated channels in locust muscle. In: Pethes G, Frenyo VL (eds) Advances in animal and comparative physiology. Pergamon, Oxford/Akadémiai Kiadó, Budapest (Adv Physiol Sci 20) p 377-383

Nicolson GL 1974 Interactions of lectins with animal cell surfaces. Int Rev Cytol 39:89-190

Poo MM, Robinson KR 1977 Electrophoresis of concanavalin A receptors along embryonic muscle cell membranes. Nature (Lond) 265:602-605

Receptor mechanisms mediating the action of 5-hydroxytryptamine

MICHAEL J. BERRIDGE and JOHN P. HESLOP

ARC Unit of Invertebrate Chemistry and Physiology, Department of Zoology, University of Cambridge, Downing Street, Cambridge, CB2 3EJ, UK

Abstract The salivary glands of the blowfly *Calliphora erythrocephala* respond to 5-hydroxytryptamine (5-HT) with a large increase in fluid secretion. Pharmacological studies with a range of 5-HT analogues have demonstrated 5-HT_1 receptors that open channels for calcium and 5-HT_2 receptors that are linked to adenylate cyclase to generate cyclic AMP. In addition to gating calcium the 5-HT_1 receptors are responsible for stimulating the hydrolysis of phosphatidylinositol. Some of the properties of this phosphatidylinositol response are discussed, particularly its possible role in calcium signalling.

The time course of cyclic AMP formation resulting from the activation of 5-HT_2 receptors is extremely complex. An initial biphasic response precedes the slower rise towards the steady state. Analysis of this response reveals that the intracellular level of cyclic AMP is not only sensitive to external agonists acting on adenylate cyclase but is also regulated by internal feedback reactions, perhaps operating on the phosphodiesterase that hydrolyses cyclic AMP.

The nervous system of vertebrates and higher invertebrates is composed of individual neurons which interact with each other to form neural circuits of bewildering complexity. In order to unravel this circuitry we need to understand not only the spatial organization of the individual neurons but also the way in which they communicate with each other. Much of this communication is achieved through neurotransmitters and neuromodulators. These chemical agents can induce rapid changes in the postsynaptic cell by opening various ion channels, or they can cause more long-lasting changes in membrane properties by inducing the formation of second messengers such as cyclic AMP and calcium. The way in which the biogenic amine 5-hydroxytryptamine acts to generate calcium and cyclic AMP has been studied in the salivary gland of the blowfly, *Calliphora erythrocephala*. This salivary gland system has proved to be an interesting model system for studying the

---

1982 Neuropharmacology of insects. Pitman, London (Ciba Foundation symposium 88) p 260-274

nature of the transducing mechanisms responsible for generating these second messengers.

5-Hydroxytryptamine (5-HT; serotonin) plays an important role as a neuromodulator in the gill withdrawal reflex of the mollusc *Aplysia* (the sea-hare) (Klein & Kandel 1978). The ability of sensory neurons to stimulate the motor neurons innervating the gills is greatly enhanced when serotoninergic neurons which terminate on the synaptic endings of the sensory neurons are activated. When 5-HT is released it interacts with receptors on the synaptic endings to increase the level of cyclic AMP, which then acts to enhance transmitter release by prolonging the inward calcium current (Klein & Kandel 1980). These studies of *Aplysia* illustrate the way in which neuromodulators such as 5-HT can profoundly influence neuronal activity. It is important, therefore, to understand the nature of the transducing mechanisms that are responsible for generating the second messengers.

Receptor diversity

Most hormones and neurotransmitters seem to act through more than one type of receptor (Fig. 1). For example, acetylcholine can act on nicotinic or muscarinic receptors. These receptors vary not only in the way in which they bind to different agonists and antagonists but also in the nature of the transducing mechanisms with which they are associated (Berridge 1980a,b,c). Nicotinic receptors are usually associated with sodium channels and agonist binding results in membrane depolarization, whereas muscarinic receptors usually open ion channels for calcium. Such receptor diversity, which seems to be a common feature, greatly enhances the range of action of individual agonists because each agonist can transmit a variety of signals, depending on the nature of the transducing mechanism associated with each receptor. For example, neurons in mammalian sympathetic ganglia have both nicotinic and muscarinic receptors. Activation of the former gives rise to a fast excitatory postsynaptic potential whereas occupation of the muscarinic receptors produces a slow excitatory postsynaptic potential (Greengard 1976). The possibility that there might be separate 5-HT receptors was first suggested by Gaddum & Picarelli (1957). The fact that 5-HT induces two distinct effects in NCB-20 neuroblastoma–brain hybrid cells (MacDermot et al 1979) is also consistent with receptor diversity. Binding studies with [^{3}H]5-hydroxytryptamine, [^{3}H]lysergic acid and [^{3}H]spiroperidol have confirmed the existence of two different 5-HT receptors in rat brain (Peroutka & Snyder 1979). One of these receptors seems to be associated with adenylate cyclase; the nature of the transducing mechanism associated with the other receptor is unknown.

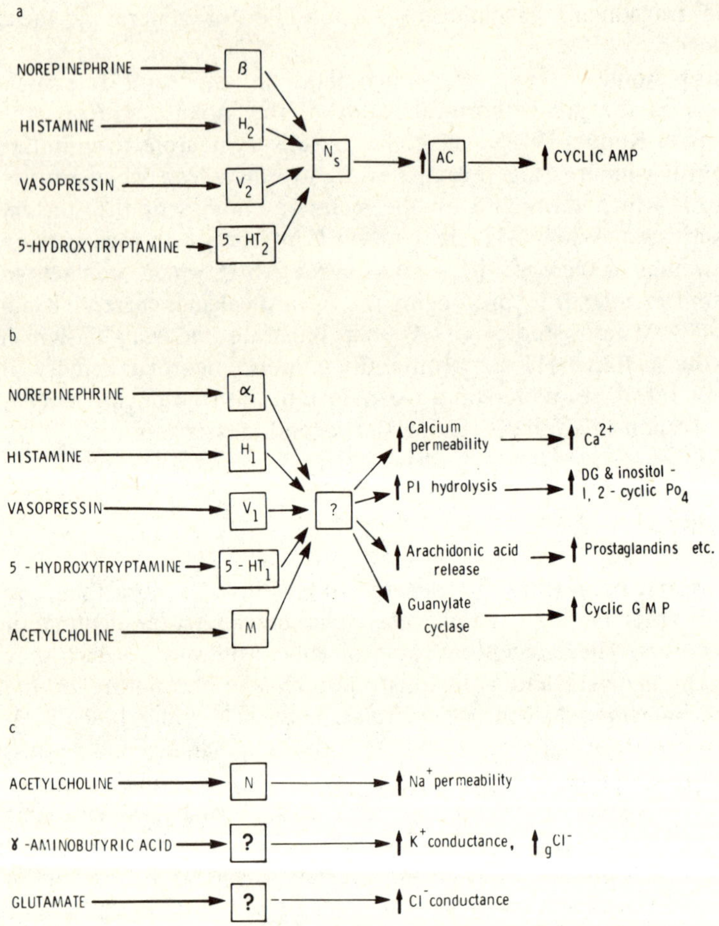

FIG. 1. A summary of receptor mechanisms responsible for the action of a variety of hormones and neurotransmitters. (a) Agonists acting through cyclic AMP. A GTP-regulatory subunit (N_s) couples receptors to adenylate cyclase (AC). (b) Multifunctional receptors responsible for generating calcium signals, for hydrolysing phosphatidylinositol (PI), for releasing arachidonic acid and for activating guanylate cyclase. (c) Receptors linked to various ion channels. M, muscarinic receptor; N, nicotinic receptor; N_s, guanine nucleotide-binding protein; AC, adenylate cyclase.

Identification of separate 5-HT receptors in Calliphora salivary glands

The salivary glands of *Calliphora* respond to 5-HT with an increase in fluid secretion (Berridge 1970). Electrophysiological studies indicate that the prime mover for fluid secretion is a cation pump located on the apical membrane which transports potassium into the lumen of the gland to create the electrical gradient for a parallel flow of chloride (Berridge & Prince 1972, Berridge et al 1975). 5-HT accelerates both these ionic pathways, activating two different second messengers: cyclic AMP stimulates potassium transport whereas the increase in chloride permeability is controlled by calcium. 5-HT thus seems to use two separate second messengers to regulate secretion. The next question is whether these two second messengers are generated by separate 5-HT receptors.

There are no recognized agonists or antagonists for distinguishing between different 5-HT receptors. We therefore attempted to identify such pharmacological agents by screening a wide range of 5-HT analogues. A rapid screening procedure was developed, based on the electrophysiological response of the gland to these analogues. Previous studies had shown that cyclic AMP caused the transepithelial potential to hyperpolarize whereas calcium had a depolarizing effect. By studying the transepithelial potential it is therefore possible to identify analogues which might act through one or other second messenger pathway (Berridge 1981). This screening procedure identified several analogues with opposing effects on transepithelial potential, suggesting that they might be acting through either cyclic AMP or calcium.

To identify the action of these analogues more precisely we had to study their effects on the second messengers. We studied the effect mediated by adenylate cyclase by measuring the intracellular level of cyclic AMP using a radioimmunoassay technique (Heslop & Berridge 1980). Receptors acting through calcium are harder to study because of the difficulty of measuring the intracellular concentration of free calcium. However, in *Calliphora* salivary glands, the activity of 5-HT$_1$ receptors can be monitored indirectly by measuring the transepithelial flux of $^{45}Ca^{2+}$ (Berridge & Lipke 1979, Fain & Berridge 1979a). Another change invariably associated with the activation of this class of receptor is the hydrolysis of phosphatidylinositol—a process often referred to as the phosphatidylinositol (PI) response (Michell 1975, Berridge 1980c). The significance of PI, which usually represents about 10% of the membrane phospholipids, is that its hydrolysis is always induced by the same receptors which act through calcium (Fig. 1). The possible relationships of this response to calcium gating will be explored later. We monitored the PI response in *Calliphora* salivary glands by measuring the release of [^{3}H]inositol from prelabelled glands (Fain & Berridge 1979a).

The effect of the parent compound, 5-HT, on fluid secretion, cyclic AMP accumulation, [^{3}H]inositol release and $^{45}Ca^{2+}$ transport is illustrated in Fig. 2. The physiological response of fluid secretion has a steeper dose–response curve than the other parameters, which are more closely linked to the receptor mechanism. In addition, a maximal increase in fluid secretion is reached with

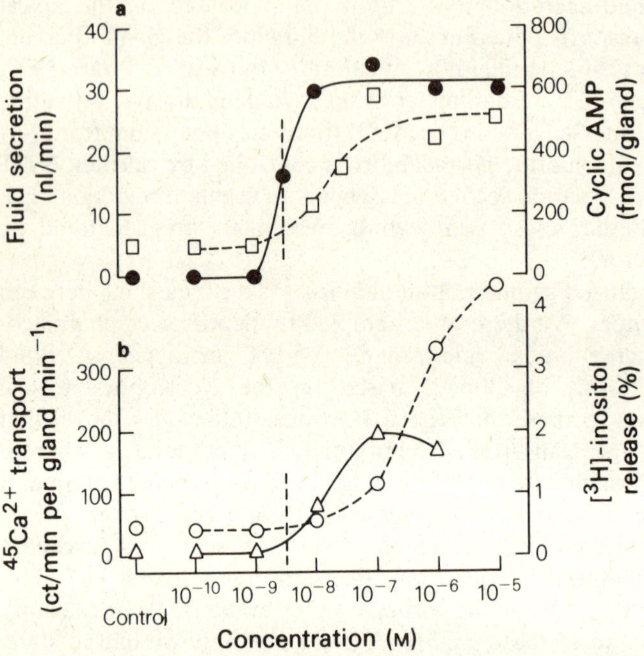

FIG. 2. The effect of 5-HT on the rate of fluid secretion (●), the concentration of cyclic AMP (□), the release of [^{3}H]inositol (○) and the transport of $^{45}Ca^{2+}$ (△) by salivary glands of *Calliphora*. (From Berridge & Heslop 1981.)

small changes in the receptor-mediated events. This relationship indicates the existence of spare receptors, in that the occupation of relatively few receptors generates enough cyclic AMP and calcium to activate fluid secretion maximally. Any further increase in 5-HT concentration leads to the occupation of more receptors and to increases in the receptor-mediated events—cyclic AMP production, [^{3}H]inositol release and $^{45}Ca^{2+}$ transport (Fig. 2). There is no further increase in the rate of fluid secretion, which is already maximal at the lower doses of 5-HT.

In similar studies with the analogue 4-fluoro-α-methyltryptamine, which gave a hyperpolarizing response in the electrophysiological study, indicating an action through cyclic AMP, a different relationship was found between these four parameters (Fig. 3). This analogue stimulated fluid secretion and cyclic AMP formation but had no effect on either [^{3}H]inositol release or the transport of $^{45}Ca^{2+}$. This analogue thus appears to act specifically on a receptor responsible for the production of cyclic AMP, but without activating the receptor mechanism which gates calcium. The electrophysiological study suggested that certain 5-substituted analogues such as 5-chlorotryptamine

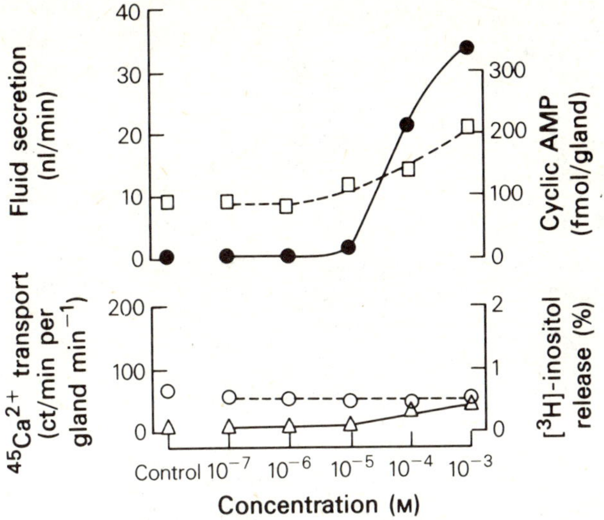

FIG. 3. The effect of 4-fluoro-α-methyltryptamine on the rate of fluid secretion (●), the concentration of cyclic AMP (□), the release of [^{3}H]inositol (○) and the transport of $^{45}Ca^{2+}$ (△) by salivary glands of *Calliphora*. (From Berridge & Heslop 1981.)

induce a rapid depolarizing response, indicating an action through calcium. In fact, 5-chlorotryptamine did increase $^{45}Ca^{2+}$ transport and [^{3}H]inositol release at concentrations far lower than those needed to stimulate the production of cyclic AMP (Fig. 4). Although 5-chlorotryptamine does not show an absolute specificity it is clearly able to discriminate between the two receptors. These pharmacological studies seem to indicate, therefore, that the *Calliphora* salivary gland has separate receptors for 5-HT—a 5-HT_1 receptor responsible for gating calcium and a 5-HT_2 receptor activating adenylate cyclase to generate cyclic AMP.

The 5-HT_1 receptor mechanism

Fig. 1 shows that a number of agonists acting on specific receptors (α_1, H_1, V_1, 5-HT_1, M) can stimulate a calcium signal. It is also apparent that calcium is not the only second messenger generated by this type of receptor. As described earlier, these receptors also stimulate the hydrolysis of PI, with the formation of diacylglycerol and inositol 1:2-cyclic phosphate. It has been proposed that in mammalian cells diacylglycerol might act as a second messenger because it has been found to activate a membrane-bound protein kinase (Kishimoto et al 1980). The activation of these receptors also seems to

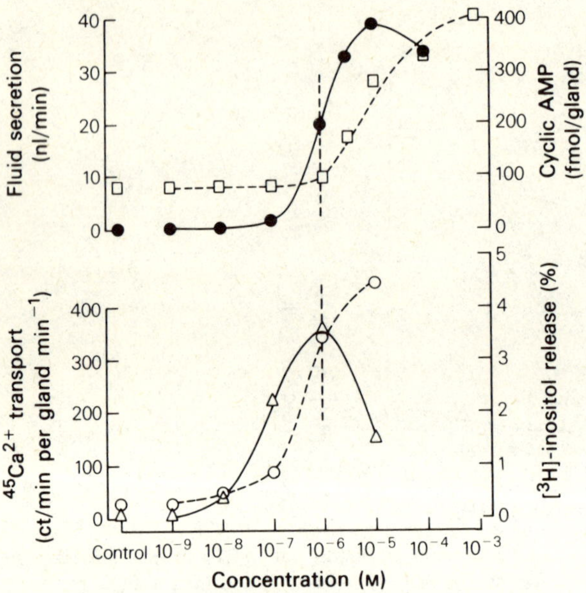

FIG. 4. The effect of 5-chlorotryptamine on the rate of fluid secretion (●), the concentration of cyclic AMP (□), the release of [^{3}H]inositol (○) and the transport of $^{45}Ca^{2+}$ (△) by salivary glands of *Calliphora*. (From Berridge & Heslop 1981.)

be important for the release of arachidonic acid, the precursor of prostaglandins and related compounds such as the thromboxanes and the leukotrienes. Finally, these receptors are responsible for the activation of guanylate cyclase, leading to the production of cyclic GMP. These receptors appear to be extremely versatile, and the obvious question is how a single receptor stimulates so many second messengers. As indicated in Fig. 1, the nature of the transducing mechanism responsible for activating all these events is unknown. There are suggestions that the hydrolysis of PI is a key event which is then responsible for mobilizing calcium (Michell 1975, Michell & Kirk 1981), for releasing arachidonic acid (Irvine et al 1981) and for activating guanylate cyclase (Nishizuka & Takai 1981). The possible involvement of the PI response in the gating of calcium has been investigated in the *Calliphora* salivary gland (Fain & Berridge 1979a,b, Berridge & Fain 1979).

A close relationship between the hydrolysis of PI and the transport of $^{45}Ca^{2+}$ is apparent from the pharmacological studies described earlier. When a wide range of analogues was tested the hydrolysis of PI was also associated with an increase in $^{45}Ca^{2+}$ transport. The problem is to discover the precise relationship between these two events. Are they parallel events, as depicted in Fig. 1, or are they organized sequentially? In the latter case, what comes first? Michell and his colleagues (Michell 1975, Michell & Kirk 1981) have proposed that the hydrolysis of PI is an early event responsible for the subsequent gating of calcium. A crucial test of this hypothesis concerns the calcium dependency of the PI response. If the hydrolysis of PI is an early receptor event responsible for calcium gating, it must be able to occur at the resting level of intracellular calcium. Numerous studies in other systems indicate that the PI response is independent of calcium (Michell & Kirk 1981). A similar conclusion has been reached for *Calliphora* salivary glands (Fain & Berridge 1979a). When calcium was removed from the bathing medium, the release of [^{3}H]inositol from prelabelled glands in response to 5-HT was not altered even though the secretory response was severely diminished. Apparently the PI response occurs in conditions in which the level of calcium falls below that required to induce maximal rates of fluid secretion.

It might be argued, however, that the PI response is sensitive to much smaller changes in calcium concentration than are necessary to induce the secretory mechanism. This possibility was tested by using the ionophore A23187 to increase the intracellular level of calcium artificially. A concentration of 1×10^{-6} M-A23187 stimulated the same amount of $^{45}Ca^{2+}$ transport as 5×10^{-8} M-5-HT, yet the former had no effect on [^{3}H]inositol release, in contrast to the large changes induced by the natural hormone. These studies seem to indicate, therefore, that the ability of 5-HT to stimulate the hydrolysis of PI is independent of any change in the intracellular level of calcium.

Another way of trying to determine the relationship between PI hydrolysis and calcium gating is by studying the effect of inhibiting the PI response. Unfortunately there are no known inhibitors that specifically block PI hydrolysis. The approach that we adopted, therefore, was to deplete the cell of PI by stimulating the cell for a prolonged period with a high concentration of 5-HT. This not only stimulates the hydrolysis of PI but simultaneously inhibits its resynthesis (Berridge & Fain 1979). Prolonged stimulation led to an almost complete inactivation of $^{45}Ca^{2+}$ transport, suggesting that this phospholipid may be necessary in order for 5-HT to stimulate the entry of calcium. Some confirmation for this view was obtained by finding that these inactivated glands could be resensitized by providing them with inositol, which enabled them to resynthesize PI (Fain & Berridge 1979b). The fact that the gating of calcium could be adjusted by altering the level of PI strongly suggests that this phospholipid is intimately connected with the 5-HT$_1$ receptor mechanism responsible for generating calcium signals.

The 5-HT$_2$ receptor mechanism

The pharmacological studies described earlier revealed the existence of a 5-HT$_2$ receptor on *Calliphora* salivary glands that is responsible for stimulating the synthesis of cyclic AMP. An increase in the intracellular level of cyclic AMP seems to be essential for increasing fluid secretion, because cyclic AMP stimulates the potassium pump that provides the driving force for the parallel flow of chloride and water. The nature of the receptor mechanism responsible for generating this cyclic AMP signal has not been studied. In other systems, a GTP-regulatory protein (N_s, Fig. 1) plays an important role as a transducer in the transfer of information from the receptor to adenylate cyclase. It remains to be shown whether a similar transducing mechanism is used by this invertebrate system.

Studies on the kinetics of the cyclic AMP response of *Calliphora* salivary glands after stimulation with 5-HT showed that the activation of adenylate cyclase may not be the only factor responsible for regulating the intracellular level of cyclic AMP (Heslop & Berridge 1980). When salivary glands were stimulated with 5-HT there was a rapid increase in the intracellular concentration of cyclic AMP, reaching a peak in 30–40 seconds (Fig. 5). After this

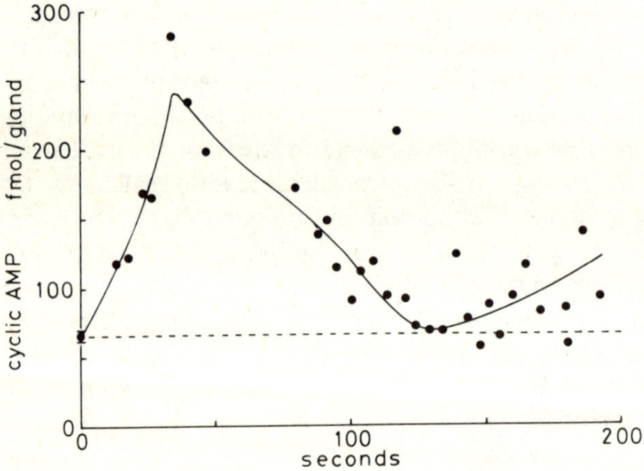

FIG. 5. The time-course of the cyclic AMP response after 5×10^{-9} M 5-HT had been added to isolated salivary glands of *Calliphora*. The horizontal broken line shows the resting level of cyclic AMP (66 ± 3.7 (5) fmol/gland) ($\bar{x}$ ± SEM (*n*)).

initial peak the level fell back close to the basal level by about 130 seconds and then began to rise again. The sudden fall in the level of cyclic AMP after the early peak is apparently not the result of a decrease in adenylate cyclase activity because if theophylline is included in the incubation medium to reduce the activity of phosphodiesterase (which hydrolyses cyclic AMP) the cyclic AMP level continues to rise, going as high as 2000 fmol/gland (Heslop & Berridge 1980). The fact that theophylline prevents the fall suggests that it is caused by the activation of phosphodiesterase. Thus the intracellular level of cyclic AMP may be determined by changes in the activity not only of adenylate cyclase but also of phosphodiesterase.

This biphasic change in cyclic AMP level after hormonal stimulation may be a general phenomenon, as it has also been described in the Malpighian tubules of *Rhodnius prolixus* (Aston 1975), in rat erythrocytes (Clayberger et al 1981) and in the amphibian heart (Flitney & Singh 1980). There seems to be general agreement that the fall in cyclic AMP level is the result of increased phosphodiesterase activity but there is less agreement about the mechanism responsible for switching on this enzyme. Clayberger et al (1981) suggest that the membrane-bound phosphodiesterase in rat erythrocytes is activated through calcium, with calmodulin as an intermediary. Activation in the amphibian heart is caused by cyclic GMP, whose level rises to a peak at the time when cyclic AMP is falling back towards its resting level. Despite this uncertainty about the mechanism responsible for activating phosphodiesterase, all these observations draw out attention to the complexity of the intracellular mechanisms responsible for generating second messengers. The activation of specific receptors is usually responsible for enhancing the production of these signals but the final level they achieve in the cell will depend on many other factors, including changes in the enzyme responsible for removing these signals.

REFERENCES

Aston RJ 1975 The role of adenosine 3′:5′-cyclic monophosphate in relation to the diuretic hormone of *Rhodnius prolixus*. J Insect Physiol 21:1873-1877

Berridge MJ 1970 The role of 5-hydroxytryptamine and cyclic AMP in the control of fluid secretion by isolated salivary glands. J Exp Biol 53:171-186

Berridge MJ 1980a Receptor coupling to adenylate cyclase and calcium gates. In: Sattelle DB et al (eds) Receptors for neurotransmitters, hormones and pheromones in insects. Elsevier/North-Holland Biomedical Press, Amsterdam, pp 27-38

Berridge MJ 1980b Hormone action—a search for transducing mechanisms. In: Locke M, Smith DS (eds) Insect biology in the future. Academic Press, London, p 463-478

Berridge MJ 1980c Receptors and calcium signalling. Trends Pharmacol Sci 1:419-424

Berridge MJ 1981 Electrophysiological evidence for the existence of separate receptors for 5-hydroxytryptamine. Mol Cell Endocrinol 24:115-140

Berridge MJ, Fain JN 1979 Inhibition of phosphatidylinositol synthesis and the inactivation of calcium entry after prolonged exposure of the blowfly salivary gland to 5-hydroxytryptamine. Biochem J 178:59-69

Berridge MJ, Heslop JP 1981 Separate 5-hydroxytryptamine receptors on the salivary gland of the blowfly are linked to the generation of either cyclic adenosine 3′:5′-monophosphate or calcium signals. Br J Pharmacol 73:729-738

Berridge MJ, Lipke H 1979 Changes in calcium transport across *Calliphora* salivary glands induced by 5-hydroxytryptamine and cyclic nucleotides. J Exp Biol 78:137-148
Berridge MJ, Prince WT 1972 Transepithelial potential changes during stimulation of isolated salivary glands with 5-hydroxytryptamine and cyclic AMP. J Exp Biol 56:139-153
Berridge MJ, Lindley BD, Prince WT 1975 Membrane permeability changes during stimulation of isolated salivary glands of *Calliphora* by 5-hydroxytrypamine. J Physiol (Lond) 244:549-567
Clayberger CA, Goodman DBP, Rasmussen H 1981 Regulation of cyclic AMP metabolism in the rat erythrocyte during chronic β-adrenergic stimulation. Evidence for calmodulin-mediated alteration of membrane bound phosphodiesterase activation. J Membr Biol 58:191-201
Fain JN, Berridge MJ 1979a Relation between hormonal activation of phosphatidylinositol hydrolysis, fluid secretion and calcium flux in the blowfly salivary gland. Biochem J 178: 45-58
Fain JN, Berridge MJ 1979b Relationship between phosphatidylinositol synthesis and recovery of 5-hydroxytryptamine-responsive Ca^{2+} flux in blowfly salivary glands. Biochem J 180:655-661
Flitney FW, Singh J 1980 Inotropic responses of the frog ventricle to adenosine triphosphate and related changes in endogenous cyclic nucleotides. J Physiol (Lond) 304:21-42
Gaddum JH, Picarelli ZP 1957 Two kinds of tryptamine receptor. Br J Pharmacol Chemother 12:323-328
Greengard P 1976 Possible role for cyclic nucleotides and phosphorylated membrane proteins in postsynaptic actions of neurotransmitters. Nature (Lond) 260:101-108
Heslop JP, Berridge MJ 1980 Changes in cyclic AMP and cyclic GMP concentrations during the action of 5-hydroxytryptamine on an insect salivary gland. Biochem J 192:247-255
Irvine RF, Dawson RMC, Freinkel N 1981 Stimulation of phosphatidylinositol turnover: a brief appraisal. In: The year in metabolism. Plenum Press, New York
Kishimoto A, Takai Y, Mori T, Kikkawa U, Nishizuka Y 1980 Activation of calcium and phospholipid-dependent protein kinase by diacylglycerol, its possible relation to phosphatidylinositol turnover. J. Biol Chem 255:2273-2276
Klein K, Kandel ER 1978 Presynaptic modulation of voltage-dependent Ca^{2+} current: mechanism for behavioural sensitization in *Aplysia californica*. Proc Natl Acad Sci USA 75:3512-3516
Klein K, Kandel ER 1980 Mechanisms of calcium current modulation underlying presynaptic facilitation and behavioural sensitization in *Aplysia*. Proc Natl Acad Sci USA 77:6912-6916
MacDermot J, Higashida H, Wilson SP, Matsuzawa H, Minna J, Nirenberg M 1979 Adenylate cyclase and acetylcholine release regulated by separate serotonin receptors of somatic cell hybrids. Proc Natl Acad Sci USA 76:1135-1139
Michell RH 1975 Inositol phospholipids and cell surface receptor function. Biochim Biophys Acta 415:81-147
Michell RH, Kirk CJ 1981 Why is phosphatidylinositol degraded in response to stimulation of certain receptors? Trends Pharmacol Sci 2:86-89
Nishizuka Y, Takai Y 1981 Calcium and phospholipid turnover in a new receptor function for protein phosphorylation. Cold Spring Harbor Symp Quant Biol, in press
Peroutka SJ, Snyder SH 1979 Multiple serotonin receptors: differential binding of [^{3}H]5-hydroxytryptamine, [^{3}H]lysergic acid diethylamide and [^{3}H]spiroperidol. Mol Pharmacol 16:687-699

DISCUSSION

Usherwood: In a calcium-free environment does hydrolysis of phosphatidylinositol (PI) still take place when you put 5-HT on the preparation?

Berridge: The ability of 5-HT to stimulate the hydrolysis of PI is independent of external calcium.

Usherwood: The calcium channel gating itself might be responsible for the PI response in the absence of calcium.

Berridge: Yes, but lanthanides (lanthanum) can be used to suppress the calcium gating and there is no effect on the PI response. It seems to be totally independent of anything that is done to calcium.

M. E. Eldefrawi: Do these cell membranes contain only monophosphoinositides?

Berridge: Preliminary measurements have shown that PI is the major inositide but there are small amounts of diphosphatidylinositide (3%) and triphosphatidylinositide (5%). These polyphosphoinositides bind calcium strongly and may thus function as a membrane calcium store.

Dudai: Are 5-HT_1 receptors modulated by GTP?

Berridge: I haven't looked at that yet.

Dudai: Muscarinic receptors in both vertebrates and flies are modulated by GTP, which means that the missing link for muscarinic receptors may also be a nucleotide-mediated effector system.

Berridge: That is possible but we must be careful about using GTP binding as part of our receptor classification. In the adrenergic system, α-receptors can be separated into α_1 and α_2. The α_1 gives a PI response while α_2 is linked through a GTP system to cause inhibition of adenylate cyclase. Muscarinic receptors may also have two output signals, one through a GTP protein which inhibits adenylate cyclase and another through some other coupling agent which increases PI breakdown. It is still an open question whether or not the GTP regulatory proteins are involved in the PI mechanism.

Lunt: What is the time scale of these changes in your system? Can you see whether release of [^{3}H]inositol precedes or follows calcium entry?

Berridge: I haven't done that yet because the technique for measuring calcium entry depends upon fluid secretion, which has a rather slow onset in response to 5-HT.

Lunt: Could you look at the change in calcium level within seconds of stimulation?

Berridge: The technique I have used would not be capable of resolving flux changes over a few seconds.

Lunt: I wondered whether you could look at the entry of labelled calcium from outside.

Berridge: Such techniques are difficult to apply to these small glands and the uptake data are difficult to interpret. Once calcium enters the cell it rapidly enters the internal pools which might also be regulated, thus making interpretation of the results very difficult.

House: Can you tell us something about your work with calcium-sensing microelectrodes?

Berridge: Calcium measurements were made using double-barrelled electrodes which were sensitive down to 10^{-7} M-calcium (Berridge 1980). I suspect that the resting level of calcium is somewhat below 10^{-7} M, and although I could record responses to 5-HT and get increases in calcium up to about 10^{-6} M, I think I was looking at the crests of the response and was probably missing the earlier events taking place at the resting level, which could be as low as 10^{-8} M. We need electrodes that are sensitive to this lower level.

Barnard: Is it necessary to assume, in the first class of receptor systems as you described them, that the calcium permeability change and the PI change are necessarily linked to a change in a nucleotide cyclase? In some of the systems well studied in vertebrates there seems to be no good evidence for any direct connection between insulin activation of its receptor, for example, and a cyclase activation. Could some of your systems involve cyclic GMP while others involve PI without mediation through the cyclase?

Berridge: That is possible, but so far there has been no separation, because whenever a receptor has been found to work through calcium there invariably is an increase in cyclic GMP. Insulin is really an agent on its own. So far as we know, the second messenger for insulin is probably a small peptide.

Barnard: But that is a case where no increase in any cyclic nucleotide is required, then.

Berridge: That would fit with the classification I proposed earlier, because insulin is not operating through the PI-arachidonic acid–prostaglandin system.

Barnard: Then the classification would have to include a further, fourth class of receptors—those that use a second messenger but do not cause an increase in a cyclic nucleotide. It may be important to keep this possibility open in looking at the actions of some of the invertebrate hormones.

Olsen: If phosphatidic acid is important it seems rather chancy to have two steps in the conversion from PI. How do you measure the breakdown? Is it possible that a direct phospholipase D activity cleaves inositol without phosphate?

Berridge: I don't think there is any evidence that such an enzyme operates in any of these systems. All the evidence points to phospholipase C being the primary enzyme because the initial products are diglyceride and inositol-1,2-cyclic phosphate.

Piek: One of your experiments seems to indicate that the messengers work independently. That is, when you block the 5-HT receptor there is still an effect of 5-HT in the other system. Does the pH change or does the outflow of potassium ions increase?

Berridge: It is possible to stimulate secretion by activating the 5-HT_2 receptor only. In this case, the cyclic AMP pathway is activated to drive the potassium pump without a change in chloride conductance, resulting in an enormous hyperpolarization of the lumen. This increased hyperpolarization helps to drag anions across the membranes. The gland is on a knife edge when we stimulate through this cyclic AMP pathway. It is just managing to get enough chloride across to satisfy the process of fluid secretion. If the external chloride concentration is reduced by being replaced with an impermeant anion, this cyclic AMP pathway cannot induce secretion by itself because it is unable to get enough chloride across the gland.

Olsen: The hydrolysis of the phospholipids is not stoichiometrically sufficient to account for a calcium flux. For example, for freeing bound calcium it would have to be a messenger rather than a stoichiometric reaction. Is that correct?

Berridge: If the substrate is a polyphosphoinositide, there is probably not enough calcium bound to give a full signal, except that they may account for an initial small surge of trigger calcium. If this is not augmented by calcium entering from the outside, all one gets is a phasic response. People working on smooth muscle are familiar with this phenomenon. If the muscle is stimulated, there is an initial phasic contraction which is replaced by a more tonic contraction caused by the entry of external calcium. Some release of calcium from the membrane may be sufficient to get things started, but this has to be augmented by calcium entering from the outside.

House: There is evidence in other secretory systems for phasic release (see Ginsborg & House 1980 for references).

Berridge: Yes, this happens in many systems, such as pancreas.

Mordue: With regard to the energetics of the gland itself, what sort of turnover of lipid is there? The gland is obviously committed to spending a lot of energy to drive all the fluid. Is it also committed to a substantial turnover of lipid?

Berridge: The energy demand may not be very large. For the sodium pump we know that one ATP molecule is needed to drive relatively few ions across the membrane. In the PI response, the opening of the channel may gate a large number of calcium ions before the channel closes.

Mordue: The number of receptors that were fired to give a good physiological response was rather small in your experiments. That must conserve the energy of the system.

Berridge: If you are trying to detect molecules that are present at a very low concentration it makes sense to put lots of receptors on the surface. It seems that relatively few of these receptors need to be occupied in order to generate a significant increase in the relevant second messenger.

Dudai: But is there now direct evidence for the concept of spare receptors?

Berridge: The old pharmacological evidence for spare receptors came from studying partial agonists. More recent evidence for the existence of spare receptors has come from studies on the β-receptor of turkey erythrocytes (Levitzki 1976).

Lunt: What is the role of prostaglandins and the leukotrienes in your system? The main function of PI turnover in some systems may be production of phosphatidic acid from which arachidonate could be released.

Berridge: We are investigating this now. Some people working on blood platelets and on neutrophils suggest that arachidonic acid metabolites, particularly the thromboxanes, are potent calcium ionophores and that this is the way the calcium ionophore is generated (Kao et al 1981, Sha'afi et al 1981). We have thought of trying to look at this in the insect system, because it might be possible to grow flies which do not have any arachidonic acid. It would be interesting to see whether these flies can still produce a calcium signal.

Lunt: Do those flies have PI?

Berridge: Yes.

Lunt: At normal levels? And what are the fatty acids on position 2?

Berridge: We have not determined the nature of the fatty acid on position 2, but it is not arachidonic acid.

REFERENCES

Berridge MJ 1980 Preliminary measurements of intracellular calcium in an insect salivary gland using a calcium-sensitive microelectrode. Cell Calcium 1:217-227

Ginsborg BL, House CR 1980 Stimulus–response coupling in gland cells. Annu Rev Biophys Bioeng 9:55-80

Kao K-J, Hagen P-O, Pizzo SV 1980 Membrane protein mediation of thromboxane A_2 induced platelet aggregation. Biochem Biophys Res Commun 97:87-93

Levitzki A 1976 Catecholamine receptors. In: Cuatrecasas P, Greaves MF (eds) Receptors and recognition, Series A. Chapman & Hall, London, vol 2:199-229

Sha'afi RI, Naccache PH, Alobaidi T, Molski TFP, Volpi M 1981 Effect of arachidonic acid and the chemotactic factor F-Met-Leu-Phe on cation transport in rabbit neutrophils. J Cell Physiol 106:215-223

Solitary wasp venoms and toxins as tools for the study of neuromuscular transmission in insects

TOM PIEK

Pharmacological Laboratory, University of Amsterdam, Polderweg 104, 1093 KP Amsterdam, The Netherlands

Abstract The venom of *Microbracon hebetor* contains two protein toxins, *Microbracon* toxin-A and toxin-B, which slowly and reversibly paralyse lepidopteran larvae, a maximum of paralysis being reached after about 20 h. Both toxins block excitatory neuromuscular transmission (not inhibitory transmission) in insects at a presynaptic site. The toxins can be used to study inhibitory transmission in insects, and to study postsynaptic effects without presynaptic interference.

The venom of *Philanthus triangulum* paralyses insects and blocks excitatory as well as inhibitory neuromuscular transmission. The effect on excitatory neuromuscular transmission in insects is presynaptic (possibly inhibition of glutamate uptake) as well as postsynaptic (ion channel block).

P. triangulum venom contains at least four active components (α-, β-, γ- and δ-philanthotoxin) of low relative molecular mass (<700 M_r). The α-PTX component blocks synaptic transmission in the insect central nervous system and contains a muscarinic agonist. β-, γ- and δ-PTX block glutamate potentials. The most active toxin, δ-PTX, blocks open glutamate channels in the junctional and extrajunctional membrane of locust muscle fibres. *P. triangulum* venom can be used to block glutamatergic transmission in skeletal and intestinal muscles so that proctolinergic effects (modulation or transmission) can be studied. As well as giving further insight into the properties of membrane ionophores, δ-PTX may prove of value in the isolation of ionophores.

This contribution reviews our present knowledge of solitary wasp venoms, including their effects on synaptic transmission in insects, especially on glutamatergic and GABAergic neuromuscular transmission. It also discusses their possible application to the study of the putative monoaminergic and peptidergic neuromodulation of insect skeletal and intestinal muscles.

As Narahashi pointed out in his review of chemicals as tools in the study of

1982 Neuropharmacology of insects. Pitman, London (Ciba Foundation symposium 88) p 275-290

excitable membranes in 1974, 'it is by no means a new idea to use natural toxins for the study of neurophysiology'. For example, many alkaloids have been used and are still being used as tools for studying cholinergic and adrenergic systems. The development of a variety of chromatographic techniques over the past 30 years has brought the use of animal toxins within the reach of students of excitable membranes.

It is not altogether surprising that animal toxins, especially those in solitary wasp venoms, have specific modes of action, as they have been evolved to enable the hymenopteran, in this case, to paralyse its prey or host. The specific properties of the various toxins developed by nature to affect nervous conduction or synaptic transmission in insects might be valuable in designing new classes of pesticides or as tools for studying the pharmacology and physiology of the insect nervous system.

A number of arthropod toxins, isolated for example from spider venoms (atraxotoxin or α-latrotoxin), from scorpion venoms (tityustoxin and leiurustoxin) or from solitary wasp venoms (the *Microbracon* toxins and the philanthotoxins) may be used for the study of transmitter release, for the study of uptake mechanisms in nerve terminals and glial cells, for studying the action of cations in relation to the function of ionophores in nerve and muscle membranes, and for unravelling the confusing number of different 'receptor' sites that may be present on ionophores (Piek 1981).

This review will be restricted to the venoms of *Microbracon hebetor* and *Philanthus triangulum*.

Microbracon venom

The female of the wasp *M. hebetor* produces a very potent venom that paralyses Lepidoptera. Beard (1952) estimated that 5 ng venom per gram of haemolymph was enough to completely paralyse the larva of the greater wax moth, *Galleria mellonella*. Visser et al (1976) estimated that the relative molecular mass of the active principle was about 61 000 by gel chromatography and about 63 000 by the sedimentation equilibrium method. The pI was 6.8. The purified toxin was inactivated by a number of proteolytic enzymes as well as by urea and sodium dodecyl sulphate, which are known to denature proteins, and by dithiothreitol and β-mercaptoethanol, which disrupt disulphide bonds. These results strongly indicate that the active principle is a protein. Later, Spanjer et al (1977) showed that the venom contained two toxins, *Microbracon* toxin-A (A-MTX), and *Microbracon* toxin-B (B-MTX), with relative molecular masses of about 42 000 and 57 000 respectively. The two toxins have a similar mode of action. The crude venom as well as A-MTX and B-MTX completely block excitatory but not inhibitory neuromuscular transmission in Lepidoptera. The onset of paralysis as well as recovery is slow, and at low doses maximum paralysis is reached about 20 h after injection. Locusts and mealworms are about 100 000 times less sensitive than Lepidoptera to the venom and to the two toxins. Despite this enormous difference, Walther & Rathmayer (1974) found that the venom of *M. hebetor* also specifically blocked excitatory neuromuscular transmission in locusts (see Fig. 1). The venom and the toxins cause a decrease in the frequency of miniature excitatory postsynaptic potentials (MEPSPs) recorded from insect muscle fibres, without a marked effect on their amplitude distribution. This indicates an effect on the spontaneous release of transmitter. Since the decrease in MEPSP frequency was always correlated with the decrease in EPSP amplitude at all venom concentrations used (Piek & Spanjer 1978), this venom may be considered to act exclusively at the presynaptic site. So far there is no indication of a postsynaptic effect.

Microbracon venom can be used to block excitatory neuromuscular transmission selectively, for example in order to study inhibitory transmission (Walther & Rathmayer 1974, Rathmayer & Walther 1976). Fig. 1 shows the

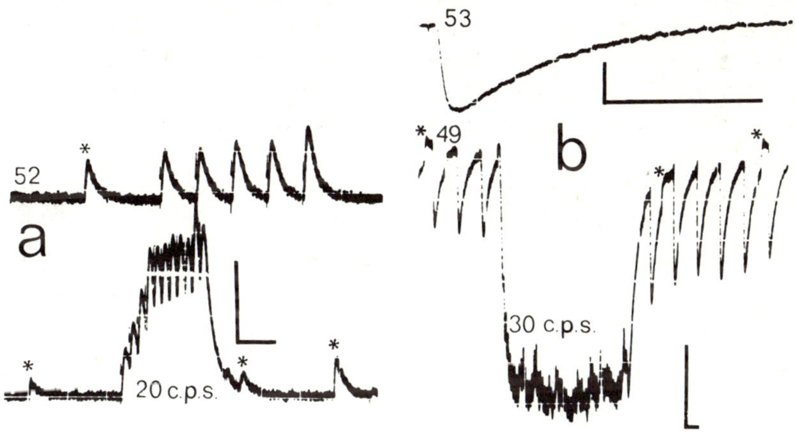

FIG. 1. Postsynaptic potentials in the extensor tibiae muscle in the metathoracic leg of *Locusta migratoria*, 2–3 h after 5 μl saline containing an extract equivalent to half of the venom organ of *Microbracon hebetor* had been injected into the leg. The numbers at the beginning of the traces are the measured resting potentials in mV.

(a) *Upper:* two hours after injection the amplitude of the EPSPs (5 Hz) has been reduced to that of the small spontaneous EPSPs (asterisks). *Lower:* effect of stimulation of the main crural nerve (20 Hz). Note facilitation (upper trace) and summation (lower trace) of the evoked potentials.

(b) Three hours after injection hyperpolarizing potentials (IPSPs) were recorded from the distal part of the extensor tibiae muscle (small bundles separated from the main part of the muscle). During stimulation of the nerve at 30 Hz the 'tetanus' is not smooth. This is probably caused by an interaction of EPSPs and IPSPs. Note the posttetanic facilitation of spontaneous IPSPs. Spontaneous EPSPs, asterisks. Calibration, 1 mV and 200 ms.

slow action of the venom on excitatory transmission in locust skeletal muscle. Two hours after injection into the jumping leg of a locust of 5 μl saline containing an extract equivalent to half a venom organ of *M. hebetor*, small EPSPs with amplitudes comparable to MEPSPs were recorded from a muscle stimulated via the main nerve containing an excitatory as well as an inhibitory axon. Inhibitory potentials (IPSPs) were recorded from the distal part of the muscle and about 3 h after injection the EPSPs were hardly visible. Walther et al (1976) used the venom in locust muscle to selectively suppress the spontaneous and nerve-evoked excitatory release of transmitter so that they could study the stimulating effect of the venom of the scorpion, *Androctonus australis*, on inhibitory transmission.

At a concentration of 3×10^6 units/ml (the unit is defined as the amount of venom activity causing a 50% paralysing effect after 2 h when injected into the haemolymph of 100 mg *Galleria mellonella* larvae) the venom of *M. hebetor* causes an initial increase in slow contraction in the extensor tibiae of *Schistocerca gregaria*, followed by a complete block (Fig. 2).

Spontaneous and stimulation-induced myogenic contractions are not affected by the venom. The intrinsic rhythm of the tonic fibres of the extensor muscle is considered to be under the control of an octopaminergic system (Evans & O'Shea 1978) and is also sensitive to proctolin (Piek & Mantel 1977, May et al 1979). The myogenic contractions can be studied by stimulating nerves after excitatory transmission has been blocked with *Microbracon* venom (Fig. 2).

Although the venom of *M. hebetor* contains labile toxins of high relative molecular mass, production of the venom is easy and it can be used in a crude form.

Philanthus venom

Like other solitary aculeate wasps the bee wolf, *Philanthus triangulum*, stings its prey (workers of the honeybee, *Apis mellifera*) into or in the direction of the main thoracic ganglion complex. Although this stinging behaviour suggests that the venom may act on the central nervous system (CNS), the venom contains at least three toxins that affect the locomotor system by a distinct peripheral effect, in addition to one (or a few?) component(s) that affect transmission in the CNS (Piek et al 1980a, 1982, Spanjer et al 1982).

The relative molecular mass of all the components of the venom is less than 700. On the basis of the separation procedure described by Spanjer et al (1982) the active components have tentatively been called α-, β-, γ- and δ-philanthotoxin (α-, β-, γ- and δ-PTX). The exact localization on thin layer chromatograms is known for β-PTX and δ-PTX, and elucidation of their chemical structure now seems possible.

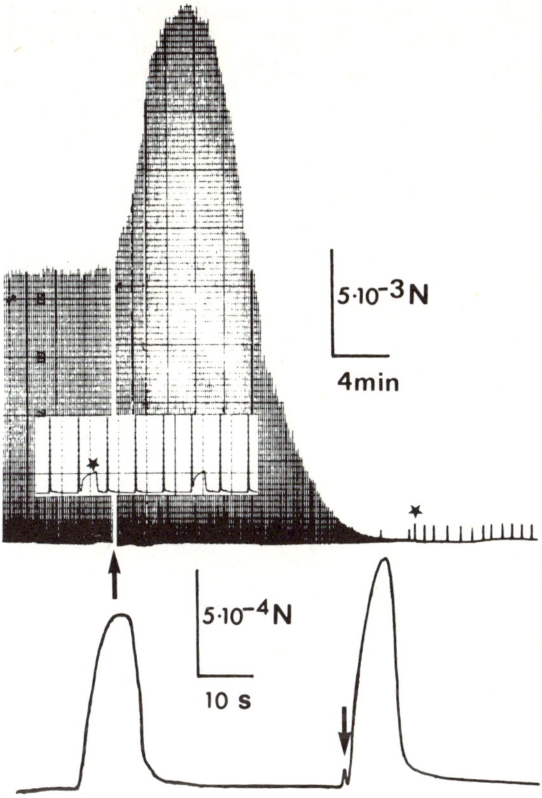

FIG. 2. Effect of the venom of *Microbracon hebetor* on contractions of the metathoracic extensor tibiae muscle of *Schistocerca gregaria*. The preparation was stimulated via nerve 3b with trains of 16 pulses at 70 Hz. *Upper:* stimulation every 10 s. After addition of venom, 3×10^6 U/ml (arrow), the contraction force initially increased and subsequently decreased to zero. Myogenic contractions are visible at the bottom line (asterisks, see also inset) and are not blocked by the venom. *Lower:* after a 2-h wash the paralysis was still present and the preparation showed spontaneous as well as stimulus-evoked myogenic contractions. In the latter the contraction is preceded by a greatly reduced slow excitatory contraction (arrow).

Studying the postsynaptic effects of β-, γ- and δ-PTX, Spanjer et al (1982) found that all three toxins cause a depression of glutamate-induced postsynaptic potentials. The relative amounts of toxins in the crude venom preparation indicated that the relative activities of β-, γ- and δ-PTX on glutamate potentials were about 1:30:100 respectively. Most of our knowledge concerns the most active toxin, δ-PTX. In locust muscles neither the crude venom nor δ-PTX seems to act as a glutamate receptor antagonist; they seem to act on glutamate-gated ion channels. This action occurs on ionophores gated by the activation of both junctional and extrajunctional glutamate receptors in the muscle fibre membrane (May & Piek 1979, Piek et al 1980b, Clark et al 1981, 1982). The action of δ-PTX on ion channels is demonstrated in Fig. 3. The decrease in amplitude of the postsynaptic

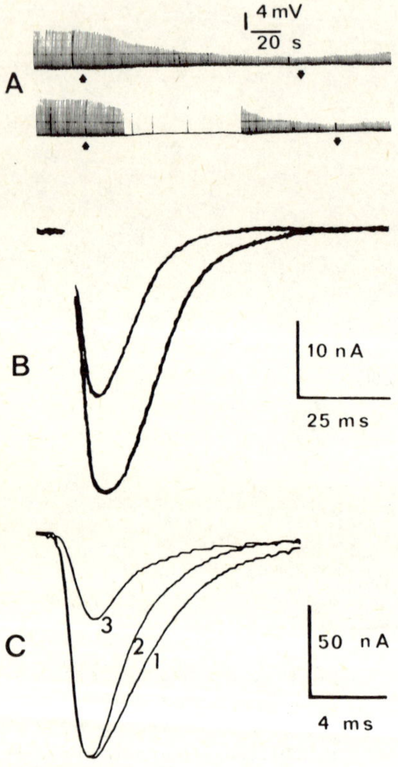

FIG. 3. Effects of δ-PTX on glutamate potentials and currents and on junctional currents recorded from locust extensor tibiae muscle fibres.

A. *Upper trace:* pulses (1 nC) of L-glutamate were applied to an excitatory nerve–muscle junction at a constant frequency of 0.74 Hz. During addition to the bath of 0.05B.U./ml of δ-PTX (between arrows) the glutamate potentials were depressed and they recovered only slowly after removal of the toxin. (The biological activity of the venom preparations is expressed in bee units, B.U. The bee unit is defined as the quantity of venom that, injected into ten bees under CO_2 'anaesthesia', causes 50% of these bees to remain paralysed for more than 1 h at 25 °C. This unit is comparable to an ED_{50}.) *Lower trace:* recording of glutamate potentials at the same junction as above. δ-PTX (0.05 B.U./ml) was applied for the period between arrows. In this experiment the pulses (1 nC, 0.74 Hz) of L-glutamate were interrupted for periods of 6, 12, 24 and 40 s. Note the slight recovery of the response amplitude after these brief periods of rest.

B. Effect of δ-PTX (0.2 B.U./ml) on the amplitude and time course of junctional glutamate currents evoked by iontophoresis. The larger of the two superimposed currents is a control response evoked during a train of iontophoretic pulses delivered in the absence of δ-PTX. The potentials in response to iontophoretically applied glutamate depends upon repeated application of glutamate to the junctional site (Fig. 3A). The postsynaptic block produced by δ-PTX is therefore activation-induced. If the iontophoretic application of glutamate is interrupted for various periods during the application of the toxin the subsequent single pulses of glutamate result in slightly increased responses but they never regain the control amplitude, even when the time between the single pulses is almost 40 s (Fig. 3A, Clark et al 1982). The block therefore seems to be semi-irreversible. If the application of glutamate is uninterrupted the block tends to reach a plateau at a level which depends mainly on the toxin concentration. This phenomenon may be explained by assuming that the unblocking is also activation-induced. According to this hypothesis unblocking should not occur without an agonist. Since there may be a slight leakage of glutamate from the iontophoresis pipette as well as spontaneous and variable release of transmitter from the nerve terminals, the time course of unblocking will be highly unpredictable in the unstimulated situation. For the junctional ionophores this problem might be solved by blocking the release of transmitter from the nerve terminals with the venom of *Microbracon hebetor*. This experiment, which we intend to do as soon as we have purified toxins available again, may be another example of how to use *Microbracon* toxin as a tool.

Experiments with glutamate-induced currents show that δ-PTX causes a slight decrease in the half-decay time of the response (Fig. 3B, Clark et al 1982). Experiments with nerve-stimulated excitatory junctional currents show that the fall in amplitude caused by δ-PTX was preceded by a decrease in half-decay time (Fig. 3C, Clark et al 1982). This confirms our earlier findings on the curtailing of glutamate potentials and of the extracellularly recorded miniature potentials by the crude venom of *P. triangulum* (May & Piek 1979, Piek et al 1980b).

Using the patch clamp technique Clark et al (1981) demonstrated that in the presence of δ-PTX the currents generated by the opening of single-ion channels in the extrajunctional membrane of the locust muscle were significantly reduced in duration and were separated either by 'normal' intervals or

---

smaller response was obtained after 20 s equilibration in δ-PTX. Stimulus frequency, 0.8 Hz; glutamate dose, 2 nC.

C. Effect of δ-PTX (1 B.U./ml) on the excitatory junctional current (EJC) recorded from a locust extensor tibiae muscle fibre under voltage clamp. EJCs were evoked by stimulating the extensor tibiae nerve at a frequency of 3 Hz. Records 1–3 are averages of 20 consecutive EJCs. Record 1 was obtained in the absence of δ-PTX. The preparation was then rested for 5 min, during which time it was 'equilibrated' in saline containing 1 U/ml of δ-PTX. Record 2 is the average of the first 20 EJCs obtained immediately after the 5-min rest period. Record 3 is the average of 20 EJCs obtained after 1 min continuous stimulation at 3 Hz in δ-PTX. Note that an increase in rate of decay of the EJC precedes the fall in amplitude of the response. Clamp potential, −60 mV; 20 °C.

by unusually long intervals. During the ‘bursts’ of activity the frequency of channel openings seen in the presence of δ-PTX was not significantly different from the control frequency recorded in the absence of the toxin. Clark et al (1981) therefore concluded that the silent intervals between the bursts represent the periods during which the individual glutamate channel is blocked. The silent intervals have a distribution showing a time constant between 500 and 1000 ms. This probably reflects the unblocking time constant for δ-PTX in the presence of an agonist.

The blocking of neuromuscular transmission by the venom of *P. triangulum* and some of its constituent toxins can be exploited as a tool for physiological and pharmacological studies of insect muscle. An example of this use of the venom is shown in Fig. 4. Spontaneous and nerve-evoked (nerve 3b) contractions of the myogenic type can be terminated at an early stage by stimulating the nerve during the myogenic contraction. This is, however, only seen when the slow excitatory contraction evoked by stimulating the slow excitatory axon to the extensor tibiae (SETi axon) is greatly reduced by the venom (3 B.U./ml, Piek & Mantel 1977). In nerve 3b the SETi axon is accompanied by the common inhibitory axon (CI), and the question arises of whether the initiation of the myogenic contraction, as well as early relaxation, are caused by SETi activity or by CI activity. In a different preparation the threshold for stimulating the SETi axon was lower (1.4 V) than for the CI axon (1.6 V). Fig. 4 (C and D) shows that a second SETi stimulus during myogenic contraction does not result in a relaxation but in a second (greatly reduced) SETi contraction. At a higher stimulus voltage the venom-reduced SETi contraction decreased further in amplitude, probably due to CI interaction. A second pulse train also resulted in a second contraction, followed, however, by an early relaxation of the myogenic contraction. At a higher concentration of venom (7–10 B.U./ml), stimulation of nerve 3b induced myogenic contraction without a preceding SETi contraction and without relaxation at a second stimulus train during the myogenic contraction (Piek & Mantel 1977). This could be explained by assuming that the CI was now completely blocked by the venom, although a block of other (neuromodulatory) axons cannot yet be excluded. According to Hoyle (1978) the intrinsic rhythm of this muscle might be regulated by at least two dorsal unpaired median neurons (DUM). One, the DUM cell to the extensor tibiae muscle (DUMETi), suppresses the frequency of the intrinsic rhythm, the other probably enhances its amplitude. The origin of the intrinsic rhythm has been found to be in the tonic bundle (Usherwood 1974), which is also innervated by SETi and CI (Cochrane et al 1972, May et al 1979). DUMETi is probably an octopaminergic neuron (Evans & O’Shea 1978).

A second example of the use of *P. triangulum* venom as a tool has been worked out by Dr S. J. Dunbar in our laboratory. As well as an apparent

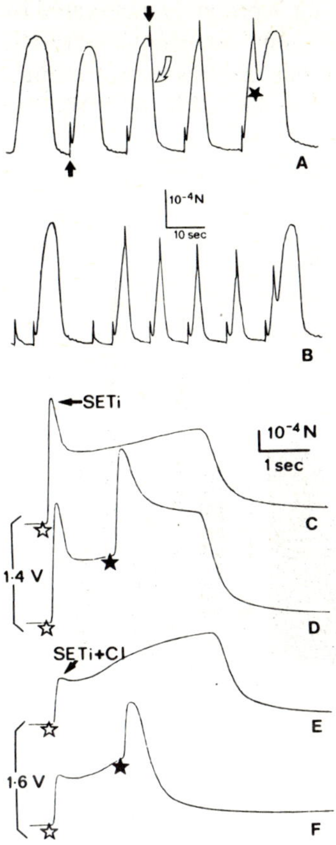

FIG. 4. Myogenic contractions of the metathoracic tibia of *Locusta migratoria* partly paralysed with 3 B.U./ml venom of *Philanthus triangulum*.

A. The first contraction is spontaneous; the subsequent contractions are induced by stimulating nerve 3b with 100 Hz (↑). Stimulation at the peak of the contraction is followed by a short contraction (reduced slow excitatory [SETi] contraction (↓), and an early relaxation (open arrow) of the myogenic contraction. Stimulation during the rising phase results in a transient relaxation followed by a renewed contraction (asterisk).

B. Stimulation at several moments during the myogenic contractions. (A and B from Piek & Mantel 1977.)

C–F. SETi-induced contractions and common inhibitor-induced (CI) relaxations (asterisks). Nerve 3b was first stimulated at 40 Hz with trains of 10 pulses and at 1.4 V (C and D) and 1.6 V (E and F) (open asterisks). At the higher stimulus voltage the SETi contraction is reduced, probably by CI firing. A second pulse train given during the myogenic contraction (closed asterisks) results in a second SETi contraction. Only at 1.6 V stimulus voltage is the second SETi contraction followed by an early relaxation.

glutamatergic neurotransmission, the insect rectum is probably innervated by a proctolinergic system (Brown 1975). There is some discussion as to the exact role of proctolin in insect visceral muscle: it may act as a neurotransmitter (Brown 1975), a neuromodulator (Cook & Holman 1979), or both. The

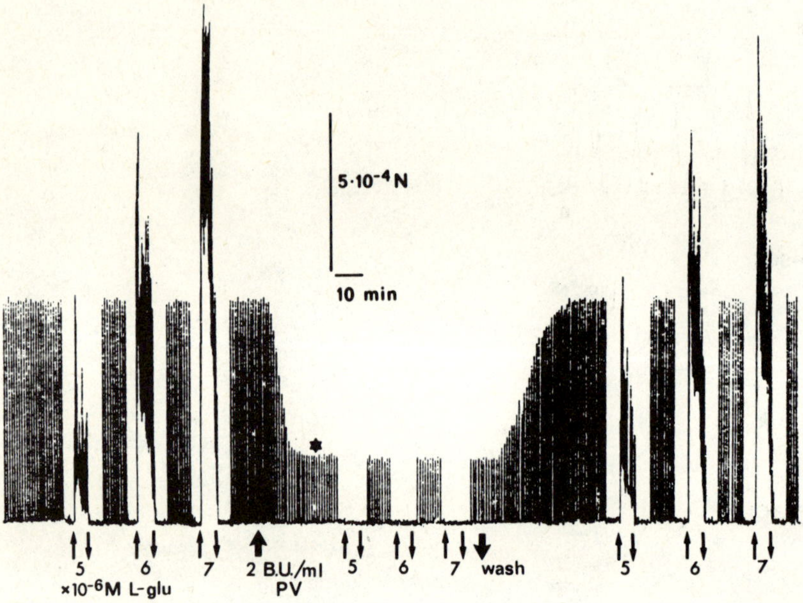

FIG. 5. Proctodeal nerve-evoked and glutamate-evoked contractions of a whole rectum preparation of *Locusta migratoria*. The proctodeal nerve was stimulated every minute for 1 s at 20 Hz, resulting in a slow phasic contraction of about 7×10^{-4} N. L-Glutamate ($5–7 \times 10^{-6}$ M) added to the bath caused a tonic contraction and a number of superimposed phasic contractions. The venom of *P. triangulum* (2 B.U./ml) reduced the nerve-evoked contraction to a plateau and completely blocked the glutamate-induced response. The remaining contractions may not be glutamatergic. (Figure kindly provided by Dr S. J. Dunbar.)

venom of *P. triangulum* blocks the glutamate-induced contraction of the rectum of *Locusta migratoria* but not the phasic proctolin-induced contraction (S. J. Dunbar, personal communication). The nerve-evoked contractions are only partially blocked (Fig. 5). Although we are not certain about the nature of the venom-resistant part of the proctodeal nerve-evoked contractions, the venom may prove to be a useful tool in unravelling this problem.

Conclusions

Although our knowledge of solitary wasp venoms is incomplete, study of these venoms has brought to light some of their biological actions. It has also opened up new fields of research with the possibility of interesting applications in various areas of agriculture, medicine and physiology. At present the significance of the isolation of these toxins from solitary wasps cannot be predicted. It is, however, evident that all the well-described natural compounds, or their synthetic analogues, that have distinct biological activity, have been or are being used as drugs, as pesticides, or as chemical tools in the study of physiology and pharmacology.

In addition to their importance as a research tool, solitary wasp venoms and their toxins may be important in the development of new principles for regulating insect populations in a specific way. We realize, however, that we are far from being able to use solitary wasp venoms as tools on a wide scale. The value of the experiments that remain to be done must not be overshadowed by the lack of spectacular practical results.

Acknowledgements

I thank Professor C. van der Meer for his stimulating interest, Dr C. Walther for enabling Fig. 1 to be prepared in his laboratory, Dr S. J. Dunbar for his critical reading of the manuscript and for permission to use Fig. 5, and Mr P. Mantel for assistance in some of the experiments.

REFERENCES

Beard RL 1952 The toxicology of *Habrobracon* venom: a study of a natural insecticide. Bull Conn Agric Exp Stn 562 (27 p)

Brown BE 1975 Proctolin: a peptide transmitter candidate in insects. Life Sci 17:1241-1252

Clark RB, Donaldson PL, Gration KAF et al 1981 Postsynaptic block at neuromuscular junctions on locust muscle by δ-philanthotoxin. J Physiol (Lond) 310:8P

Clark RB, Donaldson PL, Gration KAF et al 1982 Block of locust muscle glutamate receptors by δ-philanthotoxin occurs after receptor activation. Brain Res, in press

Cochrane DG, Elder HY, Usherwood PNR 1972 Physiology and ultrastructure of phasic and tonic skeletal muscle fibres in the locust *Schistocerca gregaria*. J Cell Sci 10:419-441

Cook BJ, Holman GM 1979 The pharmacology of insect visceral muscle. Comp Biochem Physiol 64C:183-190

Evans PD, O'Shea M 1978 The identification of an octopaminergic neurone and the modulation of a myogenic rhythm in the locust. J Exp Biol 73:235-260

Hoyle G 1978 Intrinsic rhythm and basic tonus in insect skeletal muscle. J Exp Biol 73:173-203

May TE, Piek T 1979 Neuromuscular block in locust skeletal muscle caused by a venom preparation made from the digger wasp, *Philanthus triangulum* F., from Egypt. J Insect Physiol 25:685-691

May TE, Brown BE, Clements AN 1979 Experimental studies upon a bundle of tonic fibres in the locust extensor tibialis muscle. J Insect Physiol 25:169-181

Narahashi T 1974 Chemicals as tools in the study of excitable membranes. Physiol Rev 54:813-889

Piek T 1981 Arthropod venoms as tools for the study of neuromuscular transmission. Comp Biochem Physiol 66C:75-84

Piek T, Mantel P 1977 Myogenic contraction in locust muscle induced by proctolin and by wasp, *Philanthus triangulum*, venom. J Insect Physiol 23:321-325

Piek T, Spanjer W 1978 Effects and chemical characterization of some paralysing venoms of solitary wasps. In: Shankland DL et al (eds) Pesticide and venom neurotoxicity. Plenum Press, New York

Piek T, May TE, Spanjer W 1980a Paralysis of locomotion in insects by the venom of the digger wasp *Philanthus triangulum* F. In: Insect neurobiology and pesticide action. Soc Chem Ind (Lond) Monogr, p 219-226

Piek T, Mantel P, Jas H 1980b Ion-channel block in insect muscle fibre membrane by the venom of the digger wasp, *Philanthus triangulum*. J Insect Physiol 26:345-349

Piek T, Spanjer W, Veldsema-Currie RD, Van Groen T, De Haan N, Mantel P 1982 Effect of venom of the digger wasp, *Philanthus triangulum* F., on the sixth abdominal ganglion of the cockroach. Comp Biochem Physiol, in press

Rathmayer W, Walther C 1976 Mode of action and specificity of *Habrobracon* venom (Hymenoptera, Braconidae). In: Ohsaka H et al (eds) Animal, plant, and microbial toxins, vol 2. Plenum Press, New York, p 299-307

Spanjer W, Grosu L, Piek T 1977 Two different paralysing preparations obtained from a homogenate of the wasp *Microbracon hebetor* (Say). Toxicon 15:413-421

Spanjer W, May TE, Piek T, De Haan N 1982 Partial purification of components from the paralysing venom of the digger wasp, *Philanthus triangulum* F. (Hym.Sphec.) and their action on neuromuscular transmission in the locust. Comp Biochem Physiol, in press

Usherwood PNR 1974 Nerve–muscle transmission. In: Treherne JE (ed) Insect neurobiology. Elsevier/North-Holland Biomedical Press, Amsterdam, p 245-305

Visser BJ, Spanjer W, De Klonia H, Piek T, Van der Meer C, Van der Drift ACM 1976 Isolation and some biochemical properties of a paralysing toxin from the venom of the wasp *Microbracon hebetor* (Say). Toxicon 14:357-370

Walther C, Rathmayer W 1974 Effects of *Habrobracon* venom on excitatory neuromuscular-transmission in insects. J Comp Physiol 89:23-38

Walther C, Zlotkin E, Rathmayer W 1976 Action of different toxins from the scorpion *Androctonus australis* on a locust nerve-muscle preparation. J Insect Physiol 22:1187-1194

DISCUSSION

Usherwood: If a preparation which has been paralysed by *Philanthus* toxin is not stimulated after removal of the toxin, recovery can take a long time. However, using patch clamp analysis the channel unblocking rate was obviously too fast to account for the slow recovery seen with bath application of toxin. What is the explanation for this apparent discrepancy?

Piek: When venom is added, the glutamate-induced postsynaptic potentials decrease to a plateau. If there were a quick recovery one would expect that after an interval the amplitude would be comparable to the control, but it is not. There was sometimes a small restoration, depending on the time, but it was not reproducible. It is an extremely slow unblocking. My explanation for the different time courses is that there is no recovery without activation; but there is always some activation, due to flow from the pipette or spontaneous release from the nerve terminal. The plateau or steady-state level for the peak amplitude of the glutamate potential depends on the concentration present in the bathing medium. From this steady-state situation we can calculate the unblocking time, since we can readily measure the blocking time. I found a value for blocking time which is close to what we (Clark et al 1982) observed in the patch clamp studies.

Barnard: On that question of the reversibility of the action of the *Philanthus* δ-toxin, is the block of the glutamate channel by the toxin not removable by wash-out alone, if this is given efficiently?

Piek: If you activate the glutamate receptors by glutamate iontophoresis, the block is quickly reversible, but without either this stimulation or nerve-evoked stimulation, recovery takes a very long time. Unblocking seems to require receptor activation, suggesting that only 'open' ion channels can be unblocked.

Barnard: I was interested in your suggestion that this toxin might serve as a probe for assay or labelling of a glutamate channel component. If the binding of the toxin to glutamate channel or receptor is, in fact, rapidly reversible, it would not be suitable.

Piek: Theoretically you can use *Philanthus* venom as a probe for the isolation of the ionophore if there is no stimulation.

Barnard: What is its structure?

Piek: We are not sure. We are collecting more material for nuclear magnetic resonance studies. All the components of *Philanthus* venom have an M_r of less than 700.

M. E. Eldefrawi: I gather that you are working with isolated and purified fractions and with crude whole venom?

Piek: Yes. The first experiments were with crude venom. δ-PTX has not been purified completely but it has been separated from the other toxins. For the δ-PTX preparation we used, thin layer chromatography showed only one spot, with some impurities (Spanjer et al 1982).

Narahashi: Does the δ component of *Philanthus* toxin block the open glutamate-gated ion channel, rather than the glutamate receptor?

Piek: The problem is that the venom also acts presynaptically by reducing evoked transmitter release from motor nerve terminals. This must inevitably reduce the effectiveness of the toxin in blocking postsynaptically during neural stimulation of the muscle. Initially I couldn't understand the interaction between pre- and postsynaptic effects of the toxin. Recently we found that its presynaptic action resulted from an inhibition of glutamate uptake. In the short term this would enhance the postsynaptic action of the toxin.

Desensitization of the postjunctional glutamate receptors might also contribute to the complex effects of this toxin.

Narahashi: But the venom has no effect on glutamate receptors?

Piek: Probably not. The venom does not cause a competitive block.

Narahashi: What is the approximate concentration range you use for δ-PTX?

Piek: Between 10^{-7} and 10^{-6} M. Because of the impurities we are not completely sure what the concentration is.

Gration: In our experiments on locust muscle, using the patch clamp technique we observed a small reduction, perhaps 20%, in the mean open time of channels recorded in the presence of δ-PTX (Clark et al 1982). The distribution of closed times you have shown is indicative of the type of channel activity recorded in the presence of δ-PTX, since with venom present we observe bursts of channel openings, separated by long closed intervals. These intervals presumably reflect the block time for the venom.

Usherwood: What about voltage sensitivity?

Gration: The action of δ-PTX doesn't appear to be voltage-sensitive, at least within the range of membrane potentials −50 mV to −120 mV.

M. E. Eldefrawi: In the natural predator–prey relationship when the wasp stings the bee it becomes paralysed very quickly. How long does it take to recover from the sting, or does it ever recover? If it doesn't, how do we reconcile the natural phenomenon, which appears to be irreversible paralysis, with your observation that it is a semi-reversible phenomenon? Are we failing to imitate the way the predator injects the material, or in manipulating the toxin are we not putting everything in?

Piek: If we inject more than one bee unit it takes from many hours to one day for the bees to recover, but they die shortly afterwards, probably not *directly* due to the venom. Cicadas are known to have been paralysed for more than one year (Riley 1892), so there are different venoms. Paralysis caused by *Philanthus* venom is reversible but I don't know what happens at a high concentration. There may be changes in the CNS. These wasps sting in the direction of the CNS, probably into the CNS.

Barnard: Aren't the irreversible effects in the field attributable to the presynaptic action? The irreversibility that I was asking about was of the postsynaptic action. The presynaptic action is likely to be lethal in these venoms.

Piek: Yes; we need a relatively high concentration of venom to obtain a presynaptic block. Therefore at low concentrations quick recovery may be only from the postsynaptic effect.

Hildebrand: Are there central effects if venom from either of these wasps is introduced into the insect CNS? Are any insect or arachnid venoms known to have central actions?

Piek: We have studied a number of solitary wasp venoms other than *Philanthus* venom. We were unable to find peripheral effects. All the wasps that paralyse spiders and flies put the sting inside the big nerve centre.

Hildebrand: Is anything known about the mechanism of action of those venoms?

Piek: We are working on this. We discovered a histamine-like component and a bradykinin-like substance and other kinin-like products in different species. Per venom reservoir, *Philanthus* venom has 400–500 ng acetylcholine in addition to the described toxins. I don't know what it is doing but it is possible that δ-PTX is also active in the CNS and that acetylcholine is opening all the channels.

Neumann: Can you increase the efficiency of your protein toxins (*Microbracon* toxins) with proteinase inhibitors?

Piek: Microbracon venom contains two large toxin proteins that are very labile. We can stabilize them with some proteinase inhibitors and not with others (B. J. Visser et al, unpublished results).

Neumann: Can you increase the toxicity to locusts?

Piek: No. We tried to inject locusts with a large amount of haemolymph from paralysed wax moth larvae but they were not sensitive to it. The toxic principle is therefore not a product made by the lepidopteran body. Haemolymph from paralysed locusts was extremely active in Lepidoptera so the toxicity is not destroyed in the haemolymph of locusts.

Cull-Candy: Are your patch clamp experiments done with toxin present? If they are not—that is, if you treat with toxin, wash off and then put the patch electrode down on the muscle surface—I imagine you would have a rather small number of channels under the tip of the patch pipette. In which case, if the channels do indeed unblock by activation after the receptor has been activated by glutamate in the patch pipette, the channel will presumably be seen to be unblocked. Is that what you found?

Piek: We had toxin inside the pipette, so we did not do the experiment you describe.

Penzlin: I was interested in the experiments on the possible proctolinergic system in the locust rectum. Can you achieve a total block of proctolin-induced contractions with tyramine?

Piek: This is complicated. S. J. Dunbar has now tested seven or eight different samples of tyramine: some are active and some are not in blocking the effects of proctolin. Other samples show a medium blocking activity. Normally we use 10^{-6} M-tyramine, but I think the so-called proctolin antagonistic action of tyramine might be an impurity. As an antagonist for proctolin, tyramine presents problems.

REFERENCES

Clark RB, Donaldson PL, Gration KAF et al 1982 Block of locust muscle glutamate receptors by δ-philanthotoxin occurs after receptor activation. Brain Res, in press

Riley CV 1892 The larger digger wasp. US Dep Agric Entomol Bull Insect Life 4:248-252

Spanjer W, May TE, Piek T, De Haan N 1982 Partial purification of components from the paralysing venom of the digger wasp, *Philanthus triangulum* F. (Hym. Sphec.) and their action on neuromuscular transmission in the locust. Comp Biochem Physiol, in press

Insecticides as probes for the study of ionic channels in nerve membranes

TOSHIO NARAHASHI

Department of Pharmacology, Northwestern University Medical School, 303 East Chicago Avenue, Chicago, Illinois 60611, USA

Abstract The major site of action of DDT and the pyrethroid insecticides has been identified to be the sodium channels in nerve membranes. A population of sodium channels is modified by low concentrations of insecticides to give rise to very slow opening and closing which in turn raises the depolarizing after-potential to the threshold for repetitive after-discharges. The insecticide molecule binds to the sodium channel in both closed and open states to yield a modified open state which inactivates very slowly. The pyrethroids and DDT appear to have a basically similar action. Many pyrethroids exist as two optical and two geometrical isomers, and their potencies differ markedly. For example, (+)-*trans* and (+)-*cis* tetramethrin are highly active, whereas (−)-*trans* and (−)-*cis* are almost totally inactive. Studies of antagonism among them have led to the conclusions that there are *trans* and *cis* agonistic sites and a negative allosteric site in the sodium channel which are all different from the tetrodotoxin-binding site. These insecticides are becoming useful probes for the study of the topography and gating mechanism of sodium channels in nerve membranes.

The development of a variety of synthetic insecticides since the Second World War has provided us not only with effective means to combat insect pests but also with valuable tools with which the basic physiological properties of nerve and muscle systems can be studied. Examples of these insecticides include chlorinated hydrocarbons such as dichlorodiphenyltrichloroethane (DDT) and benzene hexachloride (BHC), cyclodienes such as dieldrin, endrin and chlordane, organophosphates such as parathion and malathion, carbamates such as carbofuran and carbaryl, and pyrethroids such as allethrin, tetramethrin and decamethrin. The organophosphates and carbamates, in their original or bioactivated forms, inhibit cholinesterases. The mechanisms of action of the cyclodienes still remain largely unknown, but the available evidence suggests that they facilitate synaptic transmission by stimulating

1982 Neuropharmacology of insects. Pitman, London (Ciba Foundation symposium 88) p 291-306

transmitter release (Narahashi 1976). DDT and some of the pyrethroids stimulate the nervous system to cause repetitive discharges which in turn produce hyperexcitation, ataxia and convulsions in the poisoned animal (Narahashi 1971, 1976, Wouters & van den Bercken 1978).

This paper is mainly concerned with the pyrethroid- and DDT-type insecticides which have been demonstrated to exert unique and specific actions on nerve membrane sodium channels. The sodium channels are small pores, estimated to be 0.3 × 0.5 nm in cross-sectional area (Hille 1971), which allow sodium ions to pass through when their gates are opened by depolarizing stimulation. This generates an action potential. It is now becoming increasingly clear that both pyrethroids and DDT, despite their marked difference in chemical structure, exert very similar actions on the sodium channels.

Methods

All the experiments to be discussed here were done with giant axons of cockroach (*Periplaneta americana*), crayfish (*Procambarus clarki*), lobster (*Homarus americanus*) and squid (*Loligo pealei*) and with neuromuscular junctions of crayfish. Action potentials were recorded from the cockroach and crayfish giant axons by means of conventional intracellular glass capillary microelectrodes (Narahashi & Yamasaki 1960a,b). Sodium and potassium conductances of membranes were measured with the crayfish, lobster and squid giant axons using either axial wire or sucrose-gap voltage clamp techniques (Narahashi & Haas 1968, Wang et al 1972, Lund & Narahashi 1981a). Postjunctional potentials were recorded from the crayfish neuromuscular preparations using conventional intracellular microelectrodes. Extracellular (focal) recordings of nerve terminal potentials and postjunctional potentials were made with the crayfish neuromuscular preparations by means of glass capillary microelectrodes (Farley et al 1979).

Results and discussion

It has been demonstrated that in DDT- or pyrethroid-poisoned animals hyperactivity of the nerve and muscle systems, as represented by repetitive discharges, originates in nerve fibres, in particular in their terminals (Farley et al 1979, Evans 1976, Wouters et al 1977), and in sensory neurons (Roeder & Weiant 1948, see also Narahashi 1971). Repetitive discharges in various nerve fibres and in sensory neurons could be amplified through synapses and neuromuscular junctions to cause hyperexcitation of the entire nerve and muscle systems. Thus the next key question is how the repetitive discharges are produced by these insecticides.

When we investigated this by means of intracellular microelectrodes, it became clear that the depolarizing (negative) after-potential following the action potential was gradually increased after the application of DDT or certain pyrethroids (e.g. allethrin and tetramethrin) and eventually reached the threshold for the generation of action potentials (Fig. 1) (Narahashi &

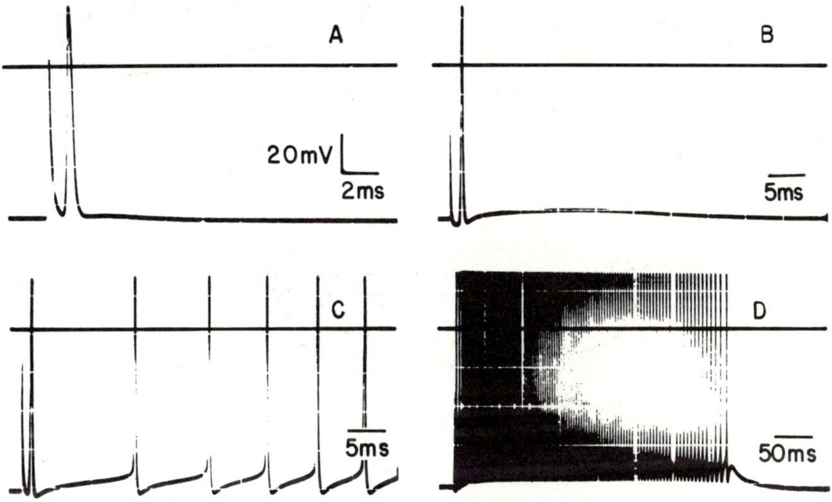

FIG. 1. Intracellular microelectrode recordings of action potentials from the crayfish giant axon before (A) and 5 min (B) and 10 min (C and D) after the application of 1×10^{-8} M-tetramethrin. The response in each trace was elicited by a single stimulus. Temperature 22 °C. (From Lund & Narahashi 1981a, with permission.)

Yamasaki 1960b, Narahashi 1962a,b, Lund & Narahashi 1981a). The same effect was also seen with 2,2-bis-(*p*-ethoxyphenyl)-3,3-dimethyloxetane (EDO), a DDT derivative (Wu ct al 1980). Certain pyrethroids, however, especially those having a cyano group in the alpha position, such as decamethrin and fenvalerate, did not cause repetitive after-discharges or increase the depolarizing after-potential as examined by a single stimulus (Narahashi 1980). The following discussion is concerned primarily with DDT, EDO, tetramethrin and allethrin, all of which cause repetitive discharges.

The next question we asked was how the depolarizing after-potential is increased by these insecticides. Since the falling phase of the action potential is produced by a delayed increase in membrane potassium conductance and a decrease in sodium conductance (sodium inactivation), the increased depolarizing after-potential will be accounted for in terms of one or both of these two parameters. Although the potassium current was decreased slightly at high concentrations of DDT or allethrin (Narahashi & Anderson 1967, Wang et al 1972, Narahashi & Haas 1968), the sodium current was found to be much more sensitive to and prolonged by the insecticides (Lund & Narahashi 1981a). We therefore tried to elucidate the mechanism by which the sodium

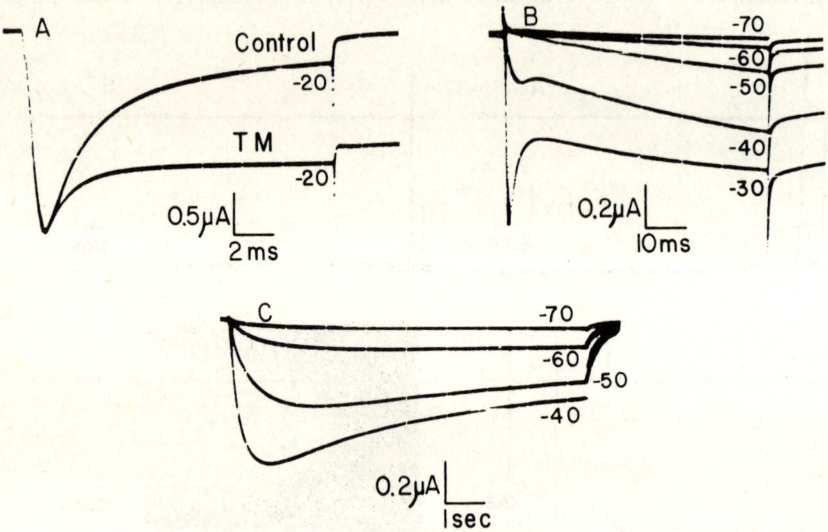

FIG. 2. A. Sodium currents associated with step depolarizations to −20 mV before and after internal treatment with 2 × 10^{-5} M-tetramethrin (TM). B and C. Sodium currents associated with step depolarizations to the levels indicated after internal treatment with TM. Note the different time scales in A, B and C. Temperature 5 °C. (From Lund & Narahashi 1981a, with permission.)

current is prolonged (Wu et al 1980, Lund & Narahashi 1981a, Narahashi & Lund 1980).

For the effects of tetramethrin on the sodium current of the crayfish giant axon to be studied, the potassium current had to be eliminated. This was done by perfusing the crayfish giant axon with K^+-free media both inside and outside. In this condition a step depolarizing pulse applied to the voltage-clamped axon produced a peak transient current which decayed exponentially to a low steady-state level (Fig. 2A, Control). This current must have been carried through the sodium channels by sodium ions, since it was completely blocked by tetrodotoxin applied externally. The current flowed in an inward direction with moderate depolarizations and in an outward direction with large depolarizations that exceeded the reversal potential for sodium. On repolarization of the membrane, a large inward surge of current appeared and decayed with a fast exponential time course (Fig. 2A, Control). This tail current was evidently also carried by sodium through the sodium channels, as it was blocked by tetrodotoxin.

Internal application of tetramethrin at a concentration of 2×10^{-5} M caused dramatic changes in the sodium current. Tetramethrin had the same effects when applied to the giant axon externally. Whereas the peak sodium current was only slightly suppressed, the falling phase of the sodium current was slowed, producing a large 'steady' current. The initial fast tail current associated with a step repolarization was followed by a large 'steady' current (Fig. 2A, TM)(tetramethrin). The steady currents during and after a depolarizing step were blocked by tetrodotoxin, indicating that they were passing through the sodium channels. These large currents were not completely maintained at a steady state, as shown with prolonged step depolarizations. With depolarizing pulses of 60 ms duration, the currents that followed the peak currents were seen to grow slowly, approaching another peak (Fig. 2B). When the depolarizing pulses were increased further, to 9 s, the slow tail currents also decayed gradually (Fig. 2C) (the initial peak sodium currents are not photographically reproduced in Fig. 2C). The fact that the tetramethrin-poisoned axon produces both the peak transient current and the secondary slow current may be interpreted by assuming two populations of sodium channels, one normal and the other modified by tetramethrin. This hypothesis was further pursued.

If tetramethrin modified a population of sodium channels to give rise to a slow sodium current, we would expect the remaining population of sodium channels to be intact, producing a normal peak sodium current. We therefore examined the characteristics of the peak sodium currents before and during application of tetramethrin to the crayfish giant axon. The time constants of sodium activation and inactivation were not changed by tetramethrin, suggesting that these values represent those of the normal, unmodified sodium channels. Similarly, the current–voltage relations for the peak sodium current were the same before and during the application of tetramethrin, except that the peak amplitude was decreased somewhat by the insecticide. The last observation can be interpreted as indicating a decrease in the population of normal channels. The time course of the initial phase of the tail current, which represents the rate at which the sodium channel closes on repolarization, was unaffected by tetramethrin.

In contrast to the normal sodium channels still present in the tetramethrin-poisoned axon, the modified slow sodium channels exhibit unique characteristics. The rates of their opening and closing and of inactivation were slowed by a factor of about 1000. Their voltage dependence was shifted in the direction of hyperpolarization, thereby allowing the modified channels to open at large negative membrane potentials, such as −60 mV. Furthermore, while the amplitude of the slow tail current increased with increases in the concentration of tetramethrin, its time constant remained unchanged. This observation is also compatible with the notion of all-or-none modification of a population of sodium channels.

The observations summarized so far gave us the kinetics and voltage dependence of the modified sodium channels. The next question was how such modification takes place. This was analysed by measuring the rate at which it occurs. The slow component of the tail current in the tetramethrin-poisoned axon was taken as a measure of modification. Thus changes in the amplitude of the slow tail current associated with various durations of depolarizing pulses indicate the time course of channel modification. It was found that the time course of the slow current development, or channel modification, could be expressed by two exponential functions, an initial short time constant in the order of 1–2 ms being followed by a slower time constant in the order of 25 ms. Although both time constants were voltage-dependent, shortening with depolarization, the difference between them was quite distinct. When the sodium inactivation was removed by internal perfusion of pronase, the slow phase disappeared.

From these and other observations, the following kinetic scheme was developed:

$$\begin{array}{ccccc} C & \rightleftharpoons & O & \rightleftharpoons & I \\ \upharpoonleft\downharpoonright & & \upharpoonleft\downharpoonright & & \\ C^* & \rightleftharpoons & O^* & \rightleftharpoons & I^* \end{array}$$

where C, O and I refer to the closed, open and inactivated states of the normal sodium channel, and C*, O* and I* refer to the corresponding states of the tetramethrin-modified channel. Thus the normal peak current is represented by O, and the slow current is represented by O*. The latter can be reached via $C \rightarrow O \rightarrow O^*$, $C \rightarrow C^* \rightarrow O^*$, or $C \rightarrow O \rightarrow I \rightarrow O \rightarrow O^*$. Pronase blocks the process $O \rightarrow I$, thereby eliminating the third route for reaching O*. Therefore the third route has a long time constant (25 ms). The first two routes seem to have the same time constant.

The time course of decay of the tail current represents the rate at which the sodium channel returns from the open state to the closed state. The time constant of the slow component of the tail current was found to vary widely, depending on the kind of insecticide used. DDT and EDO have effects on sodium channels that are similar to those of tetramethrin in lobster, crayfish and squid giant axons (Wu et al 1980, Lund & Narahashi 1981b). DDT was ineffective on squid axons when ethanol was used as the solvent (Narahashi

1971) but becomes effective with dimethyl sulphoxide as the solvent (Lund & Narahashi 1981b). However, the time constant of the slow tail current decay was shorter with DDT than with tetramethrin. Certain pyrethroids containing a cyano group in the alpha position caused an extremely slow decay of the tail current. They include decamethrin, fenvalerate and S2703. Thus all the pyrethroids and DDT compounds so far examined have basically the same effects on the sodium channels but the kinetics of the modified channels differ widely. The differences may simply be due to different association and/or dissociation rate constants at various steps of channel opening and closing.

Another important feature of the action of pyrethroids is the large difference in effectiveness among their optical and geometrical isomers. Both (−)-*trans* and (−)-*cis* forms of tetramethrin were almost completely inactive on sodium channels, whereas (+)-*trans* and (+)-*cis* forms are highly active. Furthermore, the inactive isomers prevent the action of the active isomers. Thus these inactive isomers appear to bind to channel sites without exerting the physiological action. Analyses of antagonism among these isomers have led us to the conclusion that there are *trans* and *cis* agonistic sites and a negative allosteric site in the sodium channel which are all different from the tetrodotoxin-binding site. Such highly selective stereospecificity holds the promise that these isomers could be valuable probes with which to examine the topography of the sodium channels.

Conclusions

Pyrethroid- and DDT-type insecticides modify the nerve membrane sodium channels in a unique manner, slowing their gating kinetics. This action can account for the symptoms of poisoning in the entire nerve and muscle systems. The actions of pyrethroids and DDT are remarkably similar, with only quantitative differences in reaction kinetics. Drastic differences are found among the optical isomers of pyrethroids in their ability to modify the sodium channels. Pyrethroids and DDT are becoming useful probes for the study of sodium channels.

Acknowledgements

This study was supported by grant NS14143 from the National Institutes of Health. I wish to thank Frances Novak for secretarial assistance.

REFERENCES

Evans MH 1976 End-plate potentials in frog muscle exposed to a synthetic pyrethroid. Pesticide Biochem Physiol 6:547-550

Farley JM, Narahashi T, Holan G 1979 The mechanism of action of a DDT analog on the crayfish neuromuscular junction. Neurotoxicology 1:191-207

Hill B 1971 The permeability of the sodium channel to organic cations in myelinated nerve. J Gen Physiol 58:599-619

Lund AE, Narahashi T 1981a Modification of sodium channel kinetics by the insecticide tetramethrin in crayfish giant axons. Neurotoxicology, in press

Lund AE, Narahashi T 1981b Interaction of DDT with sodium channels in squid giant axon membranes. Neuroscience, in press

Narahashi T 1962a Effect of the insecticide allethrin on membrane potentials of cockroach giant axons. J Cell Comp Physiol 59:61-65

Narahashi T 1962b Nature of the negative after-potential increased by the insecticide allethrin in cockroach giant axons. J Cell Comp Physiol 59:67-76

Narahashi T 1971 Effects of insecticides on excitable tissues. Adv Insect Physiol 8:1-93

Narahashi T 1976 Effects of insecticides on nervous conduction and synaptic transmission. In: Wilkinson CF (ed) Insecticide biochemistry and physiology. Plenum, New York, p 327-352

Narahashi T 1980 Site and types of action of pyrethroids on nerve membrane. In: Mathieu J (ed) Pyrethroids insecticides: chemistry and action. Roussel Uclaf, Romainville, France (Table Ronde Roussel Uclaf no 37) p 15-17

Narahashi T, Anderson NC 1967 Mechanism of excitation block by the insecticide allethrin applied externally and internally to squid giant axons. Toxicol Appl Pharmacol 10:529-547

Narahashi T, Haas HG 1968 Interaction of DDT with the components of lobster nerve membrane conductance. J Gen Physiol 51:177-198

Narahashi T, Lund AE 1980 Giant axons as models for the study of the mechanism of action of insecticides. In: Insect neurobiology and pesticide action. Soc Chem Ind (Lond) Monogr, p 497-505

Narahashi T, Yamasaki T 1960a Mechanism of the after-potential production in the giant axons of the cockroach. J Physiol (Lond) 151:75-88

Narahashi T, Yamasaki T 1960b Mechanism of increase in negative after-potential by dicophanum (DDT) in the giant axons of the cockroach. J Physiol (Lond) 152:122-140

Roeder KD, Weiant EA 1948 The effect of DDT on sensory and motor structures in the cockroach leg. J Cell Comp Physiol 32:175-186

Wang CM, Narahashi T, Scuka M 1972 Mechanism of negative temperature coefficient of nerve blocking action of allethrin. J Pharmacol Exp Ther 182:442-453

Wouters W, van den Bercken J 1978 Action of pyrethroids. Gen Pharmacol 9:387-398

Wouters W, van den Bercken J, van Ginneken A 1977 Presynaptic action of the pyrethroid insecticide allethrin in the frog motor end plate. Eur J Pharmacol 43:163-171

Wu CH, Oxford GS, Narahashi T, Holan G 1980 Interaction of a DDT analog with the sodium channel of lobster axon. J Pharmacol Exp Ther 212:287-293

DISCUSSION

Usherwood: What is the explanation for the non-exponential distribution of the life-times of the sodium channel, particularly the shift in the distribution of the life-times after treatment with toxin to a distribution with a mode of about 10 ms? Do the results suggest that more than one process determines the distribution of sodium channel life-times in this system?

Narahashi: This is complicated for a couple of reasons. One is a technical consideration. Since the observed single sodium channel currents are so small, being in the order of 1 pA, we are not certain whether we can reliably measure currents much smaller than that range. Very small currents may well have been within the range of noise. The second point is a consideration of modified channel kinetics. When a sodium channel opens via the normal route (C→O), it may be quickly bound by the toxin to produce the modified open state, O*. Under this condition I cannot be certain that I clearly see short open times comparable to those of control preparations before the application of toxin. However, I should emphasize that these single sodium channel experiments are preliminary ones.

Sattelle: You suggested that the pyrethroids are acting somewhere in the middle of the sodium channel. Presumably they reach the channel through the lipid phase of the membrane?

Narahashi: Yes, because they are hydrophobic compounds. Binding experiments are very difficult because of this; you might see 99% non-specific binding and only 1% specific binding.

Dudai: You suggested that pronase is acting from within the cell. Is that the case?

Narahashi: Pronase and *N*-bromoacetamide are both effective in inhibiting sodium inactivation only from inside the nerve membrane.

Usherwood: Which steps in your model should show concentration dependence?

Narahashi: Theoretically, both C→C* and O→O* should be concentration-dependent steps. We cannot measure C* because the channel is not open. The only things we can measure are O and O*, which are the peak and slow currents, respectively. The amplitude of the slow current should be concentration-dependent, and it is. The time constant is independent of the concentration; if any single channel is modified, it is modified in an all-or-none manner. Therefore if the concentration of the pyrethroid is increased, more and more channels are modified, but the time constant is not.

Hall: Binding experiments will be difficult if pyrethroids are hydrophobic and bind mainly non-specifically. Have binding inhibition studies been done? You could use the available channel probes and ask whether pyrethroids inhibit the binding of these specific probes. To take saxitoxin as an example of a probe, it wouldn't matter that the pyrethroid was binding non-specifically in other places; you could see whether it also showed inhibition at that specific site. This might give you some idea of its localization, by comparing its ability to inhibit saxitoxin versus the scorpion toxin, say.

Narahashi: That is an interesting idea and is probably feasible. Saxitoxin and tetrodotoxin (TTX) antagonize the action of pyrethroid in a non-competitive manner, so presumably they bind to different sites.

Hall: In that case, saxitoxin could serve as a control; if you say that pyrethroid binding shouldn't overlap with saxitoxin binding, then looking at that inhibition of saxitoxin binding would confirm this. You could then go on to look at effects of pyrethroids on the binding of other toxins.

Kelly: There seems to be one scorpion toxin site for each channel, whereas there seems to be more than one binding site for saxitoxin. Dr Catterall always uses veratridine to open these channels, I believe (Catterall et al 1979).

Narahashi: Yes. He says that veratridine, batrachotoxin and grayanotoxin bind to the same site, which is different from the TTX-binding site.

Perhaps I could add one comment on why we used crayfish and squid, which are not insects! This was for technical reasons. The voltage-clamp technique can be applied to the cockroach giant axon but the accuracy of measurement is a problem. For this kind of kinetic analysis we need a high precision of measurement. We also have to perfuse the axon internally to obtain flexible experimental conditions. Nobody has successfully perfused the cockroach giant axon internally yet. But the basic properties of excitation, including the action potential and conductance changes, are very similar in squid, crayfish and cockroach giant axons.

M. E. Eldefrawi: In describing the action of the pyrethroids as modulators and classifying them as similar to batrachotoxin and veratridine, are you saying that these pyrethroids can drive the sodium channel, independently of electrical stimulation?

Narahashi: Yes. We have not done many experiments on batrachotoxin yet, but according to Khodorov's results (1978) on the node of Ranvier, this toxin modulates the sodium channel in a way similar to the pyrethroids. We obtained a similar result with grayanotoxin and veratridine so, qualitatively at least, we can explain the actions of these toxins in terms of modulation of sodium channels.

M. E. Eldefrawi: So pyrethroids would activate the sodium channel?

Narahashi: To be exact, it is more complicated than that. I envisage the channel in a dynamic equilibrium, opening and closing all the time. So even when the membrane potential is −60 mV, the normal resting potential, a few channels are opening and closing. Once a channel is open, batrachotoxin or other sodium channel modulators can bind to it, so even without stimulation a depolarization is caused.

M. E. Eldefrawi: You are saying that in the resting state there are always a few open channels?

Narahashi: Yes. They are opening *and* closing, depending on the voltage, of course. At a very large negative membrane potential (e.g. −100 mV) we don't see any opening. But at the normal resting potential, channels are opening and closing at a low frequency.

Let me give another example. Consider local anaesthetics, some of which are known to block open sodium channels. If you look at the microscopic level you do not just see permanently closed channels; you see channels closing and opening, producing bursts of brief currents. In other words, the local anaesthetic molecule binds and unbinds to the channel quickly.

Olsen: You said that the cyanopyrethroids don't lead to repetitive firing. Why is that?

Narahashi: With the α-cyano-containing pyrethroid we see only a slight increase in the depolarizing after-potential because the slow sodium current following the peak sodium current is so small. The depolarizing after-potential never reaches the threshold for excitation, so no repetitive discharge is produced. However, the slow sodium current, once generated, shows little or no sign of decay; that is, the sodium channel modified by the α-cyano pyrethroid opens and remains open for a long period of time. Since the sodium channels open and close at a very low frequency, depending on the membrane potential, more and more channels gradually remain open even without stimulation, thereby causing a depolarization of the membrane. The membrane depolarization proceeds too slowly to induce repetitive discharge. Nerve conduction is eventually blocked as a result of depolarization. This is why the α-cyano-containing pyrethroids cause paralysis rather than hyper-excitation of nerves.

Usherwood: It is clear that you have made excellent use of insecticides as tools with which to dissect the sodium and potassium channels in the axonal membrane. However, can this approach contribute to the production of new chemical compounds for use as insecticides?

Narahashi: That is a difficult question! DDT, pyrethroids and their derivatives all appear to have the same basic mechanisms of action on the nerve membrane, yet the association and dissociation rate constants for each reaction step are different. Also, there are data on the metabolic detoxication of these compounds. So, if we combine these two molecules, we might be able to obtain a compound which is active on nerves with very little metabolic degradation.

Kelly: There is another feature of this. The experiments that Catterall et al (1979) do are apparently easier, and give a bigger return for the same amount of effort, so by putting these two methods together, the techniques for looking for new insecticides must be very much better than formerly.

Usherwood: This is the so-called rational approach, of course, and might provide information that would be of use to people working in the pesticide industry—but does it?

Narahashi: There are two main tests which are used widely to evaluate insecticide potency. One is the killing potency, where one obtains an LD_{50}. The other is what we call the 'knock-down' parameter. If you spray an insecticide onto, say houseflies, then depending on the dose, a certain fraction of houseflies are knocked down. This yields the KD_{50} or KD_{90} or whatever value you want. The 'knock down' value is a convenient way of estimating the *relative* potency of an insecticide. Some pyrethroids have considerable KD potency but little killing potency, or vice versa, so it is a good way to evaluate potency. But in terms of elucidating the mode of action, the knock-down parameter is not much help, because the knock-down effect could be due to many causes—to hyperactivity or paralysis of the nervous system, or various combinations of these.

Kelly: What about the economics of such research? You can now buy radiolabelled neurotoxins from commercial sources. Is that not cheaper than keeping colonies of insects, with technicians feeding them and breeding them?

Narahashi: If you have enough facilities and technicians it is not so expensive to use live insects. You can line up dishes, or whatever, and do many insecticide test experiments in one day. But electrophysiological measurements are more time-consuming.

Neumann: Experiments like those you described, Dr Narahashi, are very elegant indeed, but it may be difficult to draw conclusions from them if a specific pest problem has to be solved. Pyrethroids are a varied group of compounds; there is no one pyrethroid that is the best insecticide for all insects. The differences in activities and in structure–activity relationships depend on the species of insect. If basic research wants to contribute to finding new pesticides, it must not only show that a compound is more or less active in an *in vitro* system, but it has to give reasons for these differences in intrinsic activity. In addition, the gap between the *in vitro* system and the whole insect has to be bridged. Only a rational approach can be successful and therefore the structure–activity relationships in electrophysiological experiments must be explained and logically related to those in toxicological experiments. Acetylcholinesterase and its inhibitors represent a system that is well understood and should therefore be ideally suited to the rational approach. However, there is rarely a good correlation between inhibition of the enzyme and performance in toxicological tests.

Narahashi: A difficulty with insecticidal potency, or mammalian toxicity, is that besides the effect on the target site you have to consider the metabolism of insecticides and their penetration through a cuticle or other barrier. The final result, namely the death of an animal, is a complex product of several factors. That is why very often the nerve potency values are not in parallel with insecticidal potency.

Piek: I think we can learn from our particular kind of work on venoms how nature attacks insects, not by using a very poisonous toxin but by using a combination of two active components. When we understand how such components interact to produce a toxic effect, it may be possible to combine different insecticides and very carefully regulate insect populations without using the most toxic insecticides. But we have first to learn exactly what the different toxins are doing in the different insects!

Usherwood: Some basic research is being undertaken in the pesticide industry which is similar in nature to research being undertaken in university departments, polytechnics and research institutes. Presumably the pesticide industry feels that it is worthwhile to have people doing basic research in their laboratories.

Donnellan: Part of the job we do, as biochemists/physiologists, is to expose the chemists who will be synthesizing the new compounds to new target sites. For example, we can offer them intrinsic activity assays. If they as chemists are going to be involved in the rational design of new compounds, we are in a position to give them intrinsic activity measurements, because the problem is that if you put a compound on the outside of an insect you have no guarantee that it will reach the target. The chemists want to know if their new molecules are intrinsically active. If the molecules are active *in vitro*, the chemists can start modifying the compounds so that they will reach the target system in the insects.

Usherwood: Let us now extend the discussion and ask Professor Barnard for his opinion of the potential contributions of insect neuropharmacology and neurochemistry to neuropharmacology and neurochemistry in general.

Barnard: That is certainly a large question, and one which I should be thinking about, being here as a visiting vertebrate, as it were! That is to say, I have come here as somebody who knew virtually nothing about insect systems, and I have become impressed by the range of investigation on these which is being mounted. This is obviously going to bring much wider dividends.

I can see the work on insect neurobiology as being significant in three ways. Firstly, we have to recognize that we are inevitably anthropocentric and that we should strive not to be so, because there are intellectually important and intriguing problems in the insect world that have merit in their own right. For example, how the salivary gland of the blowfly or of the cockroach carries out its operations under the influence of a particular neurotransmitter, or how a polypeptide hormone which is unique to the insects acts, are interesting questions in their own right. There is intrinsic value in these kinds of studies, and we should try to correct the balance that our natural tendency to concentrate on vertebrate material has distorted.

Secondly, there clearly are economic merits in studying insects, in that their influence on human economic activity and health can in some cases be very great. We could usefully apply some of the advances being made currently in insect neurobiology, in view of the value to man of controlling insects. I have realized, also, that although in relation to applied studies we generally look on insects as pests, there are beneficial aspects too. One can imagine pharmacological advances which would aid in the use of insects for human benefit, as in the examples of honeybees—or of insects which control other pests—as well as in better ways of destroying unwelcome insects.

The third level seems to me to be that of neuroscience as a whole, in that it clearly has a lot to gain from studies on insects. We should of course look for favourable material for our studies in all phyla, and insects obviously have many advantages for certain operations. For example, I imagine we shall learn more about the glutamate receptor as a result of its accessibility in the insect than we shall for some time in mammalian systems. We can hope that it is essentially the same in both groups: from what I have heard the differences seem likely to be minor rather than fundamental. We should therefore be able to use the most favourable case.

There are many other examples where insect neuroscience is well placed to make a contribution through the advantages of accessibility and relative ease of experimental manipulation. I was impressed with the work on identified neurons in cockroach ganglia and similar systems. This is a topic on which the vertebrate neuroscientists are generally less advanced, for natural reasons, and where these insect systems, just as those in molluscs, can be of immediate value. I was also impressed by the genetic approach. It seems to me that genetics is going to make possible great advances in neuroscience by its application in the insect world.

If I may ask a very general kind of question, it seems to me that an important thing for us to know in neuroscience is which differences between the nervous systems in animal groups are fundamental and which are incidental. One would like to think that there are fundamental designs for receptors, for example, which will be found throughout the animal kingdom and that the apparent species-specific differences are in truth relatively minor. This is something that needs examination. Thus, we should consider the evidence marshalled by Jesse Roth (Le Roith et al 1980) that there is a phylogeny of receptors that can be found from the microorganisms through to the vertebrates, having common evolutionary origins. There could be aspects of receptor function about which we would learn much from comparative studies in different phyla.

For example, consider the following. When you look through all the known receptor types there seem to be some general principles that are valid for both vertebrate and invertebrate receptors. The fixed ion channels associated with neurotransmitter receptors are known only for the nicotinic ACh receptor and the GABA and glutamate receptors, but this association holds across all phyla where these have been looked at. Likewise the effector systems linked to the activation of cyclic AMP via a neurotransmitter seem always to go with receptors of certain classes, e.g. catecholamines, and not with the three I have just mentioned. This uniformity is probably telling us something about the evolutionary origin of the various receptor systems. It could be that multiplying such instances where a range of them is readily studied in the insect nervous system would test their generality and would give us more insight into their structural basis. What, then, are the principles of building a nervous system which are most likely to be universal, that one could learn from insect studies?

Narahashi: Dr Barnard has emphasized the importance of studying the insect nervous system. We should also stress that because insects include such a variety of forms, one can choose the most convenient material from among the vast range of insects, depending on the purpose of a measurement. Work on insect genetics is an illustration of this. Perhaps we should stress this advantage of studying insects, rather than stressing the value of comparative physiology. Insects provide the material for which we can do the best job for a given project.

Looking back over the history of insect neurophysiology, when I started using the insect nervous system about 30 years ago there were less than a dozen insect neurophysiologists in the world. Now it is a rapidly expanding field. But it takes a lot of propaganda to accomplish this. The slow reactions of people to a potentially useful yet very new approach or chemical can be illustrated by the following example. Dr C. Y. Lee of Taiwan published the first paper in English in 1963 on how α-bungarotoxin blocks nicotinic receptors (Chang & Lee 1963). Dr Lee and his colleagues then tried to persuade people to use α-bungarotoxin as a tool for studying these receptors. Nobody paid any attention until around 1970 when Ricardo Miledi, Perry Molinoff and Lincoln Potter used it for the first time as a ligand of the nicotinic receptor (Miledi et al 1971). So it took seven years before people realized the potential of this tool.

The situation is similar for insect neurophysiology. When I started using insects for neurophysiology, people tended to think that this was rather extraordinary. The same happened with channel-specific chemicals. It took several years before people started using tetrodotoxin, although that material had been available since I discovered its specific action in blocking sodium channels (Narahashi et al 1964). So we need to put a lot of effort and propaganda into telling people that insects are a useful research material, not just from the comparative point of view.

M. E. Eldefrawi: One area of growth in the neurosciences where pharmacology has benefited considerably has been tissue culture work. Insects have been the source of many successful tissue culture preparations. Not many have been used for neuropharmacological purposes yet, but in the coming years we may see this happening. I know at least three or four people who are doing excellent work on insect tissue culture, and future investigations may take advantage of this work.

REFERENCES

Catterall WA, Morrow CS, Hartshorne RP 1979 Neurotoxin binding to receptor sites associated with voltage-sensitive sodium channels in intact, lysed, and detergent-solubilized brain membranes. J Biol Chem 254:11379-11387

Chang CC, Lee CY 1963 Isolation of neurotoxins from the venom of *Bungarus multicinctus* and their modes of neuromuscular blocking action. Arch Int Pharmacodyn Ther 144:316-332

Khodorov BI 1978 Chemicals as tools to study nerve fiber sodium channels; effects of batrachotoxin and some local anesthetics. In: Tosteson DC et al (eds) Membrane transport processes. Raven Press, New York, vol 2:153-174

Le Roith DL, Shiloach J, Roth J, Lesniak MA 1980 Evolutionary origins of vertebrate hormones. Substances similar to mammalian insulins are native to unicellular eukaryotes. Proc Natl Acad Sci USA 77:6184-6188

Miledi R, Molinoff P, Potter LT 1971 Isolation of the cholinergic receptor protein of *Torpedo* electric tissue. Nature (Lond) 229:554-557

Narahashi T, Moore JW, Scott WR 1964 Tetrodotoxin blockage of sodium conductance increase in lobster giant axons. J Gen Physiol 47:965-974

Review of the symposium

P. N. R. USHERWOOD

Department of Zoology, University of Nottingham, University Park, Nottingham, NG7 2RD, UK

This meeting has been attended by experts in the field of insect neuropharmacology and by a small number of neurobiologists working with vertebrate animals. Our remit was to review the current state of insect neuropharmacology, the position that it occupies in the world of pharmacology at large, and the directions that it might profitably take in future years. Those not committed to this field of science might well ask 'What are the factors that combine to make insect neuropharmacology an interesting and important topic in science?' From a decidedly prejudiced position I can identify three reasons why the subject attracts my attention. First, I admit quite unashamedly to my affection for insects; their incredible variety of form and function not only pleases aesthetically but provides the experimentally minded with a wealth of opportunities to apply their skills and intellects. But, sadly, the day has almost passed when scientific curiosity alone guarantees support for research from the public purse. Because of the relatively high cost of much of the research undertaken by insect neuropharmacologists they readily appreciate the need to relate their endeavours to the well-being and development of mankind. For many insect neuropharmacologists this is best realized by having an interest in and perhaps even a commitment to the discovery of new poisons for controlling insect pests.

About three out of every four animal species are insects and as a group insects have been successful in colonizing most ecological niches in this planet: as our competitors for food and accommodation they must surely command our respect. Although many insect species benefit mankind others contribute to human misery on an enormous scale, especially in the Third World countries. Among the many techniques that have been devised and used commercially to control insects there is no doubt that chemical poisons or insecticides have been the most successful. Although the first insecticides were discovered by chance, the majority in current use were found by random

1982 Neuropharmacology of insects. Pitman, London (Ciba Foundation symposium 88) p 307-317

screening of potentially toxic compounds, a procedure that originated during the Second World War. Compounds such as chlorinated hydrocarbons, organophosphates and their newer brethren, the pyrethroids, perform a major role throughout the world in minimizing loss of agricultural yields and reducing the prevalence of diseases carried by insect vectors. Yet, people in the West are probably less aware of the contributions made by the pesticide industry to the maintenance and improvement of their living standards and life-style than of the damaging environmental side-effects which can result from the widespread use of insecticides. These chemicals are harmful environmentally because they are non-selective poisons. Many are neuropoisons, and because the nervous systems of insects have properties in common with those of other animals, including humans, such poisons do not discriminate between animals on the basis of leg counts. It is unlikely that cheap and effective poisons which are insect-specific will be discovered in the near future, so we must continue to accept that the use of neuropoisons developed to kill insects will also cause fatalities among other animals. Obviously a balance must be established between the need to minimize the adverse environmental impact of pesticides and the need to maximize the survival and health of human populations, but there can be no doubt that the search for new classes of insecticides must go on.

Some insect neuropharmacologists might argue that their basic research can contribute to the discovery of insect-specific poisons and may even lead to the development of insecticides which are species-specific. But these are long-term ambitions: some might say 'pie in the sky'. Others might argue that if the pesticide industry already has a powerful armoury of toxicants for use against insect pests then why is there any need, apart from environmental considerations, to search for new chemicals? The answer is frighteningly simple. Insect pests are rapidly learning to live with our present insecticides through their growing resistance to these chemicals. The pesticide industry cannot meet this challenge without resort to new poisons and the human race cannot afford to lose this battle, unless it is prepared to accept a bleak future. The discovery of toxicants by random screening and optimization of structure–activity relationships must continue to play a major role in this battle but the alternative so-called 'rational approach' must be tried and tested because it is more likely to lead to the discovery of completely new classes of insecticides. In this approach fundamental work on insect nervous systems aims to gain insight into the sites and modes of action of existing insecticides and to increase our knowledge of insect neurobiology, particularly of those neuronal systems most likely to offer target sites for novel neuropoisons.

Obviously, insect neuropharmacology is central to this approach, which is the second reason for my attraction to this field of research and my belief in its importance. I have my doubts whether, in the short term, the rational approach will produce insect toxicants capable of competing commercially with those already on the market, but novel neuropoisons may be discovered which in the longer term can be deployed in the battle against resistance. Evidence that the pesticide industry is taking this new approach seriously is to be found in a growing number of small but active interdisciplinary research groups which have been established by many companies to undertake basic research in insect neuropharmacology and to foster relationships between the pesticide industry and those universities, polytechnics and research institutes engaged in insect neurobiology research.

Most of the chemical signals found in the nervous systems of vertebrates are also found in the central and peripheral nervous systems of insects. Because of the similarities between insect neuropharmacology and that of other animals, it is not surprising that some insect preparations are providing information the significance of which extends beyond the large but restricted world of insects. At this meeting we have heard of research being undertaken on octopaminergic, glutamatergic, GABAergic and dopaminergic systems in the central and peripheral nervous systems of insects. These studies are being made on experimental material which is readily available, relatively simple in terms of its structural organization, and readily accessible. Of course, at the tissue level of organization, factors which appear to be unique to insects will hinder attempts to extrapolate data from insect nervous systems to nervous systems of other animals, but at the molecular and even cellular levels such extrapolations are often perfectly admissible and can be very profitable. The third reason, therefore, for my interest in insect neuropharmacology is my belief that it can, and undoubtedly will, make an important contribution to the advancement of neuropharmacology by giving major insights into the rules which govern chemical signalling between excitable cells—insights which will be applicable generally throughout the animal kingdom.

Perhaps not unexpectedly, our knowledge of cholinergic transmission in insects is well advanced compared with our knowledge of other chemical signalling systems in insect nervous systems. The essential components of a cholinergic system have been demonstrated in many insect species, but the criteria for identifying acetylcholine as a neurotransmitter at an insect synapse have not yet been fulfilled. Although there are cogent reasons why we need to demonstrate conclusively that acetylcholine is a neurotransmitter in insects, this should not distract us from the task of characterizing the receptors for acetylcholine that have been identified and subjecting them to the same detailed biochemical, pharmacological and physiological analyses that have been so profitably applied to vertebrate material. The central nervous systems of insects are probably well endowed with cholinergic synapses but there is growing evidence that many of the acetylcholine receptors found on insect neurons are extrajunctional. Using radiolabelled ligands Sattelle has shown that the membrane of cockroach neurons contains binding sites for certain cholinergic ligands which identify them as acetylcholine receptors. This confirms the earlier findings of Usherwood et al (1980) who mapped the acetylcholine sensitivities of neuron somata (which lack synapses) isolated from locust central nervous system. The axon terminals of locust motor neurons are also characterized by receptors for acetylcholine (Fulton & Usherwood 1977), which suggests that these receptors may be present either over the entire extrajunctional surface of these cells or in areas of neuron membrane not covered by glial cells.

Cholinergic receptors in the insect central nervous system are of interest not only from the comparative and evolutionary standpoints but also because they could provide unique experimental material for studying genetic and developmental aspects of receptor biology. One insect species, the fruitfly *Drosophila melanogaster*, has a well-established genetics; it is easy to obtain mutants of this insect and, thereby, to identify directly the genetic coding for a given macromolecule. By this means it has been possible to demonstrate that essential enzyme components of the cholinergic system are each coded by a single gene. These and other developments represent the beginnings of a new and potentially powerful approach to neuropharmacology. The argument presented by Dudai that it is the primitive state of neurochemical knowledge rather than genetic knowledge which is the factor limiting expansion of the use of genetic dissection techniques on *Drosophila* nervous systems is well taken and will, I hope, lead to a spate of applications worldwide to funding agencies for support of basic insect neurobiology. Unfortunately, whereas our knowledge of *Drosophila* genetics is well advanced, the same is not true of many insects which either have been studied intensively from the neurobiological viewpoint or are among the most harmful to humans. My impression is that the potential contribution of genetic studies of *Drosophila* to neurobiology is so great that there will be no shortage of support for fundamental research (anatomical, physiological, biochemical, etc.) on this insect, to act as a foundation for future genetically based studies. Jan's studies using a combination of electrophysiological and immunological approaches to study wild-type and mutant *Drosophila* indicate clearly the direction that these studies should take. Let us hope that the geneticists devoted to *Drosophila* will channel some of their energies into establishing the genetics of other insects, particularly those with well-established reputations for providing neurobiological preparations and those which are major pests.

Advances in our understanding of vertebrate nicotinic acetylcholine receptors are due largely to the availability of highly specific, almost irreversibly binding natural toxins such as α-bungarotoxin. Barnard has described the technical sophistication of this approach and the fine resolution of the structure and functional properties of the acetylcholine receptor that has resulted from its use on vertebrate excitable tissues. Insect neurochemists will be encouraged by his success in isolating acetylcholine receptors from vertebrate brain tissue which has the same or a lower concentration of acetylcholine receptors than insect brain. Acetylcholine receptors in insect central nervous systems also bind α-bungarotoxin and this property has been used by Donnellan and Sattelle to isolate these receptors as a first step in the pharmacological characterization *in vitro* and purification and biochemical analysis of these receptor proteins. Both Eldefrawi and Sattelle referred to the quantitative differences between the acetylcholine receptors of insects and those of vertebrates, but detailed studies of these macromolecules have been restricted so far to a few species of insect and only a small number of vertebrate excitable systems. It therefore remains to be seen whether these differences are sufficiently great to allow for the development of insect-specific neuropoisons with acetylcholine receptors as target sites.

Little is known about the channels gated by acetylcholine receptors in insect central nervous systems except that, as Sattelle pointed out, they are (mainly?) cationic (see also Usherwood et al 1980). It was therefore of interest to see the analysis by the Eldefrawis of the binding of histrionicotoxin to acetylcholine receptors in membranes from the housefly central nervous system. This channel probe not only identifies the binding sites but also signals the conformational state of the receptor-gated membrane channel. Other papers on chemically gated membrane channels highlighted the growing interdisciplinary approach to insect receptor biology. Hall used the genetic approach to study voltage-dependent Na^+ channels in axonal membrane of *Drosophila*. Her studies of the *nap*^{ts} mutant, which has a reduced population density of Na^+ channels in axon surface membrane compared with the wild-type, should lead to the identification and isolation of the gene(s) for these membrane macromolecules. Combined with the technique of gene cloning using bacterial plasmids, this approach could lead to 'large' quantities of isolated Na^+ channels being available for biochemical analysis and for biophysical studies using black lipid membranes. Both Cull-Candy and Gration described how the techniques of noise analysis and patch clamp analysis are being used to obtain quantitative information on the conductances, kinetics of opening and closing and pharmacology of receptor-gated ion channels in locust muscle membrane. Because of the large size (for insects) and accessibility of the nerve–muscle synapses they contain, locust nerve–muscle preparations have become familiar even outside insect biology circles (see also the papers by Evans and by Piek). Although they were discovered almost 20 years ago (Usherwood & Grundfest 1963), GABA (inhibitory) synapses and receptors on insect muscle have received less attention than their glutamate counterparts and for this reason alone the description by Cull-Candy of his patch clamp studies of extrajunctional receptors on locust leg muscle was especially welcome. Glutamate synapses were first discovered on locust leg muscle by Usherwood & Machili in 1966 and have since been studied extensively from the structural (e.g. Rees & Usherwood 1973), physiological (Usherwood 1981) and biochemical (Cleworth et al 1980) viewpoints. The discovery of extrajunctional receptors for glutamate on denervated locust muscle (Usherwood 1969) and, at lower population density, on innervated locust muscle (Lea & Usherwood 1973, Cull-Candy & Usherwood 1973) has allowed the new and powerful tool of patch clamp analysis to be used to investigate individual glutamate receptors (Patlak et al 1979). Both Gration and Cull-Candy referred to their use of the lectin concanavalin A (Mathers & Usherwood 1976) which has facilitated studies of the glutamate receptor–ionophore complex of locust muscle fibres by eliminating desensitization. Lentil lectin and pea lectin have a similar sugar-binding specificity (mannose/glucose) to concanavalin A (jack bean lectin) and also block desensitization, but plant lectins with different specificities do not. Furthermore, the dimeric form of concanavalin A (at physiological pH concanavalin A is tetrameric) and succinyl concanavalin A, which is also dimeric, do not influence desensitization (Anis et al 1979). Further physiological studies, aided by biochemical investigations, of the mechanisms underlying the action of these lectins at the single-receptor level may illuminate the phenomenon of desensitization of glutamate receptors on locust muscle, as well as giving information on the polysaccharide content of these receptors.

It is sometimes stated that (with the exception of cells which form electrical synapses) all neurons signal information to their target structures by releasing chemical substances which either change the conductance of the target membrane or activate secondary messengers in the target cell. These chemical messages range from molecules of low molecular weight to complex peptides. At one extreme the message is transmitted along spatially restricted pathways, e.g. at synapses, whereas at the other extreme it is released into the bloodstream; the signals range from those lasting a few milliseconds to those lasting many hours. In truth, it seems likely that most chemical signals in the nervous system exert not one but a variety of influences over their target structures. It is not surprising, therefore, that attempts to categorize chemical mediators by their structural, spatial, temporal or functional characteristics are almost immediately undermined by discoveries of a functional complexity which destroys the boundaries erected. Certainly the division of chemical signals into neurotransmitters, neurosecretors and hormones is at best arbitrary and at times decidedly misleading, and the recent introduction of the term neuromodulator into neurobiology is to be regretted. The chemical mediators at synapses evoke conductance changes of very brief duration (milliseconds) at their postsynaptic target sites and are therefore classed as neurotransmitters, to distinguish them from neuromodulators which may not increase the conductance of their target membrane and whose action may last for many hours. Yet at synapses ions moving in and out of the postsynaptic cell, through the receptor-gated membrane channels, may either activate secondary messengers in the target cells, or, as is often the case for Ca^{2+}, for example, function as secondary messengers and thereby introduce changes in the target cell which far outlast the conductance change of the postsynaptic membrane. There is growing evidence that most neurons and other excitable cells contain calcium-modulated proteins, calmodulin being the prototype, either in the cytosol or in the membrane facing the cytosol. These proteins bind free calcium and then undergo conformational changes which activate specific enzymes in their domain. Although the rise in the Ca^{2+} concentration of the cytosol after a postsynaptic event signalled by, for example, L-glutamate may be brief, the physiological result of the signal could influence events in the postsynaptic cell for hours or days.

There are other biochemical pathways for long-term modulation of target cells by chemical messengers that are normally considered as synaptic transmitters. Several of these signals, including L-glutamate, have been shown to increase cyclic GMP levels in intact preparations of vertebrate nerve and muscle tissue (Garthwaite & Balázs 1981, Beam et al 1977), apparently by promoting Ca^{2+} entry into the cells. N. L. Robinson (personal communication) has found that the application of glutamate to blowfly body-wall muscle raises cyclic AMP and cyclic GMP levels. However, the regulation of guanylate cyclase in membrane fractions of blowfly muscle by L-glutamate is complex, involving both calcium-dependent and calcium-independent pathways. Cyclic nucleotides in neurons and other excitable cells have been recognized as secondary messengers for many years and their activity has often been linked to the binding of diffusely distributed chemical signals to surface target cells. Evans described one class of octopamine receptor in insects which activates a membrane adenylate cyclase. Of the two classes of 5-HT receptor on salivary glands of blowfly, Berridge showed that one gates Ca^{2+}, which may be a secondary messenger, whereas the other is linked to adenylate cyclase to generate cyclic AMP. The dopamine receptors on cockroach salivary glands described by House gate Ca^{2+} channels and it is possible that Ca^{2+} also acts as a secondary messenger in this system. Although studies of secondary messengers in insects are in general in their infancy there are opportunities here for the applied biologist to exploit new target sites and also for insect neurobiologists to make useful and perhaps unique contributions to this new field.

The other group of chemical signals discussed at this meeting, the neuropeptides, also probably function mainly through secondary messenger systems. These chemical signals are of prime importance as regulators of physiological, biochemical, behavioural and developmental events and their influence on the insect is therefore profound and not restricted to the nervous system. Hildebrand and Mordue have drawn our attention to the wide range of peptides that have been identified in the central and peripheral nervous systems of insects, including substances known to exert profound effects on vertebrate excitable cells. The complex structure of neuropeptides compared with other chemical signals such as amino acids might relate to their high affinity binding to receptor sites on target cells. This would certainly enhance the effectiveness of low concentrations of neuropeptide acting on distant target sites. The specificity of peptide–peptide receptor interactions might also be enhanced by the structural complexity of the neuropeptide molecule which will have the effect of enhancing the signal-to-noise ratio in a signalling system that includes one of these compounds. Because of their structural complexity the neuropeptides may perhaps provide the applied biologist with opportunities to develop insect-specific poisons. However, the specificity of those insect neuropeptides so far identified remains to be established and there is already evidence that insects have many neuropeptides in common with other animals.

Piek's paper on the effects of two wasp venoms on insect peripheral and central nervous systems confirms expectations raised in many quarters that insect toxins and venoms could provide neurobiologists with valuable tools. Admittedly, the goal of discovering new pesticides by this route is a long way off, and I am not optimistic that we shall see rapid developments in this field, but this should not overshadow the important advances that have been made in elucidating the sites and modes of action of these substances. The collection of sufficient quantities of venom for experimental purposes can be a slow, labour-intensive process; the isolation, purification and identification of the active components is incomplete and much remains to be done to clarify the influence of these fractions on excitable cells. The alternative approach adopted by Narahashi has been to use synthetic neurotoxins such as insecticides to probe the membrane of excitable cells. Narahashi's studies, which have not been restricted to insects, have elucidated the properties of Na^+ channels in axonal membrane and have clearly identified this membrane component as the major site of action of pyrethroid insecticides and DDT. Since voltage-dependent Na^+ channels are found almost universally in the membranes of nerve cells it is clear why even the new pyrethroid insecticides are unlikely to live up to claims that they are insect-specific.

It is unfortunate that subcellular fractionation techniques that have been applied successfully to central nervous tissue and to other excitable systems in vertebrates have not been particularly successful when applied to insects. Nevertheless, Breer's paper on the isolation of viable synaptosomes from locust central nervous tissue demonstrates clearly that, with techniques designed specifically for use with insects, subcellular fractionation of insect central nervous systems not only is possible but may contribute significantly to our knowledge of the physiology and biochemistry of this tissue. Obviously, his studies are at an early stage and his technique will need to be refined if he is to obtain homogeneous populations of synaptosomes, but a start has been made on which more ambitious studies can be built. Although it could be argued that the isolation procedures used in his and other studies could introduce changes in the properties of the nerve terminals, subcellular fractionation has the major advantage of providing the biochemist and neuropharmacologist with seemingly functionally intact synaptic components that can be studied without the problems of access and the influence of surrounding structures which bedevil *in situ* studies of the insect nervous system.

Because of the detailed information available on their physiology, structure and pharmacology, the synapses on insect skeletal muscle should be excellent objects for the attention of the subcellular fractionators. Olsen described how he has used crayfish muscles to obtain membrane fractions containing GABA receptors and ionophores, while Lunt gave a comprehensive account of success and failures in the use of subcellular fractionation techniques for studying locust nerve–muscle junctions. Lunt also studied the uptake of L-glutamate by locust muscle vesicles and he suspects similarities between this system and areas of vertebrate brain where glutamate is thought to function as a neurotransmitter.

Having extolled the virtues of insect preparations in terms of their potential contribution to basic neurobiology, let me sound a note of caution. Because of the success of insects in colonizing our planet, insect neuropharmacologists could be readily persuaded by the vast array of material at their fingertips into believing that they are confronted by a golden harvest awaiting the combine. This is an illusion soon destroyed by an examination of our knowledge of matters anatomical. It is evident that over the past century or so biologists have been fascinated by the structural organization of the insect nervous system and the literature is replete with papers on this subject. Nevertheless, our knowledge of structure (particularly ultrastructure) seriously lags behind our knowledge of the physiology, pharmacology and biochemistry of insect nervous systems. This will inevitably retard the future growth of insect neurobiology.

While listening to the talks and participating in discussions, both formal and informal, at this meeting I have realized that even in the world of insect neuropharmacology there is more than a hint of compartmentalization of thought, tempered only by a hazy awareness of developments outside the immediate confines of subject areas. Probably this is inevitable. We are continually bombarded with verbal and written information in our own fields of study and must often rely on brief abstracts or anecdotal comments to maintain our awareness of the rest of the neuropharmacological world. Perhaps it is a mistake, therefore, to view insect neuropharmacology as a unified subject in science. Yet insect neuropharmacologists have a corporate role to play in the search for new pesticides, and it is true that insect biology has a tradition extending back over many decades. But those neuropharmacologists working on, for example, monoamines in the insect central nervous system probably have more in common with devotees of monoaminergic systems in other animals than with, say, neuropharmacologists studying insect glutamatergic synapses. Of course, superficial synthesis of results gleaned from the disparate studies which characterize contemporary insect neuropharmacology is just about possible, but in my opinion its impact on the development of our understanding of the working of these animals is at present less important than that emanating from comparative neuropharmacology. Our knowledge of insect neuropharmacology in general is still very rudimentary and it will be some years before it will be possible to start fitting the pieces together. I must add that I do not consider that the prime objective of the insect neuropharmacologist is to compare insects with other animals. By all means let us make such comparisons, for those neuropharmacologists working with animals where knowledge is more advanced may well have arrived at the right conclusions. So be it. But unprejudiced observations of insect systems will not only provide yet another test-bed for current dogma but will also lead to full exploitation of the unique opportunities that some insect preparations may offer the neuropharmacologist. Such observations will also highlight differences, both qualitative and quantitative, which will contribute to the development of neuropharmacology in general and, it is to be hoped, prove immensely beneficial to the maintenance of the dominant position of the human race on this planet.

REFERENCES

Anis N, Evans M, Usherwood PNR 1979 Plant lectins and desensitization of locust glutamate receptors. J Physiol (Lond) 291:47P

Beam KG, Nestler EJ, Greengard P 1977 Increased cyclic GMP levels associated with contraction in muscle fibres of the giant barnacle. Nature (Lond) 267:534-536

Cleworth JF, Robinson NL, Usherwood PNR 1980 Biochemical studies on glutamate receptors in insect muscle. In: Insect neurobiology and pesticide action. Soc Chem Ind (Lond) Monogr, p 279-280

Cull-Candy SG, Usherwood PNR 1973 Two populations of L-glutamate receptors on locust muscle fibres. Nature (Lond) 246:62-64

Fulton BP, Usherwood PNR 1977 Presynaptic acetylcholine action at the locust neuromuscular junction. Neuropharmacology 16:877-880

Garthwaite J, Balázs R 1981 Excitatory amino acid-induced changes in cyclic GMP levels in slices and cell suspensions from the cerebellum. In: Di Chiari G, Gessa GL (eds) Glutamate as a neurotransmitter. Raven Press, New York, p 317-326

Lea T, Usherwood PNR 1973 The site of action of ibotenic acid and the identification of two populations of glutamate receptors on insect muscle fibres. Comp Gen Pharmacol 4:333-350

Mathers D, Usherwood PNR 1976 Concanavalin A blocks desensitization of glutamate receptors on insect muscle fibres. Nature (Lond) 259:409-411

Patlak J, Gration KAF, Usherwood PNR 1979 Single glutamate-activated channels in locust muscle. Nature (Lond) 278:643-645

Rees D, Usherwood PNR 1973 Fine structure of normal and degenerating motor axons and nerve-muscle synapses in the locust, *Schistocerca gregaria*. Comp Biochem Physiol A Comp Physiol 43:83-101

Usherwood PNR 1969 Glutamate sensitivity of denervated insect muscle fibres. Nature (Lond) 223:411-413

Usherwood PNR 1981 Glutamate synapses and receptors on insect muscle. In: Di Chiari G, Gessa GL (eds) Glutamate as a neurotransmitter. Raven Press, New York, p 183-193

Usherwood PNR, Giles D, Suter C 1980 Studies of the pharmacology of insect neurones *in vitro*. In: Insect neurobiology and pesticide action. Soc Chem Ind (Lond) Monogr, p 115-128

Usherwood PNR, Grundfest H 1963 Inhibitory postsynaptic potentials in grasshopper muscle. Science (Wash DC) 143:817-818

Usherwood PNR, Machili P 1966 Clinical transmission at the insect excitatory neuromuscular synapse. Nature (Lond) 210:634-636

Index of contributors

Entries in **bold** *type indicate papers; other entries refer to discussion contributions*

Indexes compiled by John Rivers

Subject index